Marmot Biology

Sociality, Individual Fitness, and Population Dynamics

Focusing on the physiological and behavioral factors that enable a species to live in a harsh seasonal environment, this book places the social biology of marmots in an environmental context. It draws on the results of a 40-year empirical study of the population biology of the yellow-bellied marmot near the Rocky Mountain Biological Laboratory in the Upper East River Valley in Colorado, USA.

The text examines life-history features such as body-size, habitat use, environmental physiology, social dynamics, and kinship. Considerable new data analyses are integrated with material published over a 50-year period, including extensive natural history observations, providing an essential foundation for integrating social and population processes. Finally, the results of research into the yellow-bellied marmot are related to major ecological and evolutionary theories, especially inclusive fitness and population regulation, making this a valuable resource for students and researchers in animal behavior, behavioral ecology, evolutionary biology, ecology, and conservation. Color versions of many of the figures are available for download at www.cambridge.org/9781107053946.

Kenneth B. Armitage is Baumgartner Distinguished Professor Emeritus of Ecology and Evolutionary Biology at the University of Kansas. His 40-year research project on the yellow-bellied marmot in the Upper East River Valley, Colorado, is the second longest continuous study of a mammal. He is an elected Fellow of the Animal Behavior Society and the American Association for the Advancement of Science and an Honorary Member of the American Society of Mammalogists for "distinguished service to the science of mammalogy."

Marmot Biology

Sociality, Individual Fitness, and Population Dynamics

KENNETH B. ARMITAGE

University of Kansas, Lawrence, USA

CAMBRIDGE
UNIVERSITY PRESS

CAMBRIDGE
UNIVERSITY PRESS

University Printing House, Cambridge CB2 8BS, United Kingdom

One Liberty Plaza, 20th Floor, New York, NY 10006, USA

477 Williamstown Road, Port Melbourne, VIC 3207, Australia

314-321, 3rd Floor, Plot 3, Splendor Forum, Jasola District Centre, New Delhi - 110025, India

79 Anson Road, #06-04/06, Singapore 079906

Cambridge University Press is part of the University of Cambridge.

It furthers the University's mission by disseminating knowledge in the pursuit of education, learning and research at the highest international levels of excellence.

www.cambridge.org
Information on this title: www.cambridge.org/9781107656529

First published 2014
First paperback edition 2019

A catalogue record for this publication is available from the British Library

Library of Congress Cataloging in Publication data
Armitage, Kenneth B., 1925–
Marmot biology : sociality, individual fitness and population dynamics / Kenneth B. Armitage.
  pages cm
ISBN 978-1-107-05394-6 (hardback)
1. Marmots. I. Title.
QL737.R68A757 2014
599.36´6–dc23

                                   2013046725

ISBN 978-1-107-05394-6 Hardback
ISBN 978-1-107-65652-9 Paperback

Additional resources for this publication at www.cambridge.org/9781107053946

To my wife Katie and all those (spouse) who contributed to our direct fitness: Karole; Keith (Maria), Emeline, Julia, and Sophie; Kevin (Helen) and Rita.

Contents

The color plate section appears between pages 212 and 213.

Introduction

In the early years of marmot research I was frequently asked: "what are marmots good for?" When I replied that marmots had intrinsic value as a member of the earth's biodiversity, I usually received a blank look. If I replied that marmots fed golden eagles and coyotes, I received a nod of understanding. When I stated that marmots were useful organisms for studying certain biological problems, such as what regulates animal numbers, I frequently became engaged in conversation about why animal populations did not increase without limit. As I describe in the first chapter, my interest in the yellow-bellied marmot was to use it to study the role of social behavior in population dynamics. I soon came to believe that population dynamics could be understood only in the context of marmot life history. As a consequence, the yellow-bellied marmot research program expanded to explore life-history features such as body size, habitat use, environmental physiology, social organization and dynamics, individuality, and kinship. I also came to realize that characteristics of the social and population system, such as group composition and relatedness, could be understood only by following behavioral and demographic patterns over many years. Social organization and population numbers in a marmot colony in one year were a consequence of their history, which in turn affected social composition and behavior in the years to follow. Knowing the process led to understanding the structure.

This book is written with three broad goals in mind. The first goal is to describe how one species, the yellow-bellied marmot, copes with and persists in the harsh environment where it lives. The species has to solve multiple problems; e.g., finding and processing food, finding a mate, avoiding predation, in order to evolve an integrated life history. In a broad sense, the integrated system can be thought of as all the environmental and organismal factors that enable an animal to obtain and process energy in order to produce reproductive descendents. The yellow-belled marmot system can best be understood in the context of what other species of marmots do. Thus, much of the text is a comparative biology of the 15 species of marmots.

The second goal is to present considerable new data and analyses and integrate them with material published over a 50-year span in 29 journals and 17 book chapters, symposium volumes, and conference proceedings. Extensive natural history observations are included to provide the essential foundation for integrating social and population processes.

The third goal is to relate the results of the yellow-bellied marmot research to major ecological and evolutionary theories, especially inclusive fitness and population regulation.

An overview of the book's content

This book reports the results of a 41-year empirical study of the social and population biology of the yellow-bellied marmot (*Marmota flaviventris*) in the Upper East River Valley near the Rocky Mountain Biological Laboratory in Colorado, USA. The roles played by social behavior, individual behavioral phenotypes, physiological adaptations, and resource use in reproductive strategies of males and females occupy central stage. Inclusive fitness theory provides the framework for evaluating the importance of direct and indirect fitness strategies in the formation of social groups, the sharing of critical resources, and the expression of social behavior. The significance of cooperation and competition within female kin groups is related to reproductive skew and cooperative breeding. Of special significance is relating individual reproductive strategies to the demographic factors affecting population growth and decline. A feedback model of population regulation by social behavior is evaluated and the role of population density in population dynamics is analyzed. The importance of time as a limiting factor in marmot biology is developed.

The book is organized into six topical parts. Part I includes the first seven chapters. The first chapter introduces the relationship between marmots and humans that eventually resulted in the use of marmots as experimental animals. The remaining chapters describe the evolution and diversity of marmot species, how they exploit the places where they live, the importance of a harsh environment, and the consequent evolution of sociality. Because hibernation plays a central role in marmot biology, its physiology and control receive special emphasis.

The second part (Chapters 8 and 9) describes the abiotic environment, research sites, and environmental physiology of the yellow-bellied marmot. Energy conservation is emphasized as a major adaptation enabling marmots to direct as much energy as possible to reproduction.

In the third part (Chapters 10–14) kinship provides the foundation for analyzing the functions of social behavior, resource sharing, and communication. Individual behavioral phenotypes are introduced and their effects on social play are described.

The fourth part (Chapters 15 and 16) focuses on social and demographic factors influencing reproductive success of males and females and how the strategies of one sex affect the reproductive success of the other sex.

Population dynamics is the topic of Part V (Chapters 17–20). The impact of agents, such as weather and predation, on population change through time are described and followed by an analysis of dispersal and immigration, which is followed by a discussion of metapopulation dynamics. The demographic mechanisms of population growth and decline are related to reproductive strategies of yellow-bellied marmot females living in matrilines. This section concludes with an analysis of the relevance of population-regulation and population-limitation hypotheses to marmot population dynamics.

Part VI consists of Chapter 21 in which the potential effects of climate change on marmots are evaluated. The amount and duration of snow cover are emphasized as critical factors limiting marmot distribution and reproduction and survival.

The final chapter reviews, develops, and emphasizes major life-history traits that determine the successful persistence of marmots living in a harsh, seasonal environment.

Acknowledgments

The yellow-bellied marmot project and this book would not have been possible without the dedicated efforts of students, staff, and senior collaborators. Graduate students chose projects of biological interest and used the yellow-bellied marmot as their experimental animal. They also contributed to the core activities of trapping and marking marmots, recording life-history data, and assisting with data analyses. I trust that I have fairly and adequately incorporated their contributions (also see Armitage 2010) in the chapters that follow. I most gratefully thank Douglas C. Andersen, Kelly Brady, Alison K. Brody, Janice Daniel, Jerry F. Downhower, Barbara A. Frase, Scott H. Jamieson, Dennis W. Johns, Delbert L. Kilgore Jr., Jaye C. Melcher, Stephen Nowicki, Kathleen Nuckolls, Carmen M. Salsbury, Orlando A. Schwartz, Gerald E. Svendsen, Dirk H. Van Vuren, John M. Ward Jr., and Brett C. Woods.

Additional assistance in trapping, marking, and processing of marmots, and in data gathering and analysis was provided by Karole Armitage, Keith Armitage, Kevin Armitage, Erika L. Barthelmess. Martin W. Bray, Marta Chiesura Corona, David Doll, Robert R. Fleet, Regina C. Gray, Adrienne Kovach, Daniel R. Michener, Diane L. Rains, Linda S. Rayor, Milton S. Topping, Julian Trevino-Villarreal, Kathy Wallens, and Gary Worthen.

Melissa Bayouth, Sharon Green, and especially Stephanie Otey provided invaluable typing assistance and Patty Krueger, Gill Ortiz, Kathleen Nuckolls, and Sara Taliaferro prepared, scanned, and modified many of the figures for publication. Kathleen Nuckolls provided invaluable assistance in the preparation of the index.

My wife Katie provided a loving and supportive environment throughout the 50 years of marmot research and writing.

Daniel T. Blumstein, Madan K. Oli, and Arpat Ozgul contributed significantly to the marmot project by introducing new ideas and analytical skills. Our extensive collaborative efforts enriched our understanding of marmot evolution and population dynamics. Conversations and correspondence with colleagues involved in marmot research in North America and Eurasia helped clarify ideas, provided new sources of information, and stimulated my interests in marmot biology beyond the boundaries of the yellow-bellied marmot.

The yellow-bellied marmot research was supported by grants from the National Science Foundation. The Rocky Mountain Biological Laboratory provided a base of operations, housing, and laboratory space that made the field work and laboratory measurements feasible. Without this support the scope of the marmot project would have been markedly restricted.

Dan Blumstein, Madan K. Oli, Orlando Schwartz, and Dirk Van Vuren read and provided helpful comments on specific chapters. Two anonymous reviewers provided useful comments and suggestions on multiple chapters. I deeply appreciate the support and friendly advice of my editors Martin Griffiths, Megan Waddington, and Jodie Hodgson who guided me through the publication process.

Part I

The diversity and evolutionary history of marmots

1   Marmots in human culture: from folklore to research

Marmots (Fig. 1.1) are large-bodied, ground-dwelling, herbivorous squirrels (Rodentia: Sciuridae) that originated in North America, spread into Eurasia during the Pliocene and dispersed widely during the Pleistocene (Bibikov and Rumiantsev 1996). Their large size, diurnal activity, and the propensity to alarm-call at the approach of intruders would attract attention wherever humans and marmots co-existed. Their first contacts with humans were probably with nomadic hunters who populated the periglacial zone where marmots lived. Marmot fossils are not significant until the last ice age when marmots were probably an episodic game animal. The distribution and state of marmot populations during the Holocene were likely determined primarily by environmental conditions (Bibikov and Rumiantsev 1996). Evidence of marmot/human interactions dates to the Pleistocene.

Human/marmot interactions in prehistoric times in Europe are recorded in cave paintings in France (Louis *et al.* 2002). Marmots have been hunted for fur and meat in the western Alps and the southern Jura since the Late Pleistocene (Tomé and Chaix 2003). Some archaeological sites had several hundred marmot individuals indicating specialized hunting, which was uncommon because marmots are usually in low abundance. Marmot numbers and elevational range increased during cold periods and decreased during warm periods (Table 1.1). In general, hunting of marmots is clearly linked to cold periods (Tomé and Chaix 2003).

Probably all North American species (see Table 1.2 for a list of marmot species) were utilized by Native Americans. Okanagan Indians used *M. caligata* for food and clothing (Anderson 1934). A *M. flaviventris* jaw was found in Woodchuck Cave in northern Arizona associated with a burial that was dated to about AD 200 (Lange 1956). *M. vancouverensis* remains were found in four high-elevation cave sites dated from 830 to 2630 years ago (Nagorsen *et al.* 1996). Cut marks on bones and artifacts recovered at the site indicate the remains are a result of human hunting. Quite likely the aboriginal people traveled to the mountainous sites to hunt marmots.

Nagorsen *et al.* (1996) reviewed the evidence for native use of marmots along the Northwest Coast of North America. Skins of marmots, probably *M. caligata*, were reported by fur traders in the late eighteenth century as being in great abundance and they entered into the Hudson Bay Company fur trade in the mid-nineteenth century. Marmots were hunted in the fall after molt when they provided a light but finely

Table 1.1 Changes in marmot population based on information from archaeological sites (from Tomé and Chaix 2003).

Time period	Marmot status
Younger Dryas (about 14 500 to 11 500 years ago)	Marmots common during this cold period
11 000 to 10 000 BC	Marmots decreased from the Upper Paleolithic to Mesolithic because warming restricted them to higher elevations
After 7000 BC	Marmots absent from archaeological sites under 900 m

Table 1.2 The currently recognized species of *Marmota*.

Eurasia (Palearctic)		North America (Nearctic)	
M. baibacina	gray or Altai	*M. broweri*	Alaskan or arctic
M. bobak	steppe	*M. caligata*	hoary
M. camtschatica	black-capped	*M. flaviventris*	yellow-bellied
M. caudata	red or long-tailed	*M. monax*	woodchuck or groundhog
M. himalayana	Himalayan	*M. olympus*	olympic
M. kastschenkoi	forest-steppe		
M. marmota	alpine		
M. menzbieri	Menzbier's		
M. sibirica	Siberian, tarbagan or Mongolian		

Figure 1.1   A typical adult yellow-bellied marmot. (See plate section for color version.)

furred pelt prized for clothing. Marmot hides were used in potlatches (Drucker 1950). The hunters and their families went up into the mountains to their privately owned hunting grounds where they primarily used deadfall traps to capture large numbers of marmots and harvest the valuable furs. The skins were stretched and the flesh dried for winter use.

Among the Tlingit and Gitkson of the upper Skeena River, wealth was measured directly in *M. caligata* (hoary marmot) skins. Robes were made by sewing together many of the soft-furred hides (Nagorsen *et al.* 1996). Likewise, skins of *Marmota olympus* were used to make robes, bed blankets, or a seat when one had to sit on a damp or cold spot. Marmot flesh was regarded as excellent.

Humans have most certainly used marmots for food wherever humans and marmots co-existed. In *Les Miserables* Victor Hugo relates that in 1815, Jean Valjean, hungry and tired, entered an inn where he found a "fat marmot, flanked by white partridges and goose, was turning on a long spit before the fire." In Crested Butte, Colorado, during a miner's strike in 1913–14, the miners turned to hunting for meat and greatly reduced the numbers of yellow-bellied marmots (Warren 1916). The use of marmots as food continues into the twenty-first century; the old miners and trappers state that there is no better eating than a fat marmot in the fall.

Marmots and mythology

Marmots entered into mythology from early times and were often given names that were misunderstood. Animals reported as large ants dug up gold on the Dansar Plateau for the Persian Emperor 2500 years ago and these gold-digging ants were reported by Herodotus in the fifth century BC. The word in Persian for marmot is equivalent to "mountain ant" (Ramousse and Le Berre 2007). Local people collected earth from marmot burrows and sifted out the gold dust. This legend of gold-digging ants has a basis in reality, other legends, especially creation legends, represent mythological explanations for some aspects of marmot biology.

The following two legends are taken from accounts in Martin (1994). According to the Cherokee, in the old days animals talked and lived with people and marriages between humans and woodchucks were common. At that time, the woodchuck had a long tail. One day the woodchuck encountered a pack of wolves; in an attempt to gain time and perhaps escape from the wolves, the woodchuck offered to teach the wolves to dance. They all danced and as the woodchuck danced, it moved closer to a hole and dove in. The wolves quickly followed and grabbed the tail, which broke off. The woodchuck escaped but has had a short tail ever since.

Originally the Mohawk people and animals lived underground. One day a crack was noted and an individual went above ground and returned and told everyone what a beautiful place it was. All the people were happy to move to the surface and live above ground except the woodchuck, who was content to live in the ground.

A Pahute Indian legend tells how the whistler (marmot) and badger got their homes (Palmer 1973). The badger and the whistler were good friends. They traveled together;

the badger was stronger but the whistler was the better singer. Trouble arose when they both wanted the same wife and she preferred the badger's warm fur to the whistler's music.

But one day the whistler ran off with the wife. The badger was furious but could not run fast enough to catch them. In his anger and frustration he began to scratch and claw at the earth and dug a hole and sent rocks down the hillside. The hole caught fire and the fire grew bigger and threw out molten rocks. The badger fled the molten stream, but the whistler tried to stop it by blowing on it until the stream stopped. The whistler kept blowing all winter and the molten lava froze and broke up into big rocks.

Whistler believed his blowing had cooled and stopped the lava and led animals along the lava flow to show what he had done. He was very vain and the god Shinob decided only whistler liked all the black rocks and that they should be his home. From that time whistler's home has been in the rocks. He comes out on the big rocks and sits in the sun and whistles. As for the badger, he was punished for losing his temper and forced always to dig a hole in the ground.

The Mongolian people have a long relationship with the tarbagan (*M. sibirica*) that apparently began with cattle breeding on the Eurasian steppe (Bibikov and Rumiantsev 1996). The following legend shows that the people had excellent knowledge of tarbagan biology (Dimitriev *et al.* 2003). In ancient times there were seven suns. A drought set in, the soil became hot, domestic animals were exhausted, and it was impossible to sit or stand. The people requested a highly skilled hunter to shoot with his bow and arrows to decrease the number of suns in the sky. The hunter swore that if he couldn't strike the seven suns one by one that he would cut off a thumb and become a tarbagan that doesn't drink water or eat dry grass, and lives in a dark hole.

He began to shoot the suns, which were built in the sky from east to west. He shot six suns, but as he shot his seventh arrow, a swallow flew between him and the sun. The arrow cut a notch in the swallow's tail; ever since swallows have had forked tails. The sun, afraid of being shot, disappeared behind the West Mountains. The hunter became a tarbagan with only four fingers on the front legs and appears at his burrow at dawn and sunset to shoot the last sun. Because the sun hides to escape the hunter, day and night appeared.

One legendary behavior of the alpine marmot was described by several authors beginning in about AD 77 (Ramousse and Le Berre 2007). A marmot lying on its back, paws lifted up, holding herbs cut by other marmots, is dragged by its tail to its burrow. As a consequence of being dragged, the back is bare as the hair is worn off. Le Vasseur de Beauplan reported in 1660 that he watched the behavior several times. This legend persisted into the nineteenth century as writers used such cooperative behaviors among the animals as moral examples to humans.

The Sieur de Beauplan lived for many years in Ukraine and reported that some steppe marmots (*M. bobak*) were lazy and lay on their backs; the active marmots piled herbage on their bellies, which they held with their paws, and the active marmots dragged the lazy marmots to their burrows (Gudger 1935). There is a moral to the story; the lazy marmots were treated as slaves. Thus the dragging of marmots by their tails was used in two ways as moral lessons.

Figure 1.2 Yellow-bellied marmot tracks as a decoration on the window of the Tracks Restaurant in Leadville, Colorado.

Cultural use of marmots

In the last half of the twentieth century the marmot name became more widely used in human activities. In France, marmot was an infrequent surname but commonly used as a trading label for hotels, restaurants, and gifts, especially in alpine departments (Ramousse and Giboulet 2002). Similar use developed in North America. Thus, there is "La Marmotte Restaurant Francais" in Telluride, Colorado, and "Hotel Restaurant Marmotte" in Chartres, France. The "Whistle Pig Saloon" was located in Saratoga, Wyoming, and the "Tracks Restaurant" in Leadville, Colorado (Fig. 1.2). Outdoor clothing and equipment carry the name "Marmot" by Marmot Mountain Ltd. of Santa Rosa, California. And one can drink "Woodchuck Draft Cider" from Vermont.

Marmots grace fountains in Interlaken and Zermatt, Switzerland (Fig. 1.3). Marmots have been featured on stamps: the woodchuck on an United States stamp, the Vancouver Island marmot on a Canadian stamp, and the alpine marmot on a stamp from the Republic of Equatorial Guinea. Marmots are often overlooked in wildlife displays where the more charismatic megafauna, e.g., bears, wolves, deer, are featured. But the yellow-bellied marmot sits conspicuously on a rock in a mural at the Bannock County Historical Museum and Fort Hall in Pocatello, Idaho (Fig. 1.4).

Some communities have embraced marmots. In Crested Butte, Colorado, the Crested Butte Mountain Theater awarded the Golden Marmot for excellence in performance, staging, etc. One of the streets has a traffic warning that features a marmot (Fig. 1.5). And the "The Marmot Chronicles" (Zeidman 2000) takes readers on a tour of Crested Butte as the adventures of Elmo Marmot are recounted. Some animals and birds have animal characteristics, but Elmo and his other animal friends are highly anthropomorphized. Elmo and his wife Shirley live in a house with furniture, and Elmo is careful to attempt to avoid Shirley's displeasure. In another story the whistle pig wrangler rides a horse, but is rejected by the other animals and not allowed to join the roundup. He follows them anyway and spies coyotes sneaking up on the wranglers and their cattle. He gives a loud whistle, the coyotes flee, and the wranglers welcome the whistle pig into their group (Allen 1995).

Figure 1.3   Fountains in Switzerland decorated with marmot sculpture. Upper, Interlaken; Lower, Zermatt, photo courtesy of Tim Karels.

A charming set of drawings illustrate the trials and tribulations of Max, an alpine marmot (Giovannetti 1954), who attempts to do what humans do: fence, cook, ski, play the violin, smoke a pipe, all with unexpected consequences. When he decides that a mask and false nose make an inadequate costume, he paints a face on his belly, covers his head with a cone-shaped cloth, and saunters forth. Some of the scrapes Max gets into seem less

Figure 1.4   The mural in Pocatello, Idaho, with a conspicuous yellow-bellied marmot on rocks in the lower right. (See plate section for color version.)

Figure 1.5   The marmot traffic sign warning in Crested Butte, Colorado. (See plate section for color version.)

far-fetched after reading the adventures of a pet young marmot who manages to climb up on a table and plop himself in the middle of a large bowl of fruit salad prepared for a ladies bridge club (Van Wormer 1974). The book is well illustrated with the author's photographs and should be read before adopting a marmot as a pet.

Not all marmot stories are anthropomorphized. In "The Marmot Drive" (Hersey 1953), woodchucks are treated as woodchucks. The marmots form the setting in which conflicts in a small Connecticut town are played out. A young biologist, Pliny Forward, provides information as the town's people prepare to rouse the marmots from their burrows and chase them into a corral where they can be killed. The drive is held because the town selectman believes that the woodchucks have become too numerous and are damaging gardens. The drive is considered a failure, the selectman is punished, and human relationships are changed. But the woodchucks continue on as before.

Marmots have been used in advertisements. During World War II, The Travelers Insurance presented the hoary marmot as "Nature's Air Raid Wardens." The vigilant marmot sees a golden eagle and, like an air raid warden, gives the alarm and the marmots rush to their burrows. Their "homes" do not have rugs or toys to trip over nor stairs to fall down. Any one of these can cause a human accident; there is no warden to warn of the possible danger. Therefore, one should obtain accident insurance to eliminate worry and potential high costs.

In the August 1984 issue of National Geographic Magazine a Canon camera ad presents a picture of the endangered Vancouver Island marmot with the comment that the marmot could never be brought back if it completely vanished, but photography could record it for posterity. Furthermore, photography can assist in conservation efforts and provide faster, greater understanding of this marmot among people.

The traditional relationships between humans and marmots in the French Alps also had a strong degree of anthropomorphism; burrow systems resembled a real house and the marmots appeared very human (Dousset 1996). Marmots were used for meat, which was a main item in the diet. Marmot fat was used for lubrication and it was believed to be so powerful that it could penetrate and prevent a fracture from healing. Hence, fat was applied to the stomach of a pregnant woman to make delivery easier and to her breasts to maintain lactation. Captive marmots were used by itinerant musicians; dancing marmots performed as early as the mid-eighteenth century. The alpine marmot was the mascot of chimney sweepers. However, all of these activities gave way as marmots became more associated with tourism (Fourcade 1996). Decorated artifacts and traditional wood carvings of the past were supplanted with many objects: wooden toys, puzzles, glove puppets, music boxes, key chains, rubber toys, and wooden models with fur. But in some areas hand-carved marmots in the upright or vigilant position can be found. Post cards of marmots replaced live marmots, although some people still keep tame marmots. Eating marmot meat has declined and been replaced by eating pastries or chocolates shaped like or named "marmots." Finally, cakes, spirits, wine, and aromatic herbs are labeled with graphic representation or the name of the marmot, and are widely available in tourist areas.

In Mongolia marmot bones were found in archaeological sites and the tarbagan is widely utilized at the present time. Mongolian hunters developed a costume that was used during the hunting of marmots. The traditional costume consists of a white jacket, white trousers, and a white hat with ears that stick up in the air (Formozov *et al.* 1996). An important feature is a tassel or daluur made from the tail of a white horse or yak. The tassel dangles from a small wooden handle and is shaken as the hunter stalks the marmot. The aroused marmot calls but does not flee. The hunter drops on all fours as he imitates the approach of a canine predator. As I will describe in a later chapter, marmots tend to monitor the activity of a predator and may flee only when the threat is great. The daluur stimulates the marmot to watch and call until the hunter gets close enough to shoot the marmot. The hunters shoot only marmots that call, thus avoiding marmots infected with plague as they are unlikely to call. Marmots have developed responses; many do not call when they see a person but enter a burrow.

Marmot meat is baked in the skin or with noodles, meat and vegetables are baked in a cauldron, or meat is incorporated into a steamed or fried pastry (Kolesnikov *et al.* 2009a). The valuable skin is used for manufacturing various fur products. A fresh-flayed skin has medicinal use; the outer fur is placed to treat myositis. The fat (oil) may be sold locally in southern Mongolia for $1.00 to $3.00/liter. The fat is used mainly by hunters and their relatives for healing wounds, burns, and frostbite and is also used to treat saddle sores on horses. On average, a Mongolian family uses 105 marmots per season (Kolesnikov *et al.* 2009a). The Mongolians use the doctrine of similarity; i.e., they treat an organ by eating a similar organ. For example, to treat nephritis, they eat marmot kidney.

Use of marmots is influenced by religious practices; Mongolian Muslims do not eat marmot meat (Bibikov and Rumiantsev 1996, Kolesnikov *et al.* 2009a). In Transbaikalia, the Buryats, a cattle-breeding people, settled in areas inhabited by the tarbagan and used it for food. The fur was used for caps and coats and fat and gall were used in traditional medicine (Badmaev 1996). The Evenko were hunters and reindeer breeders who lived in the northern taiga area. They hunted black-capped marmots (*M. camtschatica*). Both groups gathered in special places for religious ceremonies and hunting was banned in these areas. In effect, the religious bans on hunting protected the marmots and they persist on mountains that are sacred.

Groundhog day

The best-known cultural event in North America involving marmots is Groundhog Day, February 2, which has been observed every year since 1898 at Punxsutawney, Pennsylvania (Fig. 1.6). This event raises two questions: why the groundhog and why February 2? The equinoxes and solstices marked the seasons for the Sumerians and Babylonians and this tradition was passed through the Greeks and Romans to our culture (Kalapos 2006). However, in Celtic tradition, the seasons are measured to start in the middle and not the beginning of equinoxes and solstices; thus spring starts during a major Celtic festival, Imbolc, which means "in the belly" and symbolizes that life persists in the womb of the earth. The Irish celebrated this festival as Brigid's Night.

Brigid was known as the goddess of healing and birth and bonfires were lit to symbolize the heat and warmth of the sun (Kalapos 2006). Christianity could not banish her influence

Figure 1.6   The statue of Punxsutawney Phil. Photo courtesy of John Koprowski.

and she was made a saint in the fifth century. In 1604, February 2 became a church holiday, Candlemas Day, to commemorate the purification of the Virgin Mary (Martin 1994, Kwiecinski 1998). Fire as a symbol was retained in the form of lighted candles, which were carried in a procession around the medieval churches.

Early farmers needed some indication of whether winter would continue or whether spring planting could begin. What better sign than a hibernating mammal that awakened from its slumber and emerged above ground to start a new season. But if that animal saw its shadow on a clear, cold day, winter would continue for another 6 weeks;

no shadow foretold an early spring. In Europe, hedgehogs and badgers were the prognosticators.

When the early colonists came to North America, their traditional animals were absent, but they discovered another burrow-dweller, the woodchuck or groundhog, which became the prophet. Although Punxsutawney Phil is the most famous prognosticator and the central prophet in the movie *Groundhog Day*, other prognosticating woodchucks joined the clan: Dunkirk Dave of New York, General Beauregard Lee of Georgia, Wiarton Willie of Ontario, Pee Wee from Vermont, among others. Not surprisingly, these prophets predict different scenarios; some predict an early spring and others, six more weeks of winter.

I wrote the following song to the tune of "Oh Tannenbaum."

> *Oh Murmeltier*
> Oh Murmeltier, oh Murmeltier
> We celebrate your famous day.
> Oh Murmeltier, to you we pray
> That winter soon will go away
> We like the sun and daffodils
> We've had too much of winter's chills
> Oh groundhog friend we're warning you
> If winter stays, you'll be woodchuck stew!

The threat of becoming stew meat is in keeping with a long tradition of eating woodchuck. Recipes can be found in many cookbooks; e.g., the *Pioneer Cook Book* by Ruth Stone contains directions for fried and baked woodchuck.

Why is Candlemas Day called Groundhog Day rather than Marmot Day or Woodchuck Day? Originally marmot was used for many of the ground-dwelling sciurids and probably lacked the specific designation required for a prophet. The name is most likely related to the origins of the words woodchuck and groundhog. Woodchuck has no relationship to wood or chuck, but is probably derived from the Algonquin language. The Cree word for *M. monax* is otchek and the Ojibwa called it otchig (Kwiecinski 1998). It is likely that early English speakers substituted syllables that sounded more familiar but only approximated the original sound, hence woodchuck. But woodchuck has no connotation of living in a burrow. By contrast, a groundhog or earth pig is a literal translation of the Dutch word aardvark and does refer to a burrower.

Woodchuck and groundhog are the sources of common names for mountain marmots. The yellow-bellied marmot is associated with rocks, not wood, hence the name rockchuck is often used. The groundhog is relatively quiet, whereas the yellow-bellied marmot calls frequently. Thus groundhog or earth pig is changed to whistle pig. And the designation "whistler" refers to the whistle-like call emitted by marmots.

Marmots and research

Marmots were little studied in North America prior to the second half of the twentieth century. Several species, primarily the woodchuck and yellow-bellied marmot, were used to study physiological processes during hibernation (Lyman *et al.* 1982), to describe environmental factors affecting hibernation (Davis 1967a), and to establish the role of the endogenous annual rhythm in hibernation (Davis 1976). The woodchuck serves as a model for the study of hepatitis B virus (HBV). In this species, the life cycle of HBV, virus–host interactions, virus-related carcinogenesis, and anti-viral strategies can be investigated (Cova *et al.* 2003).

Marmot research has a much longer history in Eurasia, especially in the former Soviet Union. As early as the late nineteenth century epidemics of plague (*Yersinia pestis*) were associated with marmots, especially *M. baibacina*, *M. caudata*, and *M. sibirica* (Pole 2003). Consequently, anti-plague laboratories were opened to determine the frequency and location of plague epidemics. Natural plague foci were established in the mountains of Central Asia and the widespread marmot flea, *Oropsylla silantiewi*, was implicated as a key plague vector (Ageev and Pole 1996). Because people harvested marmots for their fur, fat, and meat, plague was readily transmitted via infected fleas from marmots to humans. Marmot mortality is usually limited to local areas; only 2–3% of the marmots in a plague epizootic are infected (Bibikow 1996). However, plague epidemics and human mortalities were frequent, and major projects were undertaken to control plague by destroying marmots (see Chapter 21). Although cases of plague are uncommon, concern over a major outbreak continues. When the earthquake of April 2010 struck western China, disease specialists acted quickly to cordon off plague-infected marmot burrows and ban marmot hunting (Stone 2010).

During World War II, marmots of the French Alps were exchanged with Italians for rice (Louis *et al.* 2002). The widespread use of marmots as food kept thousands of people from starvation during years of famine, collectivization, and post-war devastation (Bibikow 1996). Local extermination of marmots often resulted. Thus, programs were established to develop sustained harvesting through regulating hunting, establishing reserves, and reintroducing marmots to areas where they had been exterminated (Bibikow and Rumiantsev 1996). However, commercial hunting of the tarbagan, *M. sibirica*, threatens the existence of this species. Tarbagans are hunted for their pelts and over a million skins were prepared in some years (Batbold 2002b). Recent surveys indicate a severe decline that led to a 2-year hunting ban being imposed throughout Mongolia (Townsend and Zahler 2006). This situation in Mongolia exemplifies a modern environmental dilemma: harvesting marmots provides a critical economic benefit to the Mongolian people, but the activity threatens the survival of the species providing the benefit.

Several characteristics of Marmotini (the spermophiles, prairie dogs, and marmots) biology made them especially useful for behavioral and population studies. They live in permanent burrows, thus the residents of a population can be easily located and readily trapped and tagged for identification, and most importantly, they are diurnal and the above-ground activities of individuals of known age and sex can be directly observed (see Murie and Michener 1984). Some of the early population studies of North American marmots investigated the effects of exploitation on woodchuck demography (Davis *et al.* 1964) and documented an 80% decline in woodchuck numbers that were attributed to a decline in the quality of food (Davis and Ludwig 1981).

I became interested in the yellow-bellied marmot as a consequence of my participation as a graduate student at the University of Wisconsin in a mammalogy and ornithology journal club. Each week we read and discussed recent publications. The paper that stimulated my interest in marmots was by John Calhoun (1952). In this paper, Calhoun observed that population dynamics are generally "treated from the demographic point of view." Population growth or decline depended on the interrelationships of mortality, migration, and natality. Food, disease, and climatic conditions were viewed as the factors controlling population density. Calhoun went on to suggest that more attention should be directed to the role of the individual or other phenomena intrinsic to the group in which every individual lives. The possible influence of sociality on population dynamics was described for a population of Norway rats (*Rattus norvegicus*) in which population growth ceased despite a super abundance of food and available space. Population growth was altered by social behavior as expressed through socially stable or unstable groups that formed the functional units within the larger population.

The idea that population dynamics might be treated from a social behavior rather than a demographic perspective was also expressed by Paul Errington (1956) who argued that "we must not ignore the role of social intolerance" as a factor limiting higher vertebrate populations. The role of social behavior in limiting population growth was developed into a model by V. C. Wynne-Edwards (1962, 1965). Briefly, the model was based on a feedback loop. Individuals tracked population density by participating in epideictic displays, which conditioned the participants to population pressure. The displays would become more intense as a local population increased. This social intensity caused a decrease in reproduction by reducing litter size and/or the number of individuals reproducing. Mortality could also increase if social behavior forced individuals to move into areas where food was inadequate and predation more likely. Population decline was followed by a decreased intensity of social behaviors and this new information permitted an increase in reproduction and population numbers to be followed eventually by a repeat of the processes causing decline. The degree of oscillation in this feedback system depended on the degree of fine-tuning of the behavioral system. Because an animal population is divided into small, local groups, those groups that best develop the self-regulating mechanism will persist and groups with poorly developed regulation will die out. This group selection model of Wynne-Edwards was quickly challenged on theoretical grounds (Williams 1966). In essence, Williams argued that individuals acted to perpetuate their genes in the population and not to benefit the species. However, a rejection of group selection does not preclude a role for social behavior in population dynamics. Whether group or individual selection occurs depends fundamentally on what individuals do. I began my research on yellow-bellied marmots to determine what individuals did.

At the time of the Calhoun publication, I was working in the summer in Yellowstone National Park. I became familiar with yellow-bellied marmots and realized that they had three characteristics that suggested they would be a good animal for exploring the relationship between social behavior and population dynamics. First, they lived in groups; thus, there was the potential that social behavior affected group size. Second, their living in permanent burrows meant that they were localized and could be easily found from day to day. Furthermore, the marmots could be readily trapped at their burrows and that

Figure 1.7   Marmot research site in Yellowstone National Park, Wyoming. Marmots lived along the bank and in the meadow along the river. Nest burrows in the meadow were under old bridge foundations or in rocky areas.

eliminated the need for establishing extensive trapping grids. Third, the marmots are diurnal; their behavior and movements could be directly observed rather than inferred from trapping records.

My initial research was conducted through the Jackson Hole Research Station adjacent to Teton National Park. I located a colony of yellow-bellied marmots in Yellowstone National Park along a river terrace on the east side of the Snake River (Fig. 1.7) at an elevation of about 2177 m. Because the river was an effective barrier against human intrusions into the colony, human disturbance of the marmots was limited to the occasional fisherman working his fly-rod from the western shore. The Yellowstone Park staff at the south entrance kindly allowed me to use their overhead chair and pulley system to cross the river a few hundred meters downstream from my study site. Thus, I was saved from the risk of wading the river.

During this initial study, we developed our basic methods. We live-trapped virtually every marmot at the site. Upon first capture, we inserted a numbered monel metal size 3 self-piercing strap tag in each ear. For each marmot we recorded body mass, sex, and reproductive condition. Each marmot was individually marked with a pattern of black stripes or blots with Nyanzol A non-toxic fur dye. The mark was lost when the marmots molted in mid-summer; thus, marmots were re-trapped for re-marking. For further details on trapping and marking, see Armitage 1982a. Because the ear tags provided each marmot with a permanent identification, we could associate young with mothers, determine age, and work out patterns of kinship, e.g., aunts, cousins, half-siblings, etc. Because we could associate young with their mother, we could record genealogies that persisted through time in a given location. We named these mother/daughter kin groups matrilines (Armitage 1984). The study of kinship and matrilines eventually initiated a change in focus from population regulation to a consideration of the strategies marmots used to attempt to maximize evolutionary success.

The initial 4-year study at the Yellowstone colony provided important basic information concerning the behavior and biology of the yellow-bellied marmot (Armitage 1962, 1965). There was never more than one adult male present in the population; thus marmots lived in a harem-polygynous social system. Thirty-one different adult females were identified; the number present in any one year varied from 8 to 17. Although many females lived singly, groups ranging from two to five shared a burrow system. Females living in groups were concentrated in the center of the colony and all of the females utilizing peripheral burrows lived singly. Most of the reproduction occurred in the centrally located burrows; only 2 of 19 litters were weaned by a female occupying a peripheral burrow. Multiple females sharing a burrow system suggested that female sub-groups form within the colony, but the significance of this grouping was not determined until years later.

Social behaviors were classified broadly as amicable or cohesive (affiliative) and agonistic or dispersive. Amicable behaviors consisted of allogrooming, when one animal chewed skin and fur, usually around the shoulders or back of another, and greeting, when two marmots approached each other head-on and sniffed at each other's cheeks. Agonistic behaviors varied in intensity. Alert behavior occurred when one marmot became tense and focused its attention on a nearby animal and relaxed only when the other animal moved away. During alert behavior, the heart rate may increase about 9.5% above resting rate when no other activity is evident (Armitage 2003b). Alert behavior often was followed by flight when the nearby animal moved closer to the alert marmot, who fled to its burrow or across the landscape. Avoidance was characterized by one marmot moving toward another, then changing directions in order to bypass the other marmot and avoid both close proximity and direct conflict. Avoidance also was evident when one animal frequented part of her home range only when another, dominant marmot was absent. In some instances, a marmot discontinued using an area where the dominant animal was frequently active. Finally, the most intense form of agonistic behavior occurred when one marmot chased another. A chase may occur when one marmot runs toward, attacks, or lunges at another causing the attackee to flee with the attacker in pursuit. Chase also may occur when two animals face each other with bodies curved and flat to the ground and tails held low and one suddenly flees and the other pursues. Frequently the fleeing marmot emitted an alarm call immediately before taking flight. Sometimes one marmot lay flat while the other stood up and stared at the recumbent individual (Fig. 1.8). Further details of marmot behavior are described in Armitage (2003a).

Behavioral interactions were often more complex than the above description implies, as illustrated by the following observation. Two females occupied the same burrow system. The smaller female had a litter and was subordinate to the larger female. One day the larger adult was chasing the smaller adult in a circle around the group of burrow openings when the smaller female suddenly darted into one of the entrances. The larger female then moved out into the meadow, laid down in a patch of clover and began feeding while facing away from the burrow area. The smaller female emerged, looked around, focused on the larger female, moved quickly to her, nipped her on her rump, and dashed back to the burrow and entered. The larger female looked up, then resumed feeding. This observation suggested that marmots might have "personalities," a topic that will be explored in a later chapter.

Other observations suggested that females recognized each other and acted in ways that indicated an expectation of seeing another animal. A dominant 5-year-old female occupied the same burrow complex, an old bridge foundation (Fig. 1.9), as a subordinate 4-year old. The subordinate female usually emerged earlier in the day and went into the meadow where she spent most of the day. When the dominant female emerged, she moved to a corner of the foundation and peered around the corner at the entrance used by the subordinate female. If the subordinate female was present, the dominant female

Figure 1.8   Two adult females in conflict postures. The one to the left lies flat in the "ready-alert" position while the other female stares at her from an upright position.

Figure 1.9   The old bridge foundation that was the central burrow area at the Yellowstone colony. Note the young marmots sitting on the top logs. Burrow entrances were dug under the bottom logs.

dashed toward her and she fled. If the subordinate female was not present, the dominant female turned away and went into the meadow and fed. When the subordinate female returned from foraging, she slowly approached the burrow area and frequently slunk on her belly. If the dominant female was present, the subordinate female turned and ran back into the meadow. If the dominant female was not in view, the subordinate female rose and dashed to her burrow entrance. On many days she was unable to return to her burrow for most of the day. As a possible consequence of this behavior, she failed to wean the young that were born in late May (Armitage 1965).

Interestingly, two other females, who copulated with the male and who were subordinate to the same dominant female, failed to wean litters. Also, the agonistic behavior of the dominant female caused two other adults to emigrate (Armitage 1962). These results suggested that social behavior played a role in population dynamics. However, there was no support for a density-dependent feedback model. The number of litters was more closely related to the number of adult females rather than the expected inverse relationship. Nor was there any relationship between the number of adult females and mean litter size, which varied among years from 4.25 to 6.0. And, contrary to my expectations, the strongest behavioral effects occurred when the number of adults was at its lowest.

In conclusion, the results of this early 4-year study supported a role for social behavior in population dynamics but did not support the density-dependent feedback model. Clearly more research was needed to determine what factors affected reproductive success when only about half of the females weaned litters each year. Were females responding to the density of marmots in their immediate vicinity? Or were females engaged in competition for reproductive success (as suggested by the conflict between the dominant female and the females subordinate to her)? And how did this competition translate into population dynamics? To answer these and other questions, I decided that several marmot colonies should be studied each year in order to account for the effects of habitat differences on social and population dynamics and to look for those factors associated with reproductive success across all females and all habitats. In order to have ready access to several marmot colonies, I switched my research in 1962 and initiated a 41-year project in the Upper East River Valley, Gunnison County, Colorado (Fig. 1.10), where I worked at the Rocky Mountain Biological Laboratory (RMBL).

Figure 1.10 The Upper East River Valley, Gunnison County, Colorado. Yellow-bellied marmots occupy rocky outcrops and talus in meadows.

RMBL is located on the site of the old mining town of Gothic (38° 57′ N, 106° 69′ W) at 2900 m elevation. RMBL provided housing, laboratory space, and proximity to marmot colonies; no colony was more than 5 km from the townsite. The following chapters describe how social, physiological, and environmental factors affect reproductive success and population dynamics in a yellow-bellied marmot metapopulation living in a harsh environment. Throughout the book, relevant information from other marmot species is included.

2 Marmots: their history and diversity

This chapter explains the diversity and relationships among the species of marmots in order to provide a historical framework for examining the differences and similarities in their social systems and the ways by which they cope with environmental stress.

The first description of a marmot (probably *M. caudata*) occurred in about 1000 BC and the second description (probably *M. bobak*) appeared in Aristotle in 350 BC under the name of white mouse of Pontus in his chapter on hibernation (Ramousse and Le Berre 2007). The alpine marmot (*M. marmota*) was described by Pliny in AD 77 as *Mus alpinus* and translated as the rat of the Alps or alpine mouse or "marmottanes" in English or "marmottanne" in French by the seventeenth century. The word "marmot" first appeared in the twelfth and thirteenth centuries in reference to itinerant Savoyards with their marmots or in medicinal recipes promoting the use of marmot fat. "Marmot" subsequently went through several spellings in the sixteenth and seventeenth centuries and finally became standardized as "marmotte" in a French dictionary in 1694. The word was applied to other species of marmots by the middle of the eighteenth century. The general use of the French word was adopted as marmot in the English language (Ramousse and Le Berre 2007).

In a late nineteenth century popular book on the animal kingdom, "marmots" was used as a general heading for chipmunks, prairie dogs, spermophiles, and marmots, which were placed in the genus *Arctomys* (Fig. 2.1; Craig 1880). The woodchuck is referred to as the pouched marmot, but no explanation is given for the use of "pouched." Some animals were classified wrongly as marmots. For example, Thunberg (1796) classified the cape dune molerat (*Bathyergus suillus*) as *M. africana*. In his Folio finished in 1848, Audubon painted Lewis's marmot (*Arctomys lewisii*); in 1915 Oldfield Thomas declared the specimen Audubon used was a species of marmot (Ford 1951). Audubon was familiar with marmots; he drew a groundhog in 1805. However, Lewis's marmot is now recognized as the white-tailed prairie dog (*Cynomys leucurus*).

Currently 15 species of marmots, restricted to the northern hemisphere (Fig. 2.2), are recognized (Brandler and Lyapunova 2009, Steppan *et al.* 2011). The most recent compilation of mammalian species of the world recognized 14 species (Hoffmann *et al.* 1993). An analysis of marmot karyotypes reported that *M. baibacina* had 38 chromosomes, but one subspecies had 36 chromosomes. This population is geographically distinct from other gray marmots, has morphological differences, and occupies different habitats including forests. This subspecies was recognized as the fifteenth marmot species, *M. kastschenkoi* (Brandler 2003, Brandler *et al.* 2008).

POUCHED MARMOT
PRAIRIE DOG   COMMON MARMOT
LEOPARD SPERMOPHILE   BABAC

PLATE LVII RODENTIA

Figure 2.1   The "marmots," squirrels that burrow in the ground. The pouched marmot or woodchuck appears to be a reversed image of the painting of a woodchuck by Audubon. It was common in the nineteenth century for popular books of this type to borrow published images. Figure from the New York Public Library digital collection courtesy of K. C. Armitage. (See plate section for color version.)

In their comparative analysis of molecular phylogenies Brandler and Lyapunova (2009) concluded that North American species are more distinct than Eurasian. Two examples illustrate speciation problems among the Eurasian marmots. The three subspecies of the black-capped marmot (*M. camtschatica*) differ in fur color, and body and skull dimensions.

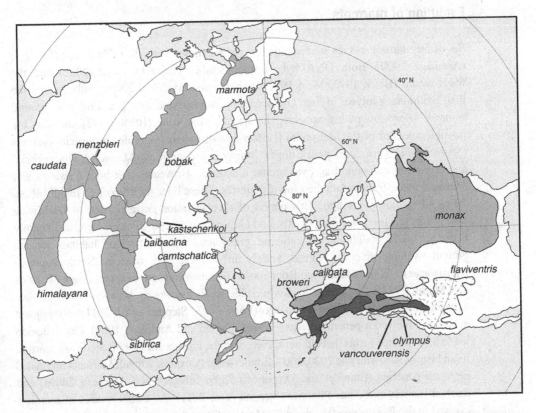

Figure 2.2 Polar-view map showing Holarctic distribution of the 15 species of *Marmota*. Darkest shading (*M. caligata*) and stippling (*M. flaviventris*) distinguish two species with large overlapping ranges in western North America. Courtesy of Scott Steppan.

All have identical karyotypes; $2N = 38$, $NF_A = 62$ (Lyapunova *et al.* 1992) but alarm calls and immunogenetic responses differ between *M. camtschatica camtschatica* and *M. c. doppelmayeri* (Zholnerovskaya *et al.* 1992). These differences led Kapitonov (1978, cited in Boyeskorov *et al.* 1996) to suggest that these two subspecies should be considered separate species. Diversification in this species is so extensive that Boyeskorov *et al.* (1996) suggested that the black-capped marmot should be considered a superspecies. Recent analysis using inter-SINE PCR data supports the independence of *M. c. camtschatica*, but *M. c. bungei* and *M. c. doppelmayeri* cluster together (Brandler *et al.* 2010).

The long-tailed or red marmot (*M. caudata*) consists of two subspecies that differ in several traits. *M. c. caudata* is larger, has a longer intestine, and is darker than *M. c. aurea*. These differences were attributed to adaptation to local conditions of food and climate (Davydov 1991). The alarm calls of the two subspecies differ (Nikol'skii *et al.* 1999). These subspecific differences were facilitated by glaciation that effectively separated the two subspecies. Distinct alarm calls were a significant factor that led to *Urocitellus richardsonii* and *U. elegans* to be considered separate species rather than conspecifics (Zegers 1984). On this basis, it seems that the two subspecies of the long-tailed marmot could be raised to species.

Evolution of marmots

Six of the marmot species are Nearctic and the remaining species are Palearctic (Fig. 2.2) (Armitage 2000). Both DNA hybridization studies (Giboulet *et al.* 2002) and the fossil record (Black 1972, Mein 1992, Erbajeva and Alexeeva 2009a) indicate that the first Sciuridae evolved in the Oligocene in North America. The first true ground squirrel, *Miospermophilus*, appeared in the late Oligocene (Black 1972) and was the possible ancestor of the Marmotini (Hafner 1984): the spermophiles or ground squirrels (*Spermophilus sensu lato*), prairie dogs (*Cynomys*), and marmots (*Marmota*). Phylogenetic analysis using mitochondrial cytochrome b sequences revealed that both *Cynomys* and *Marmota* arose from within the ground squirrels. These lineages diversified rapidly about 10 to 14 million years ago. This period of diversification corresponded to worldwide changes in sea level, tectonic events, and climate; paleo-temperatures decreased sharply between 14–12 million years ago and grasslands, which provided habitats for the ground-dwelling sciurids, spread widely in Eurasia and western North America (Harrison *et al.* 2003, Mercer and Roth 2003). Thus marmots are of recent origin (Thomas and Martin 1993); the earliest incontrovertible fossil marmot is *M. minor* from the late Miocene, about 8–10 million years ago (Polly 2003, Steppan *et al.* 2011). Subsequent evolution involved a general increase in size (Kurtén and Anderson 1980). One radiation led to large-bodied forms that did not survive into the Pleistocene (for discussion of marmot fossil history, see Armitage 2000, 2007, Erbajeva and Alexeeva 2009a). Marmots originated in North America (Brandler and Lyapunova 2009, Erbajeva and Alexeeva 2009a) and reached Eurasia in the late Pliocene (Steppan *et al.* 2011). All extant species of marmots evolved in the Pleistocene (Black 1972, Mein 1992), with one major speciation activity in North America and a second in Central Asia (Brandler 2007). Speciation in Eurasian marmots is associated with an epi-platformal orogeny area where eight of nine species are completely or partially located. An "epi-platformal orogeny is the process of mountain building on the territory of smoothed relief, which proceeded in a platform mode for a long time" (Nikol'skii and Rumiantsev 2012). The growing isolation of marmot populations as the orogeny occurred likely was a major factor in their diversification.

The marmots form a monophyletic group with two major clades (Kruckenhauser *et al.* 1998, Steppan *et al.* 1999, 2011, Brandler and Lyapunova 2009). One clade, consisting of *M. flaviventris*, *M. olympus* (Olympic marmot), *M. caligata* (hoary marmot), and *M. vancouverensis* (Vancouver Island marmot), occurs in western North America. This clade forms the subgenus *Petromarmota* (Fig. 2.3). The widespread woodchuck (*M. monax*) is a sister to the remaining species, all of whom, except for the Alaskan marmot (*M. broweri*), occur in Eurasia (Steppan *et al.* 1999) and are placed in the subgenus *Marmota*. The woodchuck may be the closest relative to the ancestral marmot and clearly is the most primitive living marmot (Brandler and Lyapunova 2009, Brandler *et al.* 2010). Presumably the woodchuck split off from the ancestral marmot and remained in North America while its sister group diversified in Eurasia. *M. broweri* in the Brooks Range of Alaska is another ancient species and did not cluster with either clade (Brandler *et al.* 2010). Its relationship to other marmot species remains obscure, but it could be a sister species of *M. monax* (Steppan *et al.* 2011).

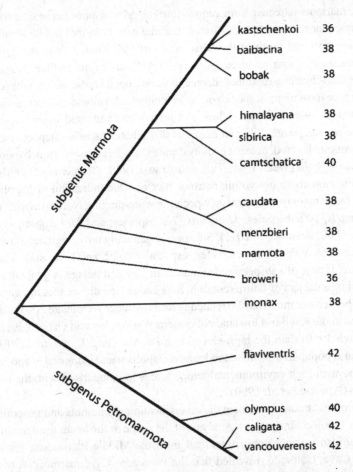

kastschenkoi	36
baibacina	38
bobak	38
himalayana	38
sibirica	38
camtschatica	40
caudata	38
menzbieri	38
marmota	38
broweri	36
monax	38
flaviventris	42
olympus	40
caligata	42
vancouverensis	42

Figure 2.3   Phylogeny of *Marmota*. The two heavy axes represent two major clades and the thin lines identify the species groups. The position of the thin lines along the major axes indicates the relative time of species diversification. The 2*n* chromosome number of each species is recorded in the right column.

Early marmots probably inhabited moist, cool habitats of the North American west (Polly 2003) and occurred as far south as Mexico (Cushing 1945). In Europe, marmots were associated with the tundra–forest–steppe fauna of the periglacial landscape (Zimina and Gerasimov 1973, Zimina 1996) and were widespread in lower mountain ranges (Kalthoff 1999a). The environment was characterized by short, warm summers and cold winters in a grassy landscape (Zimina and Gerasimov 1973). In Transbaikalia, there was a tendency for an increase in body size during the cooler Middle and Late Pleistocene; e.g., *M. nekipelovi*, and for a decrease thereafter (Erbaeva 2003, Erbajeva and Alexeeva 2009b).

Marmot distribution changed dramatically with warming and advance of the forests at the end of the last glaciation. *M. flaviventris* now lives in Colorado, where *M. monax* was present about 750 000 years ago, and retreated from low elevation southwesterly sites in California (Goodwin 1989) and other southerly locations (Polly 2003, Armitage 2007).

Similarly, marmots retreated from central Europe: *M. marmota* became restricted to the higher Alps; marmots, such as *M. bobak*, became extinct in the middle Rhine region as reforestation occurred north of the Alps (Kalthoff 1999b); and there was some shift to smaller size as warming occurred (Aimar 1992). Over several million years, the large-bodied, ground-dwelling marmots adapted to a cool, open landscape (Armitage 2007). As a consequence most marmot species occur in montane and alpine environments except for the woodchuck, which occupies low elevation woodland–field ecotones (Kwiecinski 1998), and the steppe and Siberian marmots that inhabit Eurasian steppes. Because cool environments are favored, marmot activity during the active season is much more restricted by heat than by cold (Webb 1980, Turk and Arnold 1988, Melcher *et al.* 1990).

Diversification continues within marmot species. One indication of ongoing diversi-fication is the number of described subspecies in wide-ranging North American marmots: *M. flaviventris*, 10 subspecies; *M. monax*, 7 subspecies; and *M. caligata*, 3 subspecies (Armitage 2000, Braun *et al.* 2011). Subspecies generally are designated on the basis of skull dimensions, pelage characteristics, especially color, and body size. The adaptive significance, if any, of subspecific differences in size and pelage is generally unknown and should be a subject of future research. Size and pelage differences in two subspecies of yellow-bellied marmots may be related to heat tolerance (Armitage 2005). *M. f. avara*, which lives in the semi-arid lowlands of eastern Washington and Oregon, is smaller and paler (Couch 1930) than the high-elevation mesic-dwelling *M. f. luteola* of Colorado. The semi-arid population exhibited reduced metabolism and increased evaporative water loss to cope with high environmental temperatures in comparison with the more mesic population (Armitage *et al.* 1990).

Morphological studies document the diversification of marmots and generally support the results of molecular analyses. Analysis of the shape of the lower third molar of extant North American marmot populations and from the Middle Pleistocene Pit locality in Porcupine Cave, Colorado, revealed that the Porcupine Cave marmots were related to *M. monax* and molar shape could distinguish between *M. m. monax* populations from Indiana and Virginia (Polly 2003). All woodchuck samples formed a single clade whereas the yellow-bellied, hoary, and Vancouver Island marmots formed a second clade. These results support the interpretation that the woodchuck is not the direct ancestor of the other North American marmots. The analysis of molar shape divided *M. flaviventris* into two clades, which supports the premise that diversification continues at the population level.

Cardini and his collaborators investigated patterns of morphological diversity and evolution in the marmot skull and mandible. The geometry of the marmot mandible is similar in *M. flaviventris* and *M. caligata* and differs from the other species (Cardini 2003, 2004). The mandible morphology supports monophyly of *Marmota* and the proposed subgenera. Both size and shape differed significantly across species; body mass was a good predictor of mandible size. The mandible undergoes ontogenetic changes; the morphology changes mainly in the first 2 years after birth in yellow-bellied marmots (Cardini and Tongiorgi 2003). The male mandible is larger than the female's, which is consistent with the larger body size of the male yellow-bellied marmot. Shape changes during growth produce a stronger and more efficient structure with increased surface for muscle insertion. Relationships between size and shape were examined in two species of *Petromarmota* and four species of *Marmota*$_{sg}$. Anatomical regions affected by size-related shape variation were similar among species, but allometric trajectories were divergent. The allometric patterns were produced primarily by size variation, epigenetic factors, and developmental constraints. No phylogenetic signal was evident in the allometric patterns (Cardini and O'Higgins 2005). However, traits that discriminate between the Nearctic and Paleoarctic marmots appear early in development and are conserved during post-natal ontogeny. Analyses of the size and shape of the cranium reveal highly significant species differences and clustering of species that is congruent with the two subgenera except that the woodchuck, with its relatively flat skull, is distinct among the marmots (Cardini and O'Higgins 2004, Cardini *et al.* 2005, Cardini and Thorington 2006).

M. vancouverensis illustrates how little we understand the forces producing morphological differences. It is the only insular marmot species, has low genetic diversity (Kruckenhauser *et al.* 2009), and has only minor divergence in DNA sequences from mainland *M. caligata* and *M. olympus*. The Vancouver Island marmot is a recent form (Steppan *et al.* 2011) that evolved a distinct dark pelage (Armitage 2003a), characteristic vocalizations (Blumstein 1999), and mandible and skull characteristics atypical of the genus (Cardini *et al.* 2007). An analysis of the hemimandibles from modern and Holocene subfossil *M. vancouverensis* indicates that morphological divergence occurred during a period of rapid change after isolation followed by little change in the recent past (Nagorsen and Cardini 2009). Factors contributing to the morphological divergence of this species likely included founder effects, genetic drift, or selection pressure associated with environmental change.

How do we account for the differences in cranial and mandible morphology? Some of the differences may be related to phylogeny and body size, but there is no congruence in cranial characteristics due to size similarities among the marmots (Cardini and O'Higgins 2004, Cardini *et al.* 2005). Body size and phylogeny do not account for the marked difference in the woodchuck cranium or for other differences such as the intermediate morphology of *M. broweri* and *M. olympus*. It is tempting to think that the morphological differences represent independent evolutionary events that adapted marmot species to their varied environments. But it is difficult to postulate probable selective factors. The type of food chosen by marmots appears to be quite similar among marmot species; forbs are preferred and supplemented with grasses (Armitage 2000). It also seems unlikely that predator defense can account for the diversity; marmots everywhere had to cope with canine predators as well as eagles, bears, and cougars (Bibikow 1996). The contribution of phylogenetic and environmental (local vegetation, diet, elevation, body size, temperature, precipitation) components to skull, mandible, and molar shape variation was examined in Eurasian marmots (Caumul and Polly 2005). Path analysis revealed that body size (10%), vegetation (7%), diet (25%), and mtDNA (15%) explained 57% of skull shape; by contrast, only 35% of mandible shape, which was not associated with diet or local vegetation, was explained. The low phylogenetic signal supports the interpretation that as marmots radiated into the Eurasian landscape, morphological changes occurred as environments differed. The environmental analyses are very broad and a refinement of climatic factors and diet categories could lead to a better evaluation of the importance of various environmental variables. Understanding the evolution of skull morphology remains a major problem for future research.

Marmot phylogeny

Both molecular and morphological analyses agree that the genus *Marmota* is monophyletic and that there are two major clades (Fig. 2.3). This interpretation is supported by the Nei and Li genetic distance (D_{NL}). In general, the Palearctic marmots have lower interspecific differences (D_{NL} varied from 0.07 to 0.22) than Nearctic marmots (D_{NL} varied from 0.18 to 0.29), which suggests that the Palearctic clade is younger. The mean genetic distances between the two groups are generally higher (D_{NL} varied from 0.21 to 0.35) than those within groups (Brandler *et al.* 2010).

However, different molecular analyses produce somewhat different groupings. What follows is my attempt to reconcile the differences and construct a phylogenetic diagram based on the phylogenies of Steppan *et al.* (1999, 2011) and recent reviews and analyses by Brandler *et al.* (2008, 2010) and Brandler and Lyapunova (2009). There is universal agreement that *M. monax* has the most ancestral form of the extant marmot species and is positioned at the base of the subgenus *Marmota* clade. Also, all trees agree that *M. marmota* is the oldest of the Palearctic species. *M. broweri* is now considered an ancient species and could be a sister species of *M. monax* (Steppan *et al.* 2011).

Karyotype analysis and molecular analyses provide similar groupings but also some differences. For example, *M. caudata*, *M. menzbieri*, and *M. himalayana* have similar karyotypes and comparison of G-banded sex chromosomes adds *M. bobak* to this group. By contrast, most molecular trees group *M. sibirica*, *M. himalayana*, and *M. camtschatica*. Molecular evidence indicates an ancient hybridization between *M. caudata* and *M. menzbieri*; these two species generally group together and usually are considered to be an early radiation. This grouping is strongly supported by immunogenetic studies in which *M. menzbieri* is most similar to Kazakhstan *M. caudata* and least similar to *M. sibirica* (Zholnerovskaya and Ermolaev 1996).

The *bobak* group consisting of *M. bobak*, *M. baibacina*, and *M. kastschenkoi* is clearly differentiated from the other Palearctic species and *M. kastschenkoi* is considered to be the youngest member of the cluster and is a recent derivative of *M. baibacina*. There is considerable differentiation between *M. b. centralis*, the Tien Shan population, and the Altain *M. b. bobak*, which suggests incipient speciation.

The *M. camtschatica* group consists of *M. sibirica*, *M. camtschatica*, and *M. himalayana*. Both *M. sibirica* and *M. camtschatica* have similarities with the fossil *M. tologoica*, which may be ancestral to both species. This relationship is also supported by the presence of endoparasites common to *M. sibirica* and *M. c. doppelmayeri* (Erbaeva 2003). Karyological and paleontological data suggest that *M. himalayana* is the more ancestral form and *M. camtschatica*, the youngest, but some molecular analyses group *M. sibirica* with *M. himalayana*, an interpretation supported by chromosome number (Fig. 2.3).

There is general agreement that *M. flaviventris* has more ancestral traits and is the oldest Nearctic species. The similarity in chromosome number and geographic proximity support the molecular analyses that *M. caligata* and *M. vancouverensis* are closely related and *M. olympus* diverged from the *M. caligata* group about 2.6 million years ago (Steppan *et al.* 2011).

Body size

Body size affects the expression of many morphological, physiological (Calder 1984), and life-history variables (Armitage 1981). I will first consider body-size effects in marmots in the context of their patterns in the Sciuridae (Table 2.1).

Among all squirrels, body length and body mass are closely correlated with body length (Table 2.1); tail length relative to body length varies widely (Hayssen 2008a). The short tails of the Marmotini apparently are long enough to provide support during upright vigilance and short enough to reduce drag while running. Among all squirrels, Marmotini tend to be heavier and *Marmota* are the largest members of this tribe.

Reproductive effort is affected by body size. Marmotines have larger litter masses than other sciurids, but there is no allometric relationship with body mass (Hayssen 2008b). Gestation length and duration of lactation are short in the Marmotini (Table 2.1) and they devote the least time among squirrels to the sum of gestation and lactation times. Body mass accounts for most of the variation in neonatal and litter mass across all squirrels; larger species have absolutely larger but relatively smaller litter mass at birth. Larger litters have smaller individual neonates across all squirrels. A key factor that influences reproductive effort in marmotines is that most hibernate and have a short time span in which to reproduce. Thus, the major constraint on reproduction is the length of time that resources are available (Armitage 1981). One adaptation to time limitation is that the marmotines have the fastest growth rates during gestation and lactation, measured as g/day/g adult mass (Hayssen 2008b).

The Marmotini have a specialized reproductive profile characterized by annual reproduction, small-mass offspring, fast growth rate, and many offspring (Hayssen 2008c). Not all genera follow the pattern of the tribe. For *Marmota*, there is no body-mass effect on the duration of lactation and in some species gestation may be longer than lactation.

Table 2.1 Comparison of body-size effects on marmots and other Sciuridae.

The body-mass:body-length correlation explains about 98% of the variation in the Marmotini.
Tail length relative to body length varies widely; Marmotini have shorter tails relative to body length than all other squirrels.
Other than Marmotini, gestation length increases with body size, but body mass of females accounts for only 11% of the variation in marmotines. Body mass explains only 4.7% of the variation in lactation time and the sum of gestation and lactation times is unrelated to body mass in marmotines.
Marmotini has the highest litter mass; relative litter mass is larger in smaller sciurids. Neonatal mass and litter size are positively correlated with adult female mass, $R^2 = 94\%$ for marmotines.

Although body mass explains 91% of the variation for weaning mass (individual) and litter mass at weaning across the Marmotini, the relationship is flat for *Marmota* and gestation and weaning mass are not correlated. *Marmota* produce relatively smaller young and have the smallest energetic investment of all squirrels (based on mass of young as proportion of adult female mass). Because marmots evolved in a harsh environment and large body size is highly adaptive to their life histories, marmot biology has produced similar solutions to a common problem.

The body mass of marmots changes throughout the year as a feature of an endogenous circannual rhythm (Ward and Armitage 1981a). Body mass is minimal at about the time of emergence from hibernation and is maximal at the time of immergence into hibernation (Table 2.2). Emergence mass is useful for comparing body sizes of marmots because at emergence all marmots are at a similar stage of the body-mass cycle and typically are at their lowest annual mass. Body size (measured as mass or body length) seems unrelated

Table 2.2 Body mass of adult females and time in hibernation of *Marmota* species. Classification follows that proposed by Steppan *et al.* (1999). All data from Armitage and Blumstein 2002, except as noted.

Species	Body mass (g)		Length of hibernation (months)
	Immergence	Emergence	
Subgenus *Petromarmota*			
flaviventris	3431	2422	7.5
caligata group			
caligata	6187	3283	7.5
vancouverensis	5328	3899	7.0[a]
olympus[b]	5550	4110	7.5
Subgenus *Marmota*			
monax[c]	4804	3084	4.5
marmota	3987	2825	6.5
broweri[d]	3094	2055	7.5
caudata group			
caudata	3923	2631	7.6
menzbieri	3760	2321	8.0
bobak group			
bobak	4120	2910	7.7
baibacina	5583	3978	7.0
kastschenkoi[e]	4600	3450	7.5
camtschatica group			
camtschatica	4748	2900	8.2
himalayana	6420	3445	7.5
sibirica	3960	2550	6.6

[a] Bryant and McAdie 2003
[b] Griffin, pers. com.
[c] Kwiecinski 1998
[d] Lee *et al.* 2009
[e] Taranenko, pers. com.

to phylogeny; both large and small marmot species occur in both subgenera and there can be a wide range of body sizes within a group; e.g., the *M. camtschatica* group. Although there is about a 1.8-fold difference in the body mass of the largest and smallest species at emergence, all marmot species are large and are by far the largest members of the Marmotini. The large body size of the marmots raises two questions: (1) what are the costs and benefits of large size and (2) what accounts for the differences in body size among marmots?

Large body size has three obvious benefits. First, large size enables marmots to escape small-sized predators. I observed the steppe marmot in Ukraine chase adult red foxes (*Vulpes vulpes*) and harass young foxes and drive them into their burrows. Yellow-bellied marmots vigorously chase long-tailed weasels (*Mustela frenata*) and marten (*Martes americana*) whenever they are sighted (Waring 1965, Armitage 2003g). However, large size may make marmots more attractive to large predators such as badgers (*Taxidea taxus*) (Armitage 2004a), coyotes (*Canis latrans*) (Van Vuren 2001), cougars (*Puma concolor*), and wolves (*Canis lupus*) (Armitage 2003g). Large predators cannot specialize on marmots as marmots are available as prey only during their active season of about 4.5 months, on average. Thus wolves prey on hoary marmots primarily when caribou (*Rangifer tarandus*) are scarce (Murie 1944).

Second, larger body size enables marmots to utilize a more fibrous diet. Marmots have a dentition and simple digestive tract best suited for a diet of seeds, but they are herbivorous and feed on a variety of forbs and grasses and may rarely feed on tree leaves and buds (Bibikow 1996, Armitage 2003e, Olson *et al.* 2003). Mammals that are small, less than 10 kg adult body mass, may be caecal fermenters and microbial fermentation is mainly confined to an expanded caecum. However, the main substrates used in fermentation are not plant cell walls, but cell contents that escaped digestion in the intestine (Hume 2002). Thus a critical factor is mean retention time (MRT) of food in the digestive tract. MRT increases with increasing body size; at the body size of marmots, MRT exceeds 24 h and enables marmots to expand their dietary niche.

Specializing on herbaceous vegetation coupled with large body size apparently enables marmots to hibernate for up to 8 months (Table 2.2). Prairie dogs also are herbivorous and may weigh about 1.5 kg at maximal mass in the annual cycle. However, not all species of prairie dogs hibernate and those that do hibernate for about 4 to 5 months (Hoogland 2003). Not all spermophiles hibernate and some species may hibernate as long as marmots. There is considerable diversity in the annual cycle of these relatively small Marmotini and they utilize high-energy seeds to a greater degree than marmots or prairie dogs (see Yensen and Sherman 2003, for a discussion of spermophile biology). Apparently, a complex relationship exists among body size, diet, and hibernation in this group of mammals. As we will note later, there can be high energy demands during the hibernation period of marmots which do not occur in the other groups and which are met through the advantages of large size. (Note: it is important to distinguish between hibernation, the period during which torpor occurs, and hibernation period, the time spent underground between immergence and emergence.)

Third, large body size increases energetic efficiency (French 1986). Energy (fat) stores scale directly with body mass; energy use scales to mass$^{3/4}$ at environmental temperatures within thermoneutrality and to body mass$^{1/2}$ at colder temperatures. Larger marmots accumulate more fat and use it relatively more slowly. This efficiency probably explains why body mass and body length of many Palearctic marmot species increase from south to north and with elevation above sea level (Armitage 2005).

Some of the costs of body size include greater total energy demand, which is met in part by use of forbs and grasses, and therefore affects home range area; a delay in reproductive maturity; and reproductive skipping. These topics will be discussed in the next chapter.

Fur color diversity

Fur color varies both within and among marmot species (Fig 2.4), but shades of brown and gray, often with spots, splotches, or streaks of white, black, yellow, or reddish brown predominate on the dorsal surface (Armitage 2003a). Albinism and melanism were reported for six and seven species, respectively (Armitage 2009a). Albinism was first reported for *M. flaviventris* in Colorado in 2006 (Fig. 2.5) and is the only documented instance of albinism in this species. The frequency of albinism in other species is probably rare, but was reported as frequent in some populations of *M. monax* and *M. baibacina* (Armitage 2009a). Melanism is more common and may persist in some populations of *M. monax* and *M. flaviventris* (Fig. 2.6). White marmots were previously reported for *M. monax*, *M. marmota*, *M. baibacina*, and *M. bobak* (Armitage 2009a), and for *M. flaviventris* on Mt Massive, Colorado, by an unrecorded photographer in 2009. In 2010, white marmots were reported along the south entrance road in Yellowstone National Park (Fig. 2.7) and at two locations on Mt Massive. At one location, the white young was associated with two adults and a young of normal color (Fig. 2.8) and at the other location the white marmot could be an adult male (Fig. 2.9, left). White coloration probably represents a loss of pigment and some reduction of pigment may be common in yellow-bellied marmots as light-colored individuals are frequent (Fig. 2.9, right). Four species have fur color markedly divergent from the typical shade: *M. caudata*, red; *M. vancouverensis*, dark brown; *M. caligata*, white; and *M. baibacina*, gray (Fig. 2.4).

Fur affects heat transfer between a marmot and its environment. Heat transfer through fur depends on fur structure (hair length, diameter, and density and fur depth), on fur spectral properties (reflectivity and absorptivity), and on the thermal environment (Melcher 1987). Wind speed, air temperature, and the radiant environment affect heat transfer, which is more sensitive to changes in solar radiation and far less sensitive to changes in wind velocity or ambient temperature (Melcher *et al.* 1990).

Metabolic rates calculated from predictions of heat loss based on fur properties corresponded closely with measured values for adult and young, especially when evaporative water loss was included (Melcher 1987, Armitage 2009a). The fur properties did not account for short-term physiological adjustments, but the essential points are that pelage plays a preeminent role in heat transfer between a marmot and its environment,

Figure 2.4   Fur color variation in marmots. (a) Typical *M. flaviventris.* (b) *M. olympus.* (c) *M. caligata.* (d) *M. vancouverensis.* (e) *M. monax.* (f) *M. marmota.* (g) *M. caudata.* (h) *M. baibacina.* (i) *M. camtschatica.* (j) *M. himalayana*, photo courtesy of A. Nikol'skii. (See plate section for color version.)

Figure 2.5   An albino yellow-bellied marmot. Photo courtesy of Frank Drumm.

Figure 2.6   A melanistic yellow-bellied marmot.

that heat transfer affects metabolic rate, and heat transfer is most sensitive to changes in solar radiation.

Under the intense solar radiation characteristic of montane and alpine environments, heat transfer can be toward the skin and can reduce heat required for thermoregulation or thermally stress an animal. At midday, clear with little wind, heat gain on that part of a marmot's surface receiving direct solar radiation can be up to 10 times basal heat production (Melcher 1987). Marmots can affect heat absorption by the way they orient

Figure 2.7   A young white yellow-bellied marmot. Photo courtesy of Heth Aiken and Jennifer Teti. (See plate section for color version.)

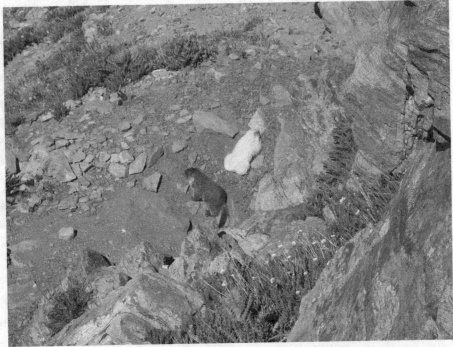

Figure 2.8   Normal and white young sibling yellow-bellied marmots. Photo courtesy of Nicholas Gianoutsos. (See plate section for color version.)

Figure 2.9   Left: An adult white yellow-bellied marmot at about 3600 m. Photo courtesy of Troy Helm. Right: A light-colored yellow-bellied marmot. Photo courtesy of Karole Armitage.

Figure 2.10   Yellow-bellied marmots with bodies angled toward the sun in the early morning. Drawing by Sara Taliaferro from author's photograph.

their bodies toward the sun. Thus, early in the morning when solar heat gain could reduce heat production for thermoregulation, yellow-bellied marmots orient their bodies perpendicular to the solar rays (Fig. 2.10). However, when solar radiation is stressful, heat absorption is reduced by orienting the body parallel to the sun's rays (Fig. 2.11).

Figure 2.11   Yellow-bellied marmot with body oriented longitudinally toward the sun. Drawing by Sara Taliaferro from author's photograph.

Also, when sitting or lying on rocks, marmots reduce the proportion of the body angled toward the sun as the thermal environment increases in intensity (Armitage 2009a).

The primary role of fur color in this heat transfer system is to affect absorption of solar energy. In general, light fur color decreases and dark fur color increases absorption of radiation primarily in the visual range. Marmots may take advantage of solar radiation to reduce metabolism; the lowest heart rate of a free-ranging yellow-bellied marmot was recorded when the marmot was lying in the sun (Armitage 2003b). Variability in fur properties is likely a source of adaptability to climatic variation (Dawson and Maloney 2008). I hypothesized that fur color functions primarily in heat transfer. This hypothesis is supported by light-colored populations of *M. flaviventris*, *M. himalayana* (Kholodna, pers. com.), and *M. c. aurea* (Davydov 1991) that occupy semi-arid environments.

How do we account for both light-colored and dark-colored species of marmots that seem to live under similar conditions; e.g., *M. vancouverensis* and *M. caligata* (Fig. 2.4) (Armitage 2000)? In brief, we know very little about the fur properties of marmots and the thermal regime in which they live. Extensive research is required to determine the factors responsible for fur color diversity.

3 Marmot habitats

Marmots evolved in cool, moist environments in western North America and in the periglacial landscape in Eurasia. Global warming in the 12 000 years since the last glaciation resulted in most species of marmots retreating to higher elevations (Chapter 2). Living in their environment profoundly affected the evolution of sociality (Chapter 5), marmot physiology (Chapter 7), and habitat selection. The characteristics of marmot habitats are the subject of this chapter.

Distribution patterns

General patterns of marmot distribution, especially of Eurasian species, were summarized by Bibikow (1996). The avoidance of woodlands by marmots produces a patchy distribution, and a preference for open terrain with "short" vegetation. Three settlement types or patterns are recognized. In the steppe or diffuse settlement, marmots are distributed rather evenly in a landscape that is uniform over a vast area. This pattern was more extensive in the past before human activity divided the landscape into agricultural and industrial uses (Rumyantsev *et al.* 2012). A species typical of this settlement type is the steppe marmot, *M. bobak* (Fig. 3.1). There is no equivalent to the marmot steppe settlement in North America. The black-tailed prairie dog (*Cynomys ludovicianus*) inhabiting the North American prairie, is the ecological equivalent of the Eurasian steppe marmot.

The second type is the belt or ravine settlement. Marmot colonies occur as strips along one of the slopes and the colonies are separated by unsuitable habitat. In the third type, local or mosaic settlement, suitable habitat occurs in small patches. For example, black-capped marmots occur in glacial cirques above timberline (Tokarsky and Vasiljev 1991). This type is especially prominent in the mountain landscape where marmots occupy alpine or subalpine meadows (Fig. 3.2); e.g., the Vancouver Island marmot occupies small and scattered habitat patches, each typically containing fewer than five adults (Bryant and Janz 1996). Marmots are absent in mountainous terrain continuously free from snow or where snow cover is maintained for a long time in the spring. Indeed, snow cover plays a prominent role in marmot biology, as will be discussed in several sections of this book.

Figure 3.1   *M. bobak* in typical steppe vegetation in Ukraine.

Figure 3.2   *M. olympus* lying on its "porch" in a small patch of meadow. The trees in the background establish the limits of the habitat (also see Fig. 3.7).

Habitat use

For the most part, marmots occupy open habitats, such as the Eurasian steppe, mountain meadows in forest zones, and alpine meadows (Armitage 2000). A striking exception is the forest-steppe marmot, *M. kastschenkoi*, which likely is the only forest-dwelling marmot. This species occupies broad-leaved and coniferous forests, tolerating limited view sheds and high vegetation cover (Ricankova *et al.* 2013). *M. kastschenkoi* is a member of the

M. baibacina complex; *M. b. centralis* occurs in a spruce forest zone but it is not known if it occupies grassland patches in the forest zone or actually occupies forests. If grassland patches are occupied, they should be of considerable size. Yellow-bellied marmots may settle on small patches with meadow vegetation in spruce–aspen forest; occupancy is brief as recruitment does not occur and local extinction follows. Old unoccupied yellow-bellied marmot burrows were found in meadows undergoing colonization by lodgepole pine (*Pinus contorta*); presumably the increasing tree cover rendered the habitat unsatisfactory.

M. b. baibacina of the southern Altai represent the population from which *M. kastschenkoi* evolved. The key factors that influenced habitat use by this subspecies were examined in an attempt to reveal the possible origin of forest dwelling in this complex. On both small and large scales, the marmots preferred grasslands and clearly avoided woodlands and tall vegetation, but were present on sites with tree cover up to 10% (Ricankova *et al.* 2013). The strong rejection of woodlands suggests that along with speciation in this complex, habitat utilization also evolved. An extensive analysis of habitat requirements of the forest zone *M. b. centralis* and the forest-steppe marmot may clarify how marmots live in this unusual habitat (Ricankova *et al.* 2013), but it is likely that research on behavioral patterns will be necessary.

Marmots do not occupy all of what appears to be suitable habitat. *M. bobak* formerly occupied flat steppes, but available habitat was reduced due to plowing of the steppe between 1950 and 1960. Currently, the steppe marmot occupies gullies (Fig. 3.3), small flat-bottom valleys, pastures, and agricultural fields and prefers south-facing slopes (Mashkin 1991, Tokarski *et al.* 1991) or open fields with good visibility and lush vegetation (Mashkin *et al.* 1994). Areas that have suitable vegetation may not support marmots for reasons such as high levels of subsoil water, subsurface bedrock, or soils too fine to support burrows (Rumiantsev 1991).

Figure 3.3   *M. bobak* habitat along a gulley in the Chuvash Republic, Russia.

The tarbagan (*M. sibirica*) also occupies steppe areas of gently sloping uplands covered with forb–grass vegetation but is not present on steppes with sagebrush or on forested slopes (Seredneva 1991). The tarbagan occurs at elevations of 1600 to 2000 m. The lower limit is set by desert steppes. At the upper limit, the species is restricted to well-drained relief in swampy river valleys and to south-facing slopes; distribution is limited by snow cover that persists beyond the time of the onset of marmot activity in the spring (Suntsov and Suntsova 1991). In some areas, tarbagan occur above 3300 m in the alpine and occupy large-stone screes (Rogovin 1992).

The other low elevation marmot is the woodchuck (*M. monax*), which prefers flat or slightly rolling land (Hamilton 1934). This species occupies a woodland/meadow eco-tone; wooded areas are preferred for hibernation and fields for foraging (Kwiecinski 1998). There is considerable plasticity in the choice of habitat; woodchucks may occupy gullies and creek banks where most dens are located on slopes of 34% or less in rocky soil (Twitchell 1939). The woodchuck expanded westward in the twentieth century along wooded riparian streams in Kansas where the river system provided the forest/grassland edge habitats this species prefers (Roehrs and Genoways 2004).

These steppe and forest-edge species do not encounter the more severe environment of the remaining mountain species. All 15 species share two environmental conditions: (1) a season in which their food plants are either unavailable or have a greatly reduced nutritional quality due to senescence of shoots and leaves; and (2) prolonged snow cover such that marmots may emerge before the snow melts; e.g., the steppe marmot (Le Berre *et al.* 1994).

Mountain environments in the northern hemisphere are highly variable but those areas inhabited by marmots can be broadly characterized as having short, often cool summers and prolonged winters with extensive snow cover. Within this broad environmental framework, marmots live under various conditions. For many species, data are minimal, but support the concept that marmots live in a harsh environment (Table 3.1). The Himalayan marmot lives at 3100 to 3400 m elevation where the mean temperature of the warmest month is 9° to 12°C (Huang *et al.* 1986). The two subspecies of the long-tailed marmot, *M. c. caudata* and *M. c. aurea*, occupy those slopes of the mountain ranges that are covered by steppe, meadow, and drought-resistant semi-savannah vegetation.

Table 3.1 Evidence for the harshness of the environment. Modified from Armitage and Blumstein (2002).

M. flaviventris	High mortality in juveniles and reproductive adults in hibernation following summer drought
M. marmota	Can reproduce in successive years if litter size no more than two
M. baibacina	25% of all embryos reabsorbed in bad years, occurs in half of the females: reproductive females accumulate 525 g, barren females, 1100 g of fat before hibernation
M. bobak	Accumulate fat more rapidly in moist years, more juvenile mortality during hibernation following a drought year
M. camtschatica	Female breeds only after a year of good feeding
M. sibirica	Reproductive females accumulate 414 g of fat, barren females 747g before hibernation
M. caudata	Up to 48% of embryos reabsorbed when emergence conditions are poor; e.g., a cold spring; no young emerged in one year in 15 social groups

Vast spaces are not occupied by marmots because of prolonged snow cover (Davydov 1991). Some of the harshest environments are encountered by the black-capped marmot, *M. camtschatica*, where typical habitats are mountain tundra or steppe tundra where permafrost occurs (Lukovtsev and Yasiliev 1992) or glacial cirques above timberline characterized by long winters and cold summers or in mountain valleys that are snow free only 3.5 to 4 months of the year (Zheleznov 1991, Mosolov and Tokarsky 1994). This species also occurs along coasts in the subalpine zone in steep cliffs mainly on the south, west, and southwest slopes (Zheleznov 1996).

Because marmots moved to higher elevations as a consequence of post-glacial warming, the elevation at which marmots live has been a focus of habitat evaluation. For example, the widespread yellow-bellied marmot (Armitage 2003b) occurs over a wide range of elevations (Table 3.2). The presence of rocky slopes or outcrops makes widespread settlement possible. Yellow-bellied marmots can colonize lower elevations, which typically are characterized by warmer, drier environments if food with sufficient water content is available. Such a situation occurs in Capitol Reef National Park in Utah. The old Mormon settlement of Fruita was established in 1880 when farms and orchards were created along the Fremont River at an elevation of about 1600 m. By 1959, the area was completely absorbed into the park. Part of the park program is to preserve some of the history of the area and orchards and irrigated lawns are maintained. Although the average precipitation is only about 180 mm and summer temperatures range from about 16°C at night to above 30°C by day, yellow-bellied marmots colonized this desert environment.

Table 3.2 The range of elevations occupied by yellow-bellied and alpine marmots.

Yellow-bellied marmot (*Marmota flaviventris*)
Great Basin of western North America (Floyd 2004)
    Mainly in rocky meadows on well-drained slopes between 2100 m and 3000 m
    Observed as low as 1500 m
Utah (Long 1940)
    A few marmots occur as low as 1500 m to 1800 m; usually occur above 2100 m
Washington and Oregon (Couch 1930, Webb 1980)
    Lowland populations in rocky locations in inland valleys as low as 800 m
Northwestern Montana (Wright and Conaway 1950)
    On rock outcroppings at about 1000 m
Colorado (Armstrong 1972)
    Occur over elevations ranging from about 1500 m to 4000 m

Alpine marmot (*Marmota marmota*)
Italy: Southeastern Alps (Chiesura Corona 1992)
    Preferred elevation is 1700 m to 2100 m
Aosta Valley (Bassano *et al.* 1992a)
    Preferred elevation is 2100 m to 2500 m
Western Alps (Macchi *et al.* 1992)
    Preferred elevation 1650 m to 2800 m; most den systems above timberline at 2200 m to 2600 m
Bavaria (Arnold *et al.* 1991)
    At 1200 m

Figure 3.4   *M. flaviventris* burrow site on the desert slope at Capitol Reef National Park, Utah.

Burrows typically occur on the desert slopes (Fig. 3.4); but marmots move down to the irrigated meadows where they forage (Fig. 3.5) or into the orchards (Fig. 3.6) where they feed on the fruit. Presumably the available plants provide sufficient water for the marmots' needs. Research is needed to determine if the Capitol Reef population has developed water balance and temperature regulation mechanisms that differ from those of montane marmots, as described for yellow-bellied marmots living under similar conditions in eastern Washington (Armitage *et al.* 1990; see Chapter 9).

As indicated for the yellow-bellied marmot, elevations where marmots occur vary in different parts of the range of the alpine marmot (Table 3.2). In these locations in Italy marmots preferred south-facing slopes (also see Panseri 1992) and slopes of 10° to 30°. In Bavaria (Table 3.2), alpine marmots occur where farmers clear areas for grazing and create meadows where several marmot families form a colony (Fig. 3.7). These habitats may disappear as trees invade the meadows if the farmers do not keep them cleared.

Vancouver Island marmots occur at an elevation of 1110 m to 1450 m (Nagorsen 1987), lower than those typical of mountain marmots. The marmots occur on steep slopes (Fig. 3.8) where avalanches and snow creep retard the invasion of trees (Milko 1984).

Elevation profoundly affects emergence times. *M. baibacina* emerge 20 to 30 days later at high elevations than at lower elevations. The young emerge and immerge later at high elevations and at the same time of the year weigh about 300 g less than the lower elevation young (Pole 1996). This shift in timing makes the higher elevation marmots more susceptible to late snow cover and probably indicates why the higher elevations are populated less frequently than expected.

Figure 3.5   *M. flaviventris* foraging on an irrigated meadow at Capitol Reef National Park. The burrow area is on the hillside at the rear.

Figure 3.6   *M. flaviventris* burrow area (foreground) with an orchard in the middle ground where marmots forage and feed on the fruit at Capitol Reef National Park.

Figure 3.7 *M. marmota* on a habitat patch in Berchtesgaden, Bavaria, Germany.

Figure 3.8 Steep slope habitat of *M. vancouverensis* on Vancouver Island.

Table 3.3 Characteristics of colonization of the Spanish Pyrenees following alpine marmot introductions.

Most colonies between 1800 m and 2400 m; range of occupancy from 1200 m to 2800 m.
  Maximum elevation of marmot colonies correlated with maximum elevation of the massif
  (Herrero *et al.* 1992).
Marmots used less area than expected below 1600 m and above 2600 m.
Marmots used more area than expected in the subalpine zone between 1800 m and 2400 m.
Marmots used areas between 1600 m and 1800 m and between 2400 m and 2600 m as expected
  (Herrero *et al.* 1994).

The colonization of the Spanish Pyrenees by alpine marmots that spread from introductions in the French Pyrenees in 1948 is instructive as marmots were free to choose where to settle, although expansion was hindered by human activity (Herrero *et al.* 1987). The lower range of occupancy was facilitated by human activity that created meadows where marmots settled. The distribution of marmots differed from the expected distribution (Table 3.3), but habitat suitability analysis accounts for the observed pattern (Barrio *et al.* 2012). Similarly, alpine marmots in the Orobic Alps were more frequent at elevations that were less available (Frigerio *et al.* 1996). Thus, elevation is a necessary but not a sufficient factor to explain marmot habitat use.

Slope affects where marmots live. When marmots have a choice, they prefer low or medium slopes even when medium to high slopes are most available (Frigerio *et al.* 1996). Finally, sun exposure is critical and may interact with slope. In general, marmots prefer sites with a southern or eastern aspect, where snow melts relatively early in the spring (Barash 1973a, Allainé *et al.* 1994). Alpine marmots emerge on average 11 days earlier on south-facing than on north-facing slopes, which provides more time for growth (Farand *et al.* 2002). The probability of finding marmots on moderate slopes with a southern aspect was 0.94 (Rodrigue *et al.* 1992). Annual phenology of *M. baibacina* may vary 10 to 15 days within the same elevation belt depending on whether burrows are on northern or southern slopes (Pole 1996). This variation in phenology is one example of the phenotypic plasticity that enables marmots to occupy diverse environments. From the previous discussion, we conclude that marmots prefer meadows at higher elevations with moderate, south-facing slopes (if not south-facing, where snow melt is early).

Marmots may occur in habitats that are not preferred or be missing from what appears to be satisfactory habitat. One important factor affecting occupancy appears to be population density. During recolonization by alpine marmots in the Eastern Italian Alps, more recent populations had a more restrictive habitat and slope selection than older populations. The more recent populations preferred habitat types that were more strongly selected (preferred) by the older population; hence, the high density of the older populations restricted settlement of more recent populations, which as a consequence, were less dense (Borgo 2003). Borgo suggested that habitat selection by marmots is in accordance with ideal free distribution theory (Fretwell 1972). Early colonizers were free to choose preferred habitat and as the population increased, marmots that could not settle in preferred habitat then settled in less-preferred sites. Similarly, yellow-bellied marmots after a population decline repopulated colony sites before eventually occupying satellite or secondary sites (Svendsen 1974, Armitage 2003c). However, yellow-bellied marmots (Armitage 1991a) and alpine marmots (Sala *et al.* 1992, Arnold 1993b, Perrin *et al.* 1993b) defend their territories. Thus, the pattern of habitat choice may be an example of ideal despotic distribution (Fretwell 1972). Regardless of which model is appropriate, clearly marmots occur in less favorable habitats because preferred habitats are occupied.

One consequence of settling in less-preferred habitat is the possibility of increased exposure to predators. An analysis of persistence of yellow-bellied marmot colonies indicated that factors associated with visibility and safety correctly classified sites that were persistently or intermittently used (Blumstein *et al.* 2006b). Where predator detection is obstructed (e.g., by shrubs), yellow-bellied marmots do not persist at or fail to settle in such a site. The importance of an open landscape in habitat selection is strongly supported by a radio-telemetric study of predation on yellow-bellied marmots. One cluster of predation events occurred where the habitat was favorable in that suitable burrows and foraging areas were present, but patches of willows and groves of spruce or aspen provided concealment for predators, such as coyotes (*Canis latrans*) (Van Vuren 2001). Observed predation of coyotes on marmots occurred where the coyote hid in tall vegetation until a marmot came close and the coyote dashed from hiding to kill the marmot (Armitage 1982b).

Habitat use is affected by apparent interspecific competition. The gray marmot and the tarbagan are sympatric in the Mongolian Altai and are segregated by habitat differences. The two species cannot be reliably distinguished in the field by external appearance, but can be identified by the difference in their alarm calls. Most calls from gray marmots come from rock outcrops with boulder screes; tarbagan calls come from all biotopes. Gray marmots tend to be localized in the upper half of slopes whereas tarbgan occur from the valley bottom to the crest of ridges. Outside the area of sympatry, gray marmots occupy various biotopes and there is no apparent difference in habitat use by the two species. Possibly in the area of sympatry the tarbagan restricts the gray marmot to screes (Brandler *et al.* 2010).

Where sympatry occurs with the yellow-bellied marmot in northern Montana, the hoary marmot inhabits the high-elevation alpine habitats and the yellow-bellied marmot occurs at lower elevations even though it occupies alpine habitats where the hoary marmot is absent. In east–central Alaska hoary marmots are restricted to talus and the sympatric woodchuck primarily occurs in arid lowland river valleys (Braun *et al.* 2011).

To summarize, marmots mitigate the potential costs of life in a harsh environment by selecting habitats that minimize the impact of prolonged snow cover primarily by settling where the period of snow cover during the active season is minimal. Although negative effects of the abiotic environment can be reduced by habitat selection, for most marmot species they cannot be eliminated with the consequence that environmental harshness has major effects on life-history traits, especially on reproduction and preparation for hibernation (Table 3.1).

Environmental harshness

Environmental harshness provides a unifying theme that integrates all life-history traits, including body size. Environmental harshness, or environmental severity, is not readily defined and can include factors such as rainfall (drought), environmental temperature, social stress, predation risk, and length of the growing season (Barash 1989). Time is fundamental to understanding environmental harshness; e.g., the time required for marmots to accumulate sufficient body mass for hibernation, dispersal, and reproduction (Barash 1974a, Armitage 2012). Time spent in hibernation may decrease reproductive success; a delay in emergence time of 15 to 20 days resulted in no juveniles in a population of *M. camtschatica* (Valentsev *et al.* 1996). Because climate records for areas where marmot species live are generally unavailable, several biological features, which provide evidence for a harsh environment, will be emphasized (Armitage and Blumstein 2002).

Marmots may be placed into one of three groups based on the frequency of reproductive skipping (Table 3.4). Of the 10 species that skip reproduction, those in group 2 are larger (mean ± SE) than those in group 1 (immergence mass (g): group 1 = 4308 ± 312, group 2 = 5120 ± 411; emergence mass: group 1 = 3005 ± 284, group 2 = 3276 ± 311). Group 2 marmots lose more mass and a larger percentage of immergence mass during hibernation than group 1 animals and have a shorter active season (Table 3.4). Only mass loss is statistically significant (t = 1.9, 0.1 > p > 0.05). The trend is consistent; marmots that suffer more frequent reproductive skipping are larger at immergence and emergence, lose more body mass and a larger percentage of body mass during hibernation, and have a shorter active season. This relationship between body mass and reproductive skipping demonstrates a trade-off to large body size; large size makes habitation of the harsh environment possible but at a cost of a reduced frequency of reproduction.

The length of the active season contributes to environmental harshness and reproductive skipping. *M. menzbieri* is one of the smallest marmots, but has one of the shortest active seasons; *M. camtschatica* is the second smallest of group 2 marmots and has the shortest active season. Both species lose considerable mass during hibernation and have less time during the active season to gain sufficient mass for hibernation and reproduction. As a consequence, two or more growing seasons may be required to accumulate sufficient fat reserves for reproduction. By contrast, of the two species that do not typically skip reproduction, *M. monax*, one of the larger species, has the longest active season, which provides sufficient time to accumulate the necessary fat.

The active season of *M. flaviventris* is about average for the *Marmota* and its mass loss is the lowest and its percentage mass loss among the lowest of the marmots (Table 3.4). Thus, in the Upper East River Valley, *M. flaviventris* typically gains sufficient mass while active to reproduce annually. However, in North Pole Basin, about 10 km to the northwest at 3400 m elevation, no adult female reproduced in successive years during a 6-year study and 14 females who were judged pregnant failed to wean litters (Johns and Armitage 1979). The major factor reducing the frequency of reproduction is snow cover. Although the marmots may tunnel up through the snow as early as mid-May, 50% snow cover on average persisted to July 1 and hibernation began in mid-September. Although the active season was about 4.5 months, foraging was not possible until snow melted, which reduced the average favorable time for the increase in mass to about 3 months. In the Upper East River Valley, 50% snowmelt typically occurred before

Table 3.4 The relationship between reproductive skipping, mass loss (g), the percentage of immergence mass lost during the hibernation period, and the length of the active season (months) for adult female *Marmota*.

Group 1 Usually 1 year				Group 2 Often 2 or more years				Group 3 No skipping			
Species	Mass loss	% Loss	Active season	Species	Mass loss	% Loss	Active season	Species	Mass loss	% Loss	Active season
M. caudata	1292	32.9	4.4	*M. menzbieri*	1439	38.3	4.0	*M. flaviventris*	1009	29.4	4.5
M. sibirica	1410	35.6	5.4	*M. camtschatica*	1848	38.9	3.8	*M. monax*	1362	28.9	7.5
M. marmota	1162	29.1	5.5	*M. vancouverensis*	1429	26.8	5.0				
M. bobak	1210	29.4	4.3	*M. baibacina*	1605	28.7	5.0				
M. olympus	1440	25.9	4.5	*M. caligata*	2904	46.9	4.5				
Mean	1303 ± 4	30.6 ± 17	4.8 ± 0.3		1845 ± 275	35.9 ± 3.7	4.5 ± 0.2				

May 30 and vegetative growth began earlier (Svendsen 1974, Van Vuren and Armitage 1991) which allowed marmots to gain mass for nearly all of the active period.

The significance of snow cover as a major component of the harsh environment is evident in the reproductive frequency of *M. olympus*. Barash (1973a) reported that female Olympic marmots usually skip one year, but recently some females reproduced in consecutive years (Griffin *et al.* 2007a). All females that successfully weaned a second litter did so in years of early spring snowmelt.

Late snowmelt reduced the number of *M. caligata* females that reproduced in the population characterized by bi- or triennial breeding (Holmes 1984a). In a *M. caligata* population near the northern extreme of its distribution where the snow-free growing season is approximately 70 days, 45% of the females were reproductively successful. Many females skipped at least one year, but some females reproduced in successive years (Kyle *et al.* 2007). In both *M. olympus* and *M. caligata*, reproduction in successive years occurred in populations more than 25 years after reproductive skipping was reported in these species. This time period coincides with the increase in global average temperature (Intergovernmental Panel on Climate Change 2007) and indicates that climate change probably affects reproductive frequency of some marmot species.

Marmots may effectively lengthen the active season for up to 60 days by mating before emerging above ground in the spring. Five species mate after emergence; seven species mate before emergence and some development or even birth occurs in four of these species (Table 3.5). Mating and development before emergence require sufficient fat reserves to sustain metabolism and development until foraging can occur. When environmental conditions are favorable, there is sufficient time in the active season for growth and fattening. In effect, these species of marmots bet that the post-hibernation season will be favorable; when it is unfavorable, embryo absorption occurs, e.g., *M. baibacina* and *M. caudata* (Table 3.1). Presumably if marmots waited to initiate reproduction after emerging above ground, time would be too short for successful reproduction.

Environmental harshness is related to the uncertainty of weather patterns. In Transbaikalia, heavy rains in the fall are associated with extensive *M. sibirica* reproduction the following spring. Fall rains probably affect vernal plant growth; because females

Table 3.5 The relationship between emergence from hibernation and the time of mating of *Marmota*. References in Armitage and Blumstein (2002), except for *M. caligata* (Kyle *et al.* 2007).

Mating time	Species
After emergence	*M. monax*, *M. flaviventris*, *M. olympus*, *M. vancouverensis*, *M. marmota*
Before emergence	
Without development	*M. caligata*, *M. baibacina*, *M. sibirica*
With development	*M. bobak* (birth may occur)
	M. broweri (young born 1–2 weeks later)
	M. camtschatica
	M. caudata (young half-developed)

can forage to sustain reproduction, the earlier growth starts, the higher is the subsequent population density (Seredneva 1991). A long, cold spring produced high juvenile mortality in *M. bobak*, where breeding occurs before emergence from hibernation. In *M. caudata*, unfavorable weather decreases the number of breeding females and increases embryo mortality (Shubin 1991). As a consequence of these variable weather patterns, age composition fluctuates widely.

Some of the strongest evidence for a harsh environment comes from measures of mass loss after emergence (Armitage and Blumstein 2002, Armitage 2007). Mass loss may occur for several weeks, especially in years of heavy or prolonged snow cover or when the spring season is characterized by cold, snowy weather (Table 3.6). In *M. sibirica*, about two-thirds of the fat accumulated in the second half of the active period is used during hibernation and one-third is used after emergence (Seredneva 1991). Many such species were categorized as those with high post-emergence costs; they emerge with more mass than expected based on their body length (Armitage 1999). If emergence conditions are unusually poor, e.g., cold weather, reproductive failure may occur and embryos are reabsorbed (Table 3.1). The harsh conditions explain reproductive skipping; reproductive females accumulate about one-half as much pre-hibernation fat as non-reproductive females. The amount of fat apparently is insufficient to meet energy requirements for hibernation, post-emergence costs, and reproduction. The quantitative relationship between body condition, including fat reserves, and reproduction requires much additional research in marmots.

Furthermore, marmots may require extensive home ranges in order to obtain the energy and nutrients required for survival and reproduction (Armitage and Blumstein 2002). There is a rough relationship between home range area and vegetation biomass (Armitage 2000). Although data are too few for statistical analysis, home-range area decreases as vegetation biomass increases. For several species, home-range area (which may be individual or family) typically is in the order of 2–3 ha and may be as great as 13 ha in *M. camtschatica* and *M. caligata* (references in Armitage 2000). By contrast, mean home range area of *M. flaviventris* in our study area typically ranges between 0.1 and 0.5 ha. It seems likely that marmot species with larger home ranges would have higher maintenance costs and increased exposure to predators than those with smaller

Table 3.6 Mass loss after emergence. Modified from Armitage and Blumstein (2002).

Species	Mass loss
M. flaviventris	Rare, usually begins gaining mass
M. monax	Lose mass (300 g in females) for 6 weeks
M. caligata, M. olympus	May lose mass for 2 or more weeks
M. caudata	Lose mass for up to 2 weeks following emergence
M. sibirica, M. camtschatica, M. marmota, M. bobak	Use fat in years of heavy snow or of snowy weather
M. baibacina	Lose mass and use fat for 2–2.5 months

home ranges. There is little information to test this suggestion; studies of field metabolic rates, activity cycles, and predation intensity are needed.

Summary

All marmot species have a set of common habitat requirements: a location that reduces the length of snow cover, e.g., south-facing slopes, and a meadow or meadow-like area for foraging. Marmots live in a harsh environment, which reduces the time available for growth and reproduction and which may cause reproductive skipping and a loss of reproduction when early active season conditions are cold and stormy.

4   Use of resources

The places where marmots live must provide essential resources, some degree of protection from predators, and a place to hibernate. The area in which a marmot travels during its daily activities is generally called the home range, and must include the essential resources of burrows and foraging areas.

Burrows

The burrows of all marmot species have three functions: (1) provide shelter from heat, predators, and agonistic conspecifics; (2) provide a place for rearing young; and (3) serve as a hibernaculum or hibernation burrow. These functions may occur in one major burrow or be distributed among numerous burrows (Table 4.1). All species use multiple burrows; *M. monax* may travel as much as 90 m between burrows on the same day (Ferron and Ouellet 1989). Yellow-bellied marmots may make their "burrows" in hollow cottonwood (*Populus trichocarpus*) trees and forego fossorial living (Garrott and Jenni 1978).

The hibernaculum and the home burrow may be identical or differ depending on local conditions. In central Oregon, yellow-bellied marmots migrated a modal distance of 1.25 km between the summer range (50% of the young remained in their natal burrow) and hibernacula on a ridge (Thompson 1979). Woodchucks may hibernate in wooded areas and move into meadows after the spring thaw (Maher 2006). Members of a colony of hoary marmots moved from their rocky hibernacula to a rich, grassy meadow, following the melting snow and occupying burrows as they became exposed (Barash 1974b). Snow-cover patterns affect hibernacula locations; they are often located where snow accumulation provides insulation (yellow-bellied marmot, Svendsen 1976; alpine marmots on Mt Cimone in Italy, Sala *et al.* 1996). In some environments the hibernaculum may be a limiting resource; monogamy in a population of hoary marmots was attributed, in part, to the distance between hibernacula being too great to allow a male to control more then one winter burrow with its associated female and juveniles (Holmes 1984a). The critical nature of the hibernaculum is evident in its use over many years. Because all members of an alpine marmot family hibernate jointly (Arnold 1993a), only one hibernaculum is needed in its territory. Successive family groups inherit the hibernaculum; new hibernacula appeared only in newly established territories. The new hibernacula disappeared the following season when they became incorporated in another

Table 4.1 Burrow types of various species of marmots.

A. Three burrow types
1. Home burrow: provides shelter and den for rearing young, usually one
2. Auxiliary or flight burrow: provides refuge when a marmot is moving about in its home range, usually numerous
3. Hibernaculum: where the individual or family hibernates

Characteristic of *M. flaviventris* (Armitage 1962), *M. camtschatica* (Solomonov *et al.* 1996), *M. marmota* (Perrin *et al.* 1992, Lenti Boero 2003a), *M. olympus* (Barash 1973a), *M. caligata* (Barash 1974b)

B. Two burrow types
1. Central principal den: where young are reared and hibernation occurs
2. Flight burrows: located near the border of the home range

Characteristic of some populations of *M. marmota* (Pigozzi 1984)

Figure 4.1   A yellow-bellied marmot female and young at their burrow in a rocky site.

group's territory (Lenti Boero 2001). Presumably the "best hibernaculum" is used; "best" is likely determined by persistent, successful survival.

Yellow-bellied marmot burrows typically are located in talus or rock outcrops, under boulders (Fig. 4.1), or under tree or shrub roots. About 75% of the burrows occur on slopes between 15° and 40°, but may occur on steep banks or cliffs where the angle is between 70° and 90° (Svendsen 1976). Similarly, *M. marmota* prefers intermediate slopes (Bassano *et al.* 1992a, Allainé *et al.* 1994, Frigerio *et al.* 1996, Gasienica Byrcyn 1997), but utilize steeper and shallower slopes, especially when the population increases and expands into previously unoccupied areas (Panseri and Frigerio 1996). The angle of the slope is important because it provides good visibility of the surroundings and drainage to prevent flooding. Burrows are dug only where soil occurs. Rocks and/or roots

Table 4.2 Variation in hibernacula depth among marmot species.

Species	Depth (m)	Reference
M. flaviventris	0.4–0.6	Svendsen 1976
M. camtschatica	0.25–0.6	Semenov *et al.* 2001
M. monax	1.0–2.0	Armitage 2003a
M. marmota	1.0–3.0	Arnold 1990a
Eurasian	average 2.0–5.0; may be up to 7.0	Bibikow 1996

are critical as they support the tunnel entrance and walls. A rocky entrance acts as a barrier to large predators, such as bears or badgers; when a protective enclosure of rocks is missing, badgers readily dig out the burrows and prey on the marmots (Andersen and Johns 1977, Armitage 2004a).

Little is known about the structure of yellow-bellied marmot burrows because most burrows are in rocky areas that cannot be excavated. Svendsen (1976) excavated five burrows outside the study sites. One or more entrances led into a main passage within 0.5 m of the entrance. The main passageways were about 3.8–4.4 m long ending in a nest chamber hollowed out beneath a large rock. Burrow depths are typically shallow (Table 4.2). The shallow depth of the burrow system emphasizes the importance of winter snow cover to insulate marmots against low, stressful burrow temperatures, which would increase metabolic rates during hibernation (Bibikow 1996). Snow cover in our study area may prevent the soil from freezing, as reported by several winter caretakers at RMBL. The shallow depth of *M. camtschatica* hibernacula is a consequence of their being located in the permafrost which thaws to a depth of only 0.1–1.0 m. The marmots compensate in part by establishing large nests and locating hibernacula on southwest facing slopes.

An essential feature of the burrow is that its microclimate, especially temperature, should not be stressful. Burrow temperature ideally should be in the thermoneutral zone. Yellow-bellied marmot mean summer burrow temperatures varied from 9 to 11.8°C (Kilgore and Armitage 1978), which corresponds roughly with the lower end of the thermoneutral zone (Armitage 2004b). Burrow temperatures lagged air temperature by about 1 month.

Thermal characteristics of burrows were extensively explored for the steppe marmot in Ukraine. Temperatures were measured daily for a week in late August for 30 burrows. Air temperature varied from 20.3 to 30.8°C; burrow temperature did not drop below 18°C nor exceed 21°C. Air temperature did not affect burrow temperature (p = 0.92), which depended on the depth of ground ($R^2 = 0.939$) (Nikol'skii and Khutorski 2001). Sensors were placed in the ground at 190 cm and soil, air, and burrow temperatures were measured four times from 08:00 to 20:00 h. Burrow temperature depended mainly on soil temperature, but air temperature had some effect (Nikol'skii 2002a). However, seasonal changes in air temperature are more likely to affect burrow temperatures than daily fluctuations. When air temperature and burrow temperature were monitored from late July to mid-October, burrow temperature was relatively constant until beginning a rapid decrease in early September (Nikol'skii and Savchenko 2002b). Burrow temperature was most sensitive to air temperature near the soil surface (Nikol'skii and Savchenko 2002c). Daily fluctuations of burrow temperature were <1°C in 85% of cases, whereas air temperature fluctuated by 15°C in 50% of cases. From mid-August until early September, the burrow temperature was very similar to the ground (at 180 cm depth) temperature at about 15°C. From early to late September, the burrow temperature decreased by about 1°C every 5 days; ground temperature decreased much more slowly. The rapid decrease in burrow temperature occurred as a result of increased convection following an abrupt decrease in air temperature. Burrow temperature declined to 14°C, almost identical to air temperature. At this time, steppe marmots began plugging their burrows (Nikol'skii and Savchenko 2002b, c). The burrows were protected from over-cooling as they were plugged before lower air temperatures (about 5°C) induced more intense convection. The authors suggest that marmots enter hibernation when the burrow temperature reaches some threshold. Clearly, plugging the burrow prevents the development of this threshold temperature early in the season reducing maintenance metabolism.

One consequence of plugging the burrow to eliminate air convection is that the primary mechanism of gas exchange is diffusion through the soil (Boggs *et al.* 1984). During hibernation, O_2 levels may be as low as 4%, but generally are about 15%, and CO_2 levels may increase to 13.5%, but usually are <4% (Williams and Rausch 1973). Consequently, marmots encounter low O_2 (hypoxia) and high CO_2 (hypercapnia) stresses. In both *M. flaviventris* (Bullard *et al.* 1966) and *M. monax* (Hall 1965) the oxygen equilibrium curve is shifted to the left, which allows hemoglobin to reach 50% saturation at lower oxygen partial pressures. In *M. monax* there is no ventilatory response to hypoxia because typical O_2 levels provide 85% saturation. The ventilatory response to hypercapnia is low in comparison with non-burrowing mammals (Boggs *et al.* 1984). Hypercapnia decreases arterial pH producing a large Bohr effect (Boggs and Birchard 1989), which facilitates oxygen delivery to the tissues.

Concentrations of O_2 and CO_2 in the nest box of a captive population of *M. broweri* cycled with a duration of 5–8 days. Low concentrations of O_2 coincided with high concentrations of CO_2. This pattern probably represents arousal periods. During torpor CO_2 can diffuse out of, and O_2 into, the burrow thus restoring typical gas concentrations.

Probably all marmot species respond similarly to burrow gas concentrations: decreased sensitivity to elevated concentrations of CO_2 and enhanced uptake of O_2 from a low O_2 environment.

I have often been asked when the marmots in our research sites dug their burrows. And my answer has been "I don't know." During the 41 years of our research, the same major burrows were used year after year. The tenants changed with time, but the "home" did not. All sites had burrows that were used only in some years, such as when a colony increased in number and adult females spread over the entire area. Sometimes a female moved her young from one burrow to another. These moves often occurred when we attempted to trap and mark the young. Several times a female moved her young at about the time of weaning from the birth burrow into a burrow that was shared with one or more other lactating females. The reason for this move is unknown, but it usually involved a younger female who moved into the burrow system with an older female who had harassed her. Infanticide was observed when an adult female killed young of a female living nearby (Armitage *et al.* 1979, Brody and Melcher 1985). These acts occurred shortly after young were weaned and appeared above ground. Possibly moving the young into the same burrow system protected them from infanticide; agonistic behavior toward young was never observed when litters emerged from the same burrow.

Because only about 20%, on average, of the available burrows were used in a summer, yellow-bellied marmots had no need to dig new home burrows and most digging involved opening and cleaning an existing burrow. Old grass, dirt and stones, and sometimes marmot skulls or lower mandibles were kicked or pushed from the burrow. As a consequence of utilizing the same major burrows, digging was always a minor component of the time budget. Digging was also rare in alpine marmots (Ferroglio and Durio 1992, Sala *et al.* 1992, Lenti Boero 2001). The time allotted to digging by yellow-bellied marmots varied from <0.1% to <0.5% of the time, which, in absolute time, varied from about 59 s/day to a maximum of 4.5 min/day (Armitage 2003d). Prolonged digging occurred only when new flight burrows were dug in or at the edge of a large foraging area (Fig. 4.2).

An activity associated with burrows is gathering grass and carrying it to the burrow for nesting material (Armitage 2003d). Gathering grass was always a minor activity and used from 0.01 to 0.6% of above-ground activity among all age/sex groups, which convert to about 1.13 min/day (Armitage *et al.* 1996). This activity, though important, probably is minor in all species of marmots, e.g., *M. marmota* (Sala *et al.* 1992). Gathering grass

Figure 4.2   A newly dug flight burrow at Marmot Meadow Colony. Note the main burrow area in the upper left corner. (See plate section for color version.)

Table 4.3 Number of individuals gathering grass.

Gathered grass	Adult females	Adult males	Yearlings
Yes	172	26	58
No	184	97	322

Adult females vs. adult males: G = 29.6, p < 0.001
Adult females vs. yearlings: G = 96.4, p < 0.001

Figure 4.3   A yellow-bellied marmot in early October gathering dried grass along Trail Ridge Road in Rocky Mountain National Park, Colorado, prior to hibernation. Photo courtesy of Kenneth M. Highfill. (See plate section for color version.)

varied seasonally; most instances were observed during gestation and lactation before young were weaned. Grass-gathering occurred in late summer prior to hibernation (Fig. 4.3); this period was the only time young were observed gathering grass. Reproductive females gathered grass more often than other groups (Table 4.3) and allocated significantly more time to gathering grass than other groups (Armitage 2003d). Similarly, adult female *M. vancouverensis* were the predominant grass-gatherers with peaks of activity in early summer and prior to hibernation (Heard 1977). Parous females were the primary grass-gatherers in *M. olympus* colonies with the peak of activity prior to weaning (Barash 1973a).

During the time of gestation and lactation, an adult female aged 3 years or older who gathers grass multiple times in one day or multiple times during these time periods is almost certain to be reproductive. This behavior was so strongly associated with reproductive activity that it predicted which females produced litters and could be used to identify those females that lost their litters before weaning. Gathering of grass by reproductive females is probably a component of maternal care as the females presumably provide fresh nesting material to replace grass soiled by the eliminative activities of the young. This activity is not costly; little time is expended and energy expenditure is probably minor; heart rate while carrying grass does not differ significantly from that during walking (Armitage 2003b).

Like yellow-bellied marmots, woodchucks clean out their burrows (Grizzell 1955), but expend more time and energy in burrow construction. Juveniles and yearlings are primarily associated with establishing new systems (de Vos and Gillespie 1960, Henderson and Gilbert 1978), in contrast to yellow-bellied marmot young and yearlings who share existing burrows with the adult females. Henderson and Gilbert reported that woodchucks reactivated 63 old systems and built 80 new systems during one active season in which 181 woodchucks were captured two or more times. Thus, there was about one major burrow activity per woodchuck, a rate far greater than any observed for yellow-bellied marmots. This high rate of woodchuck burrow construction is feasible because the burrows are dug in easily excavated soil such as sandy loam (Moss 1940); many burrows were dug in recently cultivated areas. The difference between woodchucks and yellow-bellied marmots in digging new burrows supports Bibikow's (1996) suggestion that burrows may be used more permanently by marmots living in rocky areas where burrow sites are limited.

There are numerous variations among marmot species in the location of burrows. These variations indicate considerable flexibility in their location. For example, burrows of yellow-bellied marmots may be dug under tree or shrub roots, under a fallen log, or under buildings. Woodchucks located a burrow at the base of a fir tree in a gorge where the burrow was exposed to the morning sun. The woodchucks had to travel about 100 m from the burrow to feeding grounds along stream banks (McTaggart-Cowan 1933). Other variations are reviewed elsewhere (Barash 1989, Bibikow 1996, Kwiecinski 1998, Armitage 2000, 2003a, Edelman 2003), therefore only a few of the more interesting variations that have ecological or fitness effects will be described.

By digging burrows, marmots modify their environments. Burrows of many marmots, e.g., *M. flaviventris*, *M. monax*, *M. olympus*, *M. baibacina*, and *M. bobak*, have large mounds of excavated soil. Mounds may be up to 1.5 m high and from 8 to 19 m wide. Mounds are so conspicuous that they were used to census *M. sibirica* populations in Mongolia (Townsend 2006). Yellow-bellied marmots usually defecate in an area of the mound. I suggested that the area with feces be called a defecatorium, but this term has not entered the scientific nomenclature. Marmots also deposit feces at the end of a blind passage; the feces are later used to form the winter plug (Formozov 1966). The mounds are clearly discernible examples that marmots modify their environments. Mounds may be populated by invasive, unpalatable plant species that increase vegetative diversity. The burrows and mounds also are colonized by assemblages of invertebrates and vertebrates. For example, burrows of *M. sibirica* are an important resource for the corsac fox (*Vulpes corsac*). When a marmot colony went extinct and burrows filled with sand, corsacs no longer visited the area (Murdoch *et al.* 2009). Woodchuck burrows may be used by opossums (*Didelphis virginiana*), raccoons (*Procyon lotor*), and cottontails

(*Sylviligus floridanus*) (Schmeltz and Whitaker 1977). Marmot excavations can be so extensive that landscapes are modified. Tunnels of *M. caudata* and *M. baibacina* may be 75 m and 63 m long, respectively, and the excavated dirt forms small, flat hills (Zimina 1996). As much as 150 m³/ha may be dug out. The small hills may cover an average of 8–10% of the landscape in mountainous areas and 5–10% in steppes occupied by the steppe marmot and tarbagan. These marmot-modified landscapes may persist for several thousand years. Marmot activity may also create local salt concentrations, raise the carbonate content of the soil, or destroy the soil profile in highly saline soils (Rumiantsev 1992). In general, marmot activity increases the mosaic structure of the local area of marmot activity.

   Although we associate marmot burrows with rocky areas, several species (*M. monax*, *M. bobak*, *M. sibirica*, *M. baibacina*, *M. olympus*) regularly occupy home burrows in grassy meadows. Because burrows located in rocks are believed to provide protection against predators and were observed to do so for yellow-bellied marmots evading badgers (Armitage 2004a), the location of burrows in meadows raises the question of the role of predation in the structure and location of burrows. For those species that occupy the Eurasian steppe, i.e., *M. bobak* and *M. sibirica*, there is no choice except to locate burrows in a meadow (Fig. 4.4). However, these species do not appear to be at special risk of extinction because of predation. Although the polecat (*Mustela eversmanni*) may dig up marmot burrows, especially in winter, they do not seem to threaten the elimination of marmots (Bibikow 1996). By contrast, badgers (Fig. 4.5) may exterminate some yellow-bellied marmot populations (Armitage 2004a). Thus, in western North America, badgers may have been a major selective agent for locating burrows in rocky areas.

Figure 4.4   The steppe marmot, *M. bobak*, at its burrow in the steppe vegetation. Note the excavated dirt and lack of rocks.

Table 4.4 Marmot responses to vegetation composition.

Species	Response	Reference
M. marmota	Absent from meadows where *Carex sempervirens, C. curvula, Nordus stricta*, and *Sesleria* sp. predominate	Vita 1992
	Avoid area in home range where bilberry (*Vaccinium myrtillus*) was extensive	Sala *et al.* 1992
M. sibirica	Occupies area covered with forb/grass vegetation, seldom found outside distribution limits of *Allium, Astragulus, Bupleirum, Festuca, Oxytropis, Stipa*	Suntsov and Suntsova 1991
M. olympus	Prefers subalpine meadows with three or more plant communities	Wood 1973

Figure 4.5   A badger family at its freshly excavated burrow in a meadow where yellow-bellied marmots forage.

Foraging and food choice

The foraging area is the largest part of the home range. All marmot species are herbivores and selectively eat a variety of forbs and grasses. Because marmots feed selectively (Armitage 2000, 2003e), plant community composition affects where marmots forage. Marmots may avoid or colonize meadows with a particular species composition (Table 4.4).

Marmots are generalist herbivores (Arsenault and Romig 1985, Frase and Armitage 1989). Field studies documented that marmots did not eat all of the wide variety of plant species that were available (Table 4.5A) and plant choice varied seasonally; grasses were

Table 4.5 Field studies that indicated food selectivity.

	Species	References
A	*M. camtschatica*	Solomonov *et al.* 1996
	M. monax	Hamilton 1934, Grizzell 1955, Arsenault and Romig 1985, English and Bowers 1994
	M. caligata	Hansen 1975
	M. vancouverensis	Milko 1984
	M. olympus	Del Moral 1984
	M. marmota	Bassano *et al.* 1996
B	*M. flaviventris*	Carey 1985a, Frase and Armitage 1989
	M. vancouverensis	Milko 1984, Martell and Milko 1986
	M. marmota	Bassano *et al.* 1996
	M. sibirica	Seredneva 1991

Figure 4.6   A field of dandelion flowers. The edge of the blooming area marks the extent of the foraging range of the yellow-bellied marmots that occupied the rocky area beyond the meadow. (See plate section for color version.)

preferred in the spring and forbs, from mid- to late summer (Table 4.5B). Overall, forbs were preferred over grasses; e.g., a high elevation (3800 m) population of *M. flaviventris* overwhelmingly preferred forbs, especially clover, *Trifolium andersonii* (Stallman and Holmes 2002).

Marmots often choose flowers. *M. olympus* selectively ate flower heads when meadow flowers were particularly abundant (Barash 1973a, Wood 1973). *M. marmota* utilized flowers heavily in June and July, and vegetative parts in May and August and September. Seed ingestion peaked in August (Massemin *et al.* 1996). Yellow-bellied marmots prefer flowers of dandelions (*Taraxacum officinale*); all flowers in foraging areas may be consumed whereas flowers are abundant in the meadow outside the foraging area (Fig. 4.6).

Food selectivity

Various selectivity experiments support a preference for forbs in woodchucks (Twitchell 1939, Swihart 1990), steppe and long-tailed marmots (Ronkin and Tokarsky 1993), and yellow-bellied marmots (Woods 2009). In the latter cafeteria-style feeding trial, dandelions (*Taraxacum officinale*) were selected significantly more often than expected by a random distribution, whereas grasses (*Poa* and *Bromus*) were avoided. In woodchuck cafeteria-style feeding, grasses ranked lower than expected by chance. Even when preferred species were removed, consumption of orchard grass remained low (Swihart 1990).

Marmots will utilize other plants such as leaves of trees (Fig. 4.7). One yellow-bellied marmot had bark and wood from pine and fleshy roots of a weed in its stomach (McTaggart-Cowan 1929). In a spring with late snow cover, yellow-bellied marmots foraged on aspen and spruce by climbing trees and eating buds and branches; this behavior declined abruptly following snowmelt (Olson *et al*. 2003). Marmots are opportunistic; a woodchuck was observed dipping a paw into the water and pulling aquatic vegetation of at least four species into its mouth (Fraser 1979). They may feed on insects; remains of June bugs were found in the feces of a woodchuck (Gianini 1925).

Woodchucks may climb trees and feed on the leaves of hackberry, Norway maple, peach, and red mulberry (Swihart and Picone 1991a). Red mulberry was highly palatable and Norway maple was unpalatable. Woodchucks feed on woody plants near their burrows; sugar maple was the most severely gnawed species (Swihart and Picone

Figure 4.7   A hoary marmot, *M. caligata*, feeding on leaves of cottonwood.

1991b). When woodchuck burrows are located near orchards, their gnawing may kill young trees. Most gnawing occurs in the spring and on trees within 1.5 m of a burrow and seems to be associated with scent-marking behavior (Swihart 1991a, Swihart and Picone 1994).

Field observations of yellow-bellied marmots document how food selectivity occurs. In early summer in a meadow where lupine (*Lupinus floribundus*) is abundant, the meadow develops a blue cast as the plants produce flowers on terminal stalks. Marmots sit upright at a clump of lupines and ingest the flowers while eating few or no leaves. After about a week, the blue cast to the meadow, along with most of the lupine flowers, is gone. I observed marmots reject other plant species such as blue flax (*Linum lewisii*), yellow paintbrush (*Castilleja surphurea*, Fig. 4.8), mustard (*Draba* sp.), the little sunflower (*Helianthella quinquenervis*), and aster (*Aster foliaceus*; Fig. 4.9) after touching them to the mouth or lips. By contrast, marmots readily eat the flowers, but not the leaves, of the tall larkspur (*Delphinium barbeyi*) and the blue columbine (*Aquilegia coerula*) (Fig. 4.10). In late summer marmots of all ages ingest the seed heads of grasses (Armitage 2003e). Marmots were never observed to feed on *Veratrum californicum* (Fig. 4.11) or *Frasera speciosa* (Fig. 4.12), even though both species may be highly abundant. However, the Vancouver Island marmot was observed to feed on green false hellebore (*Veratrum viride*), which is plentiful and also contains toxic secondary compounds (Werner 2012). The costs and benefits of this food choice are unknown.

Figure 4.8   Yellow paintbrush, *Castilleja surphurea*, a plant often "tasted" but rejected by yellow-bellied marmots.

Figure 4.9   Aster, *Aster foliaceus*, may be abundant in late summer but are not eaten by yellow-bellied marmots.

Figure 4.10   Blue columbine, *Aquilegia coerula* (left), and tall larkspur, *Delphinium barbeyi* (right), are abundant in some marmot colonies, but only the flowers are eaten.

Figure 4.11 *Veratrum californicum* may form extensive patches at marmot colonies, such as these at Picnic Colony. Marmots may enter *Veratrum* patches to feed on low growing plants, but do not forage on *Veratrum*.

Figure 4.12 The green gentian, *Frasera speciosa*, plants may be dense at some marmot sites, but are never used as food.

Food choice: seasonal patterns

Food is restricted during two seasons; spring, when during early snowmelt only a few plant species are abundant, and late summer, when plant biomass declines rapidly and many plant species become senescent, turn brown, and have markedly decreased water content (Kilgore and Armitage 1978, Frase and Armitage 1989). Among the earliest plants available in the spring are the larkspur (*D. nelsoni*) and little bluebells (*Mertensia fusiformis*) (Inouye *et al.* 2000) along with spring beauty (*Claytonia lanceolata*) and the glacier lily (*Erythronium grandiflorum*) (Fig. 4.13). Bluebells (Fig. 4.14) are eaten when encountered, but larkspur is rejected (Armitage 2003e). Yellow-bellied marmots readily forage on spring beauty and continue to select it long after many other food items become available. By contrast, the glacier lily is rejected at a time when food is scarce. This rejection is surprising as we often made a fresh salad of glacier lily leaves with no negative consequences.

Yellow-bellied marmots partially compensate for the decline in leafy vegetation in late summer by eating seeds of grasses and fruits of gooseberry (*Ribes inerme*) and elderberry (*Sambucus pubens*) (Fig. 4.15). Foraging time increases as marmots spend considerable time moving among and manipulating branches to reach the fruit (Armitage 2003e). Many abundant fall-blooming species, such as goldeneye (*Viguiera multiflora*, Fig. 4.16) aster, gentian (*Gentiana calycosa*, Fig. 4.17) and fireweed (*Chamerion angustifolium*) are not eaten. The decline in palatable food sources in late summer may be one reason why marmots complete most of their mass gain for hibernation in early to mid-August.

Figure 4.13 The spring beauty, *Claytonia lanceolata*, and the glacier lily, *Erythronium grandiflorum*, are locally abundant as the snow melts, but only the spring beauty is selected as food. (See plate section for color version.)

Figure 4.14   Bluebells, *Mertensia ciliata*, grow in moist areas or along streams, and are utilized when encountered by yellow-bellied marmots, but are not present in many habitats. *M. fusiformis*, by contrast, early in the season grows in the meadows frequented by marmots and is readily eaten.

Quite possibly the decline in food plants and the widespread senescence of plants stimulate the initiation of hibernation.

Food choice and plant secondary compounds

The widespread evidence for selective feeding raises the question of what is the basis of food choice. There are two issues: why are some species tested and rejected and why are some species, such as cow parsnip (*Heracleum lanatum*), highly selected to the extent that marmots may extend their home ranges to areas where the plant is abundant (Armitage 1979)? In feeding trials with captive marmots, four plant species were rejected, except some flowers were eaten. Three of the rejected species, lupine, columbine, and tall larkspur, contain alkaloids and the fourth species, fireweed, contains tannins. Thus, the presence of toxic secondary compounds accounts for the rejection of plant species that are often abundant in marmot habitats (Armitage 1979).

Food choice and nutrition

Yellow-bellied marmots (Carey 1985a, Frase and Armitage 1989) and Olympic marmots (Wood 1973) select plants with higher protein content. Selected plants also have higher

Figure 4.15   Gooseberry, *Ribes inerme* (top), and elderberry, *Sambucus pubens* (bottom), fruits are eaten when present in yellow-bellied marmot habitats.

concentrations of required minerals; two principal forbs eaten by yellow-bellied marmots in the White Mountains of California had a calcium content 2–3 times greater than any other plant species and forbs had higher content of phosphorus, calcium, and sodium than graminoids (Carey 1985a). High cellulose content of plants reduced consumption and assimilation by *M. sibirica*, which prefers new plant growth (Seredneva 1991).

Figure 4.16   Goldeneye, *Viguiera multiflora*, may form extensive populations in late summer, but marmots do not feed on them.

Figure 4.17   The gentian, *Gentiana calycosa*, may be common in meadows in late summer; yellow-bellied marmots may forage around clumps of this plant, but do not eat it.

Woodchucks use small mineral licks and lick road surface residues of winter-applied NaCl (Weeks and Kirkpatrick 1978). I observed yellow-bellied marmots licking the mud surface where water seeped to the surface at the foot of a slope. These seeps were visited by other species; water was not the critical factor as deer were observed to cross the East River and move to the seep and lick the surface (Armitage 2000). The importance of Na was demonstrated when woodchucks were provided with wooden pegs soaked in water and various salt solutions. Pegs containing Na salts were most highly gnawed (Weeks and Kirkpatrick 1978).

Diet selectivity is also related to digestibility. Marmots are caecal fermenters; the caecum is large, approximately 8–10 cm long and 2–3 cm in diameter in woodchucks (Bezuidenhout and Evans 2005). The MRT of food in the digestive tract is all important. MRT is longer for low quality foods which require a longer, more complex digestive system (Hume 2002). MRT increases with increasing body size; larger body size results in a larger absolute gut capacity relative to metabolic rate. *M. caudata* that feed on less nutritious plants are larger and have a longer intestine whose content relative to body mass is larger than that of conspecifics that feed on more nutritious vegetation (Davydov 1991). Digestibility is greater in marmots than in voles for a low-fiber diet; marmots can use a more fibrous diet because of the longer retention of digesta in the gut (Hume *et al.* 1993), which provides more time for microbial digestion of fiber (Hume 2003). In effect, the large size of marmots allows them to utilize a wide variety of plants including high fiber grasses. Marmots feed on forbs with a lower fiber content to obtain essential nutrients and on grasses to fill out their energy requirements. The long MRT may explain why marmots spend much more time sitting and lying than feeding (Armitage and Chiesura Corona 1994). The large amount of time spent inactive was characterized as laziness (Herbers 1981) and probably represents time needed for digestion (Weiner 1989, 1992).

Food choice and polyunsaturated fatty acids (PUFAs)

Marmots and ground-dwelling squirrels require a high level of polyunsaturated fatty acids (PUFAs) for successful hibernation. PUFAs are not synthesized by mammals and must be obtained from their food (Frank *et al.* 1998). Two PUFAs, linoleic and linolenic, are critical. Yellow-bellied marmots fed an essential fatty acid deficient diet had significantly higher metabolic rates and shorter torpor bout lengths than controls during hibernation, but metabolic rates did not differ between the two groups during the summer (Thorp *et al.* 1994, Florant 1998). Marmots are highly lipolytic during winter as they rely on white adipose tissue depots as their source of energy. The release of fatty acids was not random; linoleate was significantly underrepresented (Florant *et al.* 1993). The critical factor may be not PUFAs in general, but the ratio of n-6 to n-3 PUFAs (linoleic acid to linolenic acid) (Ruf and Arnold 2008). Free-living alpine marmots, and probably all marmot species, remodel the membrane phospholipids, into which PUFAs are incorporated, independently of direct dietary intake. Remodeling occurred during hibernation when neither food nor external cues were available. There appears to be endogenous control of the process involving fatty acid transfer (Arnold *et al.* 2011). Possibly these processes are related to the circannual rhythm.

Concentrations of fatty acids vary significantly among plant species and between parts of the same plant. Several plant species (*Taraxacum officinale, Heracleum lanatum, Potentilla gracilis*) widely eaten by yellow-bellied marmots have high concentration of PUFAs (Hill and Florant 1999). Alpine marmots fed selectively on plants and plant parts of high digestibility and energy content but during the period of fattening preferred plants with high contents of linoleic acid (18: 2n-6) and avoided plants with high concentration of linolenic acid (18: 3n-3) (Arnold *et al.* 2003). Both the distribution and quality of food plants vary on a small geographical scale and could affect hibernation body-mass loss in different habitats or different years (Bruns *et al.* 2000). Much additional research is needed to determine the significance of the availability of PUFAs in food plants and marmot habitat use and survival.

Food choice and patch use

One consequence of selective feeding is that marmots spend more time in some foraging areas than in others. For example, hoary marmots foraging on a talus slope preferred old growth areas (Tyser and Moermond 1983). A preferred foraging area may be marked by a trail through the meadow, often with a flight burrow at the end of the trail (Fig. 4.18). The marmot may travel on the trail to the burrow, spend some time vigilant and then begin foraging. All age groups of yellow-bellied marmots in California had similar preferences; areas with high food biomass, the presence of the clover *Trifolium monoense* and the grass *Koleria cristata*, and low vegetation (Carey 1985b). Another population concentrated foraging on areas where clumps of *T. andersonii* were concentrated (Stallman and Holmes 2002). Hoary marmots also concentrated their feeding in areas of greatest food abundance. When feeding areas were augmented by adding nitrogen, use of the fertilized plots was 590% greater, but there was no difference between years in use of the non-fertilized patches (Holmes 1984a, b). Both availability of forage and risk of predation affected patch use by hoary marmots; patch use was positively correlated with the frequency of selected plants and the number of refuge burrows and negatively correlated with the distance of the patch from the talus containing home burrows (Holmes 1984b). In summer, tarbagan feed more in micro-depressions where moisture is higher (Seredneva 1991). I frequently observed yellow-bellied marmots in dry years shift foraging areas in mid- to late summer to low, moist areas that maintain palatable vegetation (Armitage 2003e).

Although predation risk and plant distribution influenced the location of foraging areas in two colonies of yellow-bellied marmots in Colorado, kinship was a critical factor in the determination of the amount of foraging area shared among individual marmots (Frase and Armitage 1984). Spatial overlap was greater among close kin, but the degree of overlap was modified by individual behavioral phenotypes, age, and reproductive status. For example, mothers and juveniles, and young and yearling littermates had almost identical foraging areas. By contrast, adult female littermates in conflict had reduced sharing of foraging areas.

Figure 4.18 Marmots may make a trail through the vegetation between a home burrow and a flight burrow (at the top). They then forage in the meadow beyond the burrow.

Foraging and modification of plant communities

Another consequence of food selectivity and differential patch use is that marmots modify plant communities. Grass biomass decreased and forb biomass increased as a function of distance from woodchuck burrows. Over an entire field, woodchucks caused an increase in biomass of orchard grass and a decrease in biomass of alfalfa, which was selected 10 times more than orchard grass in feeding trials (Swihart 1991b). Plant biomass (*M. caligata*, Holmes 1984a) and plant cover (*M. monax*, English and Bowers 1994) increase with distance from burrows as these species tend to forage near their burrows. Some plant species increased and others decreased with distance from wood-chuck burrows; the strongest effects were limited to a 4 m radius around the burrows. Species of plants close to burrows tended to be mostly unpalatable, early successional and early colonizing annual and biennial species (Armitage 2000). The disturbed areas around yellow-bellied marmot burrows often are colonized by species typical of dis-turbed areas and are generally unpalatable; e.g., fireweed, nettle (*Urtica dioica*), Rocky Mountain pentstemon (*Pentstemon strictus*), and composites such as *Happlopappus* sp. (Fig. 4.19) (Armitage 2003e). Similar colonization by unpalatable plants occurs on the

Figure 4.19 An unpalatable vegetation, such as *Happlopappus*, often develops on the dirt mounds formed by excavating burrows.

mounds around *M. olympus* burrows and species richness was greater in the meadow than on the mounds (Del Moral 1984). Diversity, species richness, and equitability were low in plots heavily impacted by black-capped marmots (*M. c. bungei*) (Semenov *et al.* 2001). Plots lightly impacted or marmot-free had high floristic homogeneity in contrast to the high heterogeneity of the marmot-impacted plots. Similar to the burrow areas of other marmot species, some plant species occurred only around the main burrows and latrine areas. Latrine areas are associated with lush green zones next to woodchuck burrows and soil nitrogen is greatly increased (Merriam and Merriam 1965).

Marmot disturbances can produce complex responses in the vegetation. The tarbagan creates mounds that differ greatly in their vegetative cover. Three types of mounds named for the species with the dominant cover, were identified: *Artemesia*, *Leymus*, and *Stipa* (Van Staalduinen and Werger 2007). The study was conducted on a *Stipa* steppe, thus *Stipa* dominated the off-mound vegetation. Species richness was highest on the *Stipa* mounds, but nitrogen and phosphorus concentrations were higher in the vegetation on the *Artemesia* and *Leymus* mounds. Disturbance by the marmots decreased species richness, which seems to be a general result of marmot disturbance. The mounds probably represent stages in succession: *Artemesia* invades bare mounds followed by *Leymus* and eventually the *Stipa*-dominated steppe vegetation develops. The authors suggested that the disturbance that preceded the *Artemesia* and *Leymus* mound vegetation increased forage quality (Van Staalduinen and Werger 2007), which contrasts sharply with the unpalatable vegetation that develops on the mounds of other marmot species. However, utilization by tarbagans of the mound vegetation was not determined.

Burrowing, grazing, urination, and defecation by tarbagan increase species diversity and modify the soil's physical properties. These effects are evident at the landscape scale; mountain areas had lower sensitivity to marmot disturbance than plains areas (Yoshihara *et al.* 2010a). Factors such as the pre-existing abundance of water, seed source, and slope aspect and steepness affected the responses to marmot disturbance. The authors considered the tarbagan to be a keystone species acting as an ecosystem engineer.

Foraging and time budgets

The need for marmots to meet their energy requirements and to seek food items to meet specific nutritional requirements while avoiding unpalatable plants suggests that marmots spend a considerable portion of their above-ground activity foraging. Time budgets indicate that marmots are not constrained by time, all species spend more of above-ground activity sitting or lying or retreating to their burrows than foraging (Barash 1973a, 1980, Sala *et al.* 1992, Bonesi *et al.* 1996, Armitage *et al.* 1996).

Overall, yellow-bellied marmots spend about 12–23%, 37–94 min, of their daily above-ground activity foraging. Hoary marmots spend 25–34% of their time foraging (Barash 1980); Olympic marmots, up to 40% (Barash 1973a), and alpine marmots, about 130 min daily (Sala *et al.* 1992). Time spent foraging decreases around midday for alpine (Perrin *et al.* 1992, Sala *et al.* 1992), Olympic (Barash 1973a), and yellow-bellied (Armitage *et al.* 1996) marmots. Foraging time varies seasonally; the percentage of time foraging is high during the reproductive period and declines in late summer prior to hibernation (alpine marmots, Sala *et al.* 1992, Perrin *et al.* 1993a, Bonesi *et al.* 1996, Lenti Boero 2003a; hoary marmots, Barash 1976a; Olympic marmots, Barash 1973a; woodchucks, Arsenault and Romig 1985; yellow-bellied marmots, Armitage *et al.* 1996). Contrary to popular belief, foraging does not increase prior to hibernation. Marmots prepare for hibernation in advance of immergence (Chapter 7). The variation in foraging time among marmot species probably reflects differences in plant abundance, which could require more time to find and consume preferred plants.

Marmot foraging time is not significantly affected by vigilance. In California, yellow-bellied marmots spent 10.2% of their foraging time in vigilance (Carey and Moore 1986). In Colorado, yellow-bellied marmots, over all age groups and seasons, allocate 10–75 minutes per day (up to 16% of above-ground time) to vigilant behaviors (Armitage *et al.* 1996). During foraging, marmots frequently raise their head or stand and look around; these combined behaviors add a maximum of about 5% or 20 min/day to foraging time. During this alertness, marmots typically chew their food, thus it is possible that alertness during foraging does not impose a cost on foraging. Wariness during foraging can be thought of as a sampling process to look for approaching predators or agonistic conspecifics which follows a prolonged period of wariness at the burrow area before the marmot ventures forth. There is no doubt that predator concerns affect wariness during foraging. Thus marmots are more wary when foraging near the periphery of a colony (Armitage 1962), and less wary when feeding in groups (Barash 1973a, Holmes 1984b, Carey and Moore 1986), and when feeding near talus and burrows (Holmes 1984b).

Home range

Finally, how large should marmot home ranges be? Obviously, home ranges must be large enough to provide energy and required nutrients, burrow sites, and opportunities for mating. Home range area varies considerably among marmot species and among sites within species (Table 4.6). For example, home range differs significantly among four yellow-bellied marmot colonies (Armitage 2009b). The comparison of home ranges among species is confounded by the use of different methods for calculating home range areas as different methods can produce markedly different results (Swihart 1992, Maher 2004); some authors do not report their methods. Nevertheless, two factors that affect home range areas are evident. First, larger marmot species have larger home ranges ($r_s = 0.573$, $p = 0.05$, $N = 11$). However, the rank correlation explains only about 33% of the variation in the rankings. Second, limited data indicate that home range increases as vegetation biomass decreases (Armitage 2000, Armitage and Blumstein 2002). This relationship is supported by the variation in home range of yellow-bellied marmots, whose home range is larger in North Pole Basin (Table 4.6) where plant biomass is much less than in East River Valley (Armitage 2009b). Woodchucks had smaller home ranges where food resources were greater (Armitage 2003a) and vegetation cover and home

Table 4.6 Home range areas of marmots. Species are listed in order of increasing body mass. ERV = East River Valley, NPB = North Pole Basin.

Species	Area (ha)	References
M. flaviventris		
ERV	0.369 (adults)	Armitage 2009b
NPB	0.62	Armitage 2009b
M. caudata aurea	2.9–3.1	Blumstein and Arnold 1998
M. sibirica	2.0–3.6	Suntsov and Suntsova 1991
	1.7 (some over 3.0)	Seredneva 1991
M. marmota	1.52–1.92	Lenti Boero 2003a
	1.05 ± 0.22	Labriola *et al.* 2008
	1.2–2.5	Sala *et al.* 1992
	2.5	Allainé *et al.* 1994
	2.3–2.8	Perrin *et al.* 1993a
	2.1–7.2	Gasienica Byrcyn 1997
M. bobak	3.2	Mashkin 1991
M. monax	0.25(♀)–1.6(♂)	Meier 1992
	0.97	Bronson 1963
	0.88(♀)–1.32(♂)	Swihart 1992
M. camtschatica	13.0	Tokarsky 1996
	10.0–15.0	Semenov *et al.* 2001
M. vancouverensis	2.0	Heard 1977
M. olympus	3.59(♀)–12.06(♂)	Griffin 2007
M. baibacina	3.0	Dudkin, pers. com.
M. caligata	9.2	Holmes 1984b

range size were significantly correlated (r = −0.96) in the alpine marmot (Gasienica Byrcyn 1997).

There is a little information on variation in home range among individuals occupying the same habitat. For species characterized as living in restricted or extended family groups (Table 5.2) all age/sex groups occupy the family territory; e.g., Barash 1973a, 1974b, Holmes 1984b, Nagorsen 1987, Mann *et al.* 1993, Kyle *et al.* 2007. Home range is individualistic in woodchucks where that of males is generally larger than that of females (Table 4.6, Ferron and Ouellet 1989), which is expected for a polygynous species that typically does not form social groups.

Home range variation in yellow-bellied marmots

Home range area varies extensively in yellow-bellied marmots. Mean home range across all years, ages, sexes, and social status was 2302 ± 86 m^2, N = 793 (Armitage 2009b). This mean home range is much smaller than the area predicted from body-mass equations (Calder 1984: 291). As discussed above, the small home range is likely the consequence of the high vegetation biomass in the habitats occupied by marmots in the East River Valley. Multiple factors affect the variation in home range: year (p < 0.001), colony (p < 0.001), age (p < 0.001), and residency (p < 0.001). Neither sex (p = 0.373) nor the propensity to make excursions (p = 0.312) significantly affected home range size.

The difference among years was partly a function of sample size and the social structure of the colony. When the territory of a single adult male included the entire colony site or when wide-ranging yearlings were present, high yearly means occurred. Lower means occurred when multiple males had small territories. Difference in home range area among colonies is related to topography and the location of home burrows relative to foraging areas. For example, at Marmot Meadow (Fig. 8.8), the colony with the smallest mean home range, burrow sites are adjacent to the meadow and marmots need to travel only a short distance for foraging. By contrast, at Picnic Colony (Fig. 8.9), the colony with the largest mean home range, home burrows may be located in the midst of a large talus slope such that marmots must traverse the talus and travel farther to the meadows. Topography also affects the shape of home ranges (see Fig. 1, Armitage 2009b) and the routes marmots use to reach foraging areas.

Adults had the largest home range (0.359 ha) and young, the smallest (0.117 ha). The means of the four age groups differed significantly from each other: adults > 2-year olds > yearlings > young.

Some individuals make excursions from the home range into other parts of the habitat. Although making excursions did not significantly affect the overall mean home range size, marmots that made excursions had a significantly larger home range than those that did not. Because of this difference, I examined the factors associated with making excursions. Young made fewer and adults, more excursions and excursions were more likely to occur in the two larger colonies than in the two smaller colonies (Armitage 2009b). Residency status also affected the occurrence of excursions. Those marmots that were present throughout the active season made more excursions then those, such as dispersers, who were present for only part of the active season.

Residency status not only affects those who makes excursions, it also affects home range area. Territorial (colonial) males had home ranges (0.562 ha) that were significantly larger than those of residents (0.222 ha) and dispersers (0.169 ha); the latter was significantly smaller than that of residents (adult and yearling females and young). Non-colonial males had the largest and most variable home ranges (5.25 ± 2.18 ha); the mean is significantly larger than that of colonial males ($t = 2.06$, $p = 0.05$). Home range size of males is determined primarily by the distribution of females; males with large territories defended widely spaced females (Salsbury and Armitage 1994a). Transients, marmots that live outside a colony, but often travel through a colony, presumably seeking residency, had home range areas that did not differ from those of dispersers. Thus, transients could explore only a relatively small area of a colony site. Most transients were present for only a few days and generally avoided social contact with residents.

The most important factors affecting residency are age and sex. All ages older than young were more likely to be dispersers or transients than were young. Females were more likely to be residents than were males. Sex ratios of young and yearlings did not differ and fitted a 1:1 ratio. Sex ratios of adults and 2-year olds did not differ but were female-biased and differed significantly from the sex ratios of young and yearlings. This shift in sex ratio is a consequence of male-biased dispersal, which produces the polygynous mating system. Patterns of residency differed among colonies, e.g., there were more dispersers and fewer residents at North Picnic than at Picnic.

Yearling home ranges are smaller than those of adults in part because many yearlings have a small home range and disperse early in the active season. The home range of yearling dispersers is significantly lower than that of yearling residents. When dispersers are excluded from home range calculations, the mean home ranges of adult, 2-year old, and yearling females do not differ (Armitage 2009b). This similarity in home range area suggests that all three age groups require about the same amount of space to fulfill their resource requirements.

Yellow-bellied marmots may shift their home ranges in the early weeks following emergence from hibernation (Armitage 1965). Dispersing females may move among several small habitat patches in an area before eventually settling in one (Wallens 1970). Many shifts are by eventual dispersers (Armitage 2009b); others involved females moving their young from one burrow to another. Reproductive females apparently shift the area used after young emerge; for five females the similarity in the area used between pre-emergence and post-emergence of young averaged only 29.8%. Home range shifts occur between years (compare Figs. 10.3 and 10.4). One adult female at Picnic concentrated her activity in one area; the similarity between two successive years was 54%. In a later year, she modified the shape of her home range and the similarity between successive years was 27%. When 6-years old, she expanded her activity over most of Lower Picnic; the similarity with the previous year was only 18%. A female may shift her home range between years to a completely new area as a consequence of agonistic behavior with another female (Armitage 1992). Solitary females may shift home ranges between years and use several small habitat patches within a year. This behavior especially occurs in females seeking a place to settle and reproduce (Wallens 1970). Some reproductive females shift home ranges by occupying a colony site in one year and a nearby satellite site in another year.

Home range shifts also occur in woodchucks (de Vos and Gillespie 1960, Ferron and Ouellet 1989). Reasons for these shifts are not clear but in some cases probably involve the avoidance of a dominant female by a subordinate animal. However, these shifts in home range indicate that home range is behaviorally plastic and that marmots modify home range based on social and environmental factors. For example, urban woodchucks had smaller home ranges, fewer burrows per individual, and shorter distances between burrows than their rural neighbors (Lehrer and Schooley 2010).

In yellow-bellied marmots, home range area is not significantly related to the density of adult females, but the degree of overlap among home ranges increases with increasing density of females (Armitage 1975). In Chapter 10 we will explore who shares resources when home ranges overlap.

Resource competition

Although marmots typically face little or no competition from ungulates for food resources, grazing by ungulates can seriously reduce forage available for marmots. Grazing by caribou can reduce mass gain in black-capped marmots by up to 50%; as a consequence marmot populations decreased from 240 to 35 individuals in two colonies and the number of juveniles was reduced in the year following caribou grazing (Valentsev et al. 1996).

Summary

Fundamental resource use is similar in all marmots, i.e., the acquisition of burrows and foraging areas. All species prefer forbs over grasses and feed selectively in that plants with required nutrients are chosen and plants with defensive compounds are rejected. Digging burrows and food selectivity modify plant community structure. Home range area varies; essentially its size is sufficient to include adequate food resources and burrow sites.

5   Evolution of sociality

Only a minority of animal species live in groups. The rarity of group living has posed special problems because the potential costs of group living seem to overwhelm benefits (Armitage 1996b). Costs include attractiveness to predators, increased likelihood of parasite and disease transmission, competition for resources, cuckoldry and misdirected parental care, and loss of reproductive opportunities. Probable benefits include predator defense and exploitation of resources, especially when they are clumped (Alexander 1974). Obviously, both individuals and groups must cope with potential predators, acquire energy and nutrients, and also solve other environmental problems; e.g., weather factors.

A useful way of thinking about sociality is to consider it in its life-history context (Armitage 1981). In essence the model considers sociality as a critical reproductive strategy. Eight hypotheses were proposed to explain the evolution of group living in rodents; six focus on life-history and ecological constraints and two rely on new fitness benefits to individuals (Ebensperger 2001). In reality a group cannot persist if individuals do not gain net benefits; in other words, the benefits of living in the group must exceed the benefits of living individualistically. Whatever life-history and ecological constraints lead to group living, individuals must benefit.

It is now generally agreed that sociality has costs and benefits. Among the benefits are increased awareness of predators through increased vigilance (Armitage *et al.* 1996, Blumstein 1996) and increased reproductive success (Armitage and Schwartz 2000); a major cost is loss of reproduction (Armitage 2007). Thus marmot societies are characterized by cooperation (benefits) and competition (costs). These will be described in terms of reproductive success in Chapters 15 and 16. In this chapter, I present a historical analysis of the probable evolutionary events that led to marmot sociality.

Coping with winter

A major environmental problem that marmots had to solve (in the evolutionary sense) was how to cope with the long period of harshness associated with winter when forbs and grasses are no longer palatable. Marmots are both too small and too big to remain active. Their small size, in comparison with caribou (*Rangifer tarandus*), wapiti (*Cervus elaphus*), and muskox (*Ovibos moschatus*), does not allow marmots to migrate or to develop sufficient insulation and fat during summer feeding to withstand prolonged periods of starvation in a cold environment (see Speakman 2000). Marmots are too large, in comparison with voles (*Microtus montanus*) and lemmings (*Lemmus sibiricus*), to forage beneath the snow and would require large amounts of dried forbs and grasses, which marmots do not eat, for maintenance.

The remaining alternative is hibernation, for which their large body size, relative to other ground-dwelling squirrels, is adaptive (see Chapter 2). Hibernation did not originate with marmots. Because marmots rose from within the ground squirrels (Harrison et al. 2003, Mercer and Roth 2003, Steppan et al. 2004), hibernation likely evolved in an early ancestor of the Marmotini. Most species of ground squirrels are either asocial, have a low sociality index (Armitage 1981, Michener 1983) or have social complexity scores lower than those of marmots (Blumstein and Armitage 1998a). These species reach maturity and reproduce when 1-year old in their second summer of life.

Hibernators have relatively slow life histories; they reproduce at lower rates and mature at older ages than non-hibernators of similar size (Turbill et al. 2011). Age of maturity in marmots is strongly affected by body size and the length of their active season.

Age of maturity

All species of marmots except the woodchuck delay maturity to age 2 years or older (Table 5.1). This delay in maturity is the major consequence of large body size; time is insufficient during the relatively short active season for young to mature before their first hibernation. The importance of time is illustrated by the first record in 2010 of a yearling yellow-bellied marmot, tagged as a young, weaning a litter (Blumstein, pers. com.). Her body mass in June was comparable to that of a 2-year-old. She moved from her natal

Table 5.1 Maturity index and ages of dispersal and first reproduction for female marmots. Values for *M. flaviventris* based on the Upper East River population.

	Maturity index (MI)				Age (years)	
Species[a]	Juvenile (first summer)	Yearling (second summer)	Two-year-old (third summer)	Dispersal	Reproductive maturity	Population mean
M. monax	0.67	0.87		< 1	1	c.1.5[b]
M. flaviventris	0.39	0.76	1.0	1	2	3.02
M. vancouverensis	0.38	0.86	1.0	2	3	3.63
M. olympus	0.42	0.77		2	3	≥ 3.6[b]
M. caligata	0.37	0.71		2	3	> 3
M. caudata aurea	0.37	0.66	0.88		3	> 3
M. c. caudata					4	
M. sibirica	0.26			3 or older	3	
M. marmota	0.39	0.79		3 or 4	2	> 3
M. bobak	0.41			3 or older	3	> 3
M. baibacina				2	3	> 4

[a] Species are listed in order of increasing sociality (Table 5.2).
[b] Author's calculation.
References in Armitage (2007), Blumstein and Armitage (1999).

burrow to a peripheral burrow (to escape reproductive suppression?). Her growth to the body mass characteristic of a 2-year-old occurred during a period of environmental warming associated with earlier snowmelt and earlier weaning that effectively increased the length of the growing season (Ozgul *et al.* 2010). The woodchuck, *M. monax*, has an active season of about 7.5 months (the active season varies with latitude, Zervanos *et al.* 2010) and is 2–4 months longer than that of any other marmot. The long growing season enables woodchucks to become independent in their first summer. As a result of the combination of large size and a short active season, the young of all other species remain in their natal environment for their first hibernation or longer (Table 5.1).

Where the active season is longer, *M. flaviventris* disperses as young. Webb (1980) developed "available foraging time" based on heat stress, day length, and the length of the growing season to serve as an index of the time actually available for mass gain. Young *M. flaviventris* at low elevation sites (<1000 m) with more available foraging time dispersed after adults hibernated whereas at higher elevation sites (>1400 m) with less available foraging time dispersal was delayed until the yearling age class (Webb 1981). Young also dispersed in late summer from the low elevation *M. flaviventris* colony in Yellowstone National Park, but disperse as yearlings in the Upper East River Valley (Table 5.1). This pattern of delayed dispersal at higher elevations was first suggested by Barash (1974a) as a major factor in the evolution of marmot societies.

Marmots likely delay dispersal until after their first hibernation to provide favorable conditions for maturation (Blumstein and Armitage 1999). The age of maturity may be estimated by calculating a maturity index (MI, sensu Barash 1989) by dividing the body mass at a specified age by the body mass of an adult, preferably at the time of emergence. Dispersal can occur when MI ≥ 0.5, but only *M. monax* reaches that value as a juvenile (Table 5.1). All other species exceed that value as yearlings, but only high-elevation *M. flaviventris* typically disperses at that age. All other species have delayed dispersal; that is dispersal occurs one or more years beyond the year in which the MI indicates that dispersal can occur.

As a consequence of delayed dispersal social groups of varying complexity are formed (Blumstein and Armitage 1998a). The formation of social groups of varied age structure leads to delayed reproduction. Marmots are capable of reproducing the year after an MI of about 0.65 is reached. Only *M. monax* reaches that value as juveniles and this species can reproduce when 1-year old. The MI of all other species exceeds this threshold as yearlings; thus all species should be capable of reproducing as 2-year olds. However, many species do not reproduce until age 3 years and the realized age of first reproduction (population mean) is always greater than the age of physiological maturity (Table 5.1).

Social systems

The retention of offspring in their natal area is expressed in four social systems (Table 5.2). These social systems are based on the ages of dispersal and reproduction. Some authors recognize only three social systems by combining my third and fourth groups into one

Table 5.2 Social systems of marmots.

Social system	Species	Characteristics
Asocial	*M. monax*	Mating system polygynous, little overlap of female home ranges, disperse as young; typically solitary hibernation
Female kin group	*M. flaviventris*	Mother:daughter:sister kin groups persist through time as matrilines, territorial male defends one or more matrilines; mating system mainly polygynous, yearling dispersal; group or solitary hibernation
Restricted family	*M. caligata, M. olympus, M. vancouverensis*	Adult male typically with one to three females, yearlings, and young; mating within the family; disperse as 2-year olds, group hibernation
Extended family	*M. baibacina, M. bobak, M. broweri, M. camtschatica, M. caudata, M. marmota, M. sibirica*	Typical family of adult territorial pair; subordinate adults, yearlings, and young; mating monogamous or polyandrous; disperse at age 3 years or older; group hibernation and social thermoregulation

group of high sociality (Allainé 2000). However, there are important differences in ages of dispersal. Marmots in the restricted family group disperse at age 2, the year before reaching reproductive maturity. Species in the extended family group disperse at age 3 or older; dispersal may occur after the age of reproductive maturity is reached (Table 5.1). One consequence of delayed dispersal is that members of the extended family group have the capacity for social thermoregulation in which non-reproductive group members warm the young during hibernation (Arnold 1988). The non-reproductive age groups responsible for providing heat to young during hibernation are not present in species with restricted families.

Assigning a species of marmot to one of the four social systems implies that the social system is fixed and obligate. However, each of these social systems is characterized by variability that implies that they are to some degree facultative. *M. monax* is the only species considered to be asocial because all young were reported to disperse (de Vos and Gillespie 1960, Anthony 1962), and there appeared to be no opportunity for the retention of young to form social groups (Table 5.2). However, in an Ohio study, 25% of the young remained with their mother through their first hibernation and moved to different burrows as yearlings (Meier 1992). In Connecticut, some juvenile females remained close to their natal home range; males were less likely to do so (Swihart 1992). In a high-density population in Maine, up to 50% of the juveniles, both males and females, delayed dispersal until their second summer. Some juveniles hibernated with their mother and/or siblings. Several females settled in home ranges adjacent to their mother. Males were more likely to disperse in their second summer and females were more likely to be recruits (Maher 2004, 2006). These philopatric young delayed dispersal even though their MI indicated that dispersal was possible. Of the females that were philopatric, 60% were recruited into the population adjacent to or within their natal home ranges (Maher 2006). Thus, some populations of woodchucks have dispersal/recruitment patterns characteristic of yellow-bellied marmots except there is no report that mothers and daughters remain and hibernate together as adults to form female kin groups.

Why does delayed dispersal occur in some populations of woodchucks in which a large percentage settle in or near their natal home range? Because woodchucks are facultatively philopatric, it is most likely that dispersal and philopatry are strongly influenced by ecological conditions (Maher 2006). The occurrence of a relatively high level of philopatry in the high-density Maine population suggests that quality living sites were limited and that remaining philopatric provided a better opportunity for occupying a quality site than did dispersing. One consequence of delayed dispersal was that only 15.8% of the yearling females reproduced, whereas in populations in Pennsylvania where all young dispersed, typically 40–60% of yearling females reproduced (Snyder and Christian 1960, Snyder 1962). This difference in the percentage of yearling woodchucks reproducing may be a consequence of reproductive suppression (Armitage 1996b) or failure to reach reproductive maturity. Reproductive suppression in marmots will be discussed in Chapter 16.

The mating system of *M. monax* is polygynous. Adult female home ranges do not overlap or overlap slightly, especially in good habitat (de Vos and Gillespie 1960). Home ranges of adult males in many populations do not overlap, but the home range of a male may overlap the home range of several females (Ferron and Ouellet 1989, Meier 1992). In Connecticut, mean overlap among reproductive females was 13.5%; among adult males, 31%; and between males and females, 36% (Swihart 1992). There is considerable seasonal variation in overlap and size of home ranges and the degree of overlap varies among woodchuck populations (details reviewed in Kwiecinski 1998). None of the variation indicates any departure from a basic polygynous mating system in which males attempt to include several adult, reproductive, asocial females in their home ranges.

Although solitary hibernation characterizes woodchucks, females in a Canadian population occasionally hibernated in pairs. Females that hibernated with another female had a similar loss of mass during hibernation as females hibernating singly (Ferron 1996); thus there was no evidence for a physiological advantage of group hibernation.

The yellow-bellied marmot social system is also facultative. Over 41 years at all sites in the Upper East River Valley, 70 males were socially polygynous, 28 males were socially monogamous, and 24 were monogamous or polygynous in different years. When males were present for more than 1 year, they formed monogamous groups 73 times and polygynous, 186 times.

The major factor that accounts for the variation in the male mating system is the number of females with whom a male mates which is determined in large part by female demography. For example, only one female may be present for several years at a colony, thus the resident male is monogamous. At the satellite sites, the mean number of resident adult females is always less than two (Table 8.1), thus in many years only one female is present and the resident male is monogamous. A male may be polygynous in one year and because of female mortality, be monogamous the following year. On the other hand, when multiple-male territories occurred at a colony, two males were monogamous and one was polygynous. The death of the polygynous male coupled with the recruitment of additional adult females resulted in both of the surviving males becoming polygynous in the next year, each associated with a matriline. The one surviving male the following year was associated with two matrilines, each with two females. These examples indicate that the mating system is dynamic and facultative and influenced by demography.

Although the yellow-bellied marmot is characterized as forming social groups of highly related females (Fig. 5.1), known as matrilines (Table 5.2), most of the time (72.6%) an adult female lives in a matriline of one (Table 5.3). Any particular female during her residency may live in a matriline with at least one additional adult; the most

Figure 5.1   A social group of yellow-bellied marmots consisting of two adult sisters and their young at Aspen Burrow at Marmot Meadow.

Figure 5.2   A social group of the steppe marmot in Ukraine. Two adults and two young are present.

Table 5.3 Matriline size of yellow-bellied marmots, 1962–2003. Data from six colony sites and 12 satellite sites.

Matriline size		Number of matrilines
1		524
2		142
3		31
4		17
5		7
Mean	1.39	
Total		721

common multi-female group is two; the overall mean is 1.39 (Table 5.3). Variation in the size of matrilines is a result of demographic processes. Mortality during hibernation and from predation reduces the size of a matriline and recruitment of daughters increases the size of matrilines (Armitage 1996b).

Although members of a matriline may hibernate together (Johns and Armitage 1979, Blumstein *et al.* 2004a) and littermate young hibernate in the same hibernaculum (Lenihan and Van Vuren 1996a), social thermoregulation has not evolved in this species. Litter size of hibernating young was unrelated to their percent survival and young hibernating together do not have lower metabolic rates than young hibernating singly; asynchrony in the torpor cycles negated possible beneficial effects of group hibernation by increasing energy expenditure (Armitage and Woods 2003). Disturbance of one young by another in the group reduced the amount of time torpid and the grouped young had a mean daily mass loss (DML; as mg d^{-1} g^{-1} immergence mass) greater than that of single young. There is no evidence that females in different matrilines share the same hibernaculum. Furthermore, there is no difference in the hibernation mortality of yearling females that disperse and hibernate singly and the philopatric female yearlings that may be group hibernators (Van Vuren and Armitage 1994a). Thus, most of the time females hibernate alone or with juvenile (young and yearlings) offspring. The subordinate adults responsible for social thermoregulation in the alpine marmot (Arnold 1990b) do not occur in the yellow-bellied marmot. Thus group hibernation in yellow-bellied marmots may function to maintain social cohesion among matriline members and juveniles rather than as a mechanism to provide social thermoregulation benefits.

The major difference between the female kin group social system and the restricted family is that dispersal occurs at age 2 in the restricted family species. Dispersal in the latter group occurs when the MI is 1.0 (Table 5.1); therefore, some factor other than MI must account for delayed dispersal. The probable reason for delayed dispersal will be discussed later.

The social system and mating system are identical in restricted family species. The Vancouver Island marmot is primarily monogamous, but three instances of polygyny occurred (Bryant 1996a, b). The social adult pair is associated with non-reproductive 2-year olds, yearlings, and young (Heard 1977, Bryant 1990). By contrast, the typical

Olympic marmot colony consists of one adult male, two adult females, occasionally with 2-year olds, and a litter of yearlings and a litter of young (Barash 1973a). More recently, of seven social groups in which a female reproduced in consecutive years, five groups consisted of one female and two groups contained two females, both reproduced (Griffin *et al*. 2007). Barash also reported that 10 of 35 colony-years had only one adult female and four had three adult females. Several of the social groups had one adult female when the other member of a pair failed to return in the next year. The number of adult females increased from zero or one to two or three in four social groups when non-reproductive 2-year olds were recruited (my calculation from data in Barash 1973a). These patterns of increase and decrease in Olympic social groups support the role of demography in determining social group structure and size.

The social system of the hoary marmot in Glacier National Park was either monogamous or polygynous; males associated with one to three adult females; yearlings and/or 2-year olds and young were usually present. An adult female was associated with yearlings or young, but not both (Barash 1974b, 1975a). In an Alaskan population, all social groups were monogamous (Holmes 1984a). In a Canadian population, both monogamous and polygynous social groups were present (Kyle *et al*. 2007). The composition of social groups consisted of one reproductive male and between one and four reproductive females. Typically up to three sexually mature, non-reproductive female and male individuals aged 3 years or older were also present in the social groups; most of these individuals were recruited from offspring within the group. The number of juveniles, yearlings, and 2-year olds varied widely among social groups. The mating system varied within groups; both monogamy and polygyny occurred in the same social group in different years. The shift between monogamy and polygyny apparently was a consequence of demographic patterns of recruitment and mortality, but this pattern was not documented for all social groups (Kyle *et al*. 2007).

The social systems categorized as female kin group and restricted family occur in one major clade of marmot phylogeny and this group of species belongs to the same subgenus (Table 5.2 and Fig. 2.3). The mating systems are similar in that they are facultative; both monogamy and polygyny occur, but there is no evidence for polyandry. Available evidence suggests that adult females are recruited from within the social group (Barash 1973a, Armitage 1998, Bryant 1998), but long-term genealogical studies are needed to verify this evidence. Males apparently are immigrants to the social group. Genetic analysis failed to detect inbreeding in *M. vancouverensis* (Bryant 1990, Kruckenhauser *et al*. 2009), *M. caligata* (Kyle *et al*. 2007), or *M. flaviventris* (Schwartz and Armitage 1980) and this is likely because males normally do not remain in their natal social group whereas females may do so.

All the remaining species of marmots are broadly characterized as living in family groups (Bibikow 1996) or in groups of high sociality (Allainé 2000) or in an extended family and I use the latter term for all of them. The major characteristic of this social group is the presence of non-reproductive, subordinate adults associated with a dominant, reproductive, territorial pair (Table 5.2). Although usually only the dominant female weans offspring; multiple litters may occur rarely in *M. caudata* (Rymalov 1994), *M. baibacina* (Mikhailuta 1991), *M. marmota* (Barash 1976b), and *M. vancouverensis* (Bryant 1996a).

These instances probably represent facultative polygyny. In the alpine marmot, the family group can include over 10 individuals (Mann *et al.* 1993); the median group composition consists of a territorial pair, three infants, three yearlings, and two subordinate adults (Arnold 1990a, b). Although copulations usually occur in burrows, females were observed to copulate in rapid succession with all the males present (Müller-Using 1957, Arnold 1990b). Genetic analysis verified that the territorial male did not sire all the young born in the family (Arnold *et al.* 1994); this situation typically occurred in multiple-male groups. Polyandry probably represents a direct fitness cost for the territorial male and a direct fitness benefit for the subordinate males (Allainé *et al.* 2000).

Why should the territorial male tolerate the presence of subordinate males if in so doing he incurs a fitness loss? One answer to this question is that overall the subordinate males increase fitness of the territorial male. Subordinate males act as helpers during hibernation (Allainé and Theuriau 2004). The presence of helpers increases survival of young and the territorial pair (Arnold 1990b). Juvenile survival increased precisely with the number and proportion of subordinate males in the hibernating group (Allainé *et al.* 2000). Sex ratio at weaning is male-biased and this bias is maintained in the subordinate males (age 2 or older) in some populations where males disperse later (Arnold 1990b), but in other populations there is a high dispersal rate of 2-year-old males (Allainé *et al.* 2000). The sex ratio of offspring was biased toward males when helpers were absent, but not when helpers were present (Allainé 2004). Although groups with older, subordinate adult males experienced significantly reduced mortality than groups without subordinates (Arnold 1990a), there is a cost to the territorial male. The probability of monopolizing reproduction decreases as the number of sexually mature male subordinates increases and the dominant male is likely to lose dominance (Lardy *et al.* 2012). There is an apparent energetic cost; dominant males lose body mass and this loss is associated with the loss of dominance. Males that lost paternity had a dominance tenure of <5 years; those males able to secure paternity had dominance tenures as long as 11 years. The fitness of individual males varies considerably, but despite short tenure for some individuals, there are higher fitness payoffs with helpers than without.

The presence of helpers indicates that alpine marmots are cooperative breeders (Blumstein and Armitage 1999); increased survival of the adults and offspring provides direct benefits to the territorial pair. Increased survival provided by helpers more than compensates for the low percentage of young fathered by subordinate adults. We can conclude that the territorial male gains direct fitness from the presence of subordinate males and that subordinate males serve as helpers because they gain indirect fitness by increasing the survival of the young and may gain direct fitness by fathering some of the young and by achieving dominance in their natal territory (Lardy *et al.* 2012). Also, by serving as helpers, dispersal is delayed until males are of a larger, more competitive size and have a higher probability of obtaining residency in a family.

An optimality model developed by Allainé and Theuriau (2004) indicates that the optimal number of helpers for males is 1.6 and, for females, 2.0; the mean number of helpers found in families was 1.7. This mean of 1.7 subordinate adults is similar to the median value of 2.0 subordinate adults reported by Arnold (1990a). There was no further reduction in winter mortality of juveniles when more than two subordinate adults were

Table 5.4 Variation in the social structure of extended family groups.

Species	Location	Social structure	Reference
M. marmota	Mount Cimone, Italy	Over 7-year period, social group consisted of a territorial pair and yearlings or young. None of four 2-year olds returned as 3-year olds.	Sala *et al.* 1993, 1996
	France	Only one of four families had a 2-year-old female resident, no 2-year-old male resident, yearlings present in two and young in three.	Perrin *et al.* 1993a
	Gran Paradiso National Park, Italy	Both monogamous and polyandrous groups occurred in 8 years in 27 sexually mature reproductive groups in three family areas. Group size varied from 2 to 11 (excluding infants), mean was 4.7. Family territories stable and hosted different social groups.	Lenti Boero 1999, 2003a
M. bobak	Central Kazakhstan	One mature male and female plus old non-breeding males and 1–2-year olds and young. Subordinate adults not reported.	Mashkin 1991
	Ukraine	Adult female and two or three adult males, no subordinate adult females. Dispersal apparently at age 2.	Nikol'skii and Savchenko 1999
M. c. camtschatica	Kamchatka	Average family group: 1.8 adult males, 1.7 adult females, 4.3 young.	Mosolov and Tokarsky 1994
M. c. bungei	Yakutia	Family composition: adults and cubs 53.8%; adults, yearlings and cub 7.7%; adults, 2–3-year olds and cubs 15.4%; adults and yearlings 15.4%; adults and 2–3-year olds 7.7%.	Yakovlev and Shadrina 1996

present (Arnold 1993b). The social structure of many alpine marmot families differs from the optimum (Table 5.4). The differences likely are a consequence of demographic changes. Small group size in the Gran Paradiso population resulted from the lack of recruitment of subordinate adults; 7 of 14 2-year olds were present as 3-year olds if the parents were still present, only one was a male. There was considerable population turnover; residents were evicted by immigrants, some of whom were from adjacent families; mortality of pairs hibernating with young was high; satellite males replaced resident males; some predation occurred; some 3-year olds were chased from the colony; and six animals dispersed (Lenti Boero 1994, 1999, 2003a). The high mortality and subsequent turnover in this colony probably is related to its suboptimal location; it had a northeast exposure where snow cover would be expected to last longer in the spring; mortality of alpine marmots is higher when snowmelt is late (Arnold 1990b). Detailed, long-term studies of other members of the extended family social system (Table 5.2) are unavailable. For example, the family structure of *M. baibacina* is simpler in less favorable habitat than in more favorable habitat (Mikhailuta 1991). This report is consistent with the general interpretation that extended families do not develop or persist in suboptimal habitat, but what is lacking, as in most studies of this social group, is a detailed age/sex structure of the family group and its variation over time.

Either there is more variability in these social groups or possibly there may be an undescribed fifth social system (Armitage 2007). The number of adult males in families of the steppe marmot (Fig. 5.2) varied between central Kazakhstan and Ukraine (Table 5.4). In contrast to the typical alpine marmot family, subordinate adult females were not reported at either location (Mashkin 1991, Nikol'skii and Savchenko 1999).

Group structure varies between black-capped marmot subspecies. In *M. c. bungei* 42–57% of the adults lived alone or in pairs (Yakovlev and Shadrina 1996), no such groups were reported for *M. c. camtschatica* (Mosolov and Tokarsky 1994). The considerable variation in family structure in *M. c. bungei* is likely a consequence of a demography in which less than half of the families weaned young in any given year; thus the age composition among families would be asynchronous.

Social systems versus mating systems

It was generally accepted that social systems and mating systems were identical. For example, in monogamous species, it was assumed that reproduction was confined to the social pair. However, the widespread occurrence of multiple mating in rodents indicates that social and mating systems may not be congruent (Waterman 2007). Who the parents are is critically important for determining the reproductive success of individual males and females.

Because *M. monax* and *M. flaviventris* males and females do not form social groups, the basic question is whether an adult male fathers all the offspring in his territory. Isozyme analysis did not detect any instance in which the territorial male did not father all the offspring (Schwartz and Armitage 1980). However, multiple paternity occurred in 63% of woodchuck litters in a high-density Maine population (Maher and Duron 2010). Up to four males sired offspring within a litter and the average was 1.8 males/litter. Although the males were generally not closely related to the females, some pairs included close kin, e.g., mother:son, father:daughter, non-littermate sibs. Females and their mates typically lived near each other and their home range overlap was three times greater than that of randomly chosen pairs. In contrast, there was very little overlap among putative fathers. As a consequence of this mating activity, average *r* of littermates varied widely. There is little opportunity for mating outside the family in those species that mate in the hibernation burrow before emerging above ground (see Table 3.5).

Extra-pair paternity (EPP) occurs in the alpine marmot when a subordinate adult male mates with the territorial female (Arnold 1990a). Because the adult, territorial male does not exclusively mate with the territorial female, EPP must occur in a polyandrous system. However, the mating system is generally reported as monogamous (Table 5.2); the critical question is does mating occur outside the family? All patterns of mating, i.e., monogamy, polyandry, promiscuity (multiple mating by adult females), occurred in a French population (Goossens *et al.* 1998). Eight of 35 litters were sired by males from outside the family group; these matings accounted for 17.2% of the young. In the one instance of polyandry, the subordinate male was a brother of the dominant male and unrelated to the dominant female.

Under what conditions is EPP likely? Two possibilities come readily to mind: (1) the dominant male is unable to prevent other males from mating with dominant females; (2) the dominant female chooses to mate with another male for fitness benefits. In alpine marmots, EPP increased with the number of subordinate males; no EPP occurred when subordinate males were absent (Cohas *et al.* 2006). Thus, possibility number one is supported. Females may gain genetic benefits; females that mated with dissimilar extra-pair males simultaneously increased offspring heterozygosity and promoted genetic diversity within litters (Cohas *et al.* 2007a). Although heterozygous young survived better than homozygous young over their first year, extra-pair young (EPY) were not more heterozygous than within-pair young (Cohas *et al.* 2007b). However, EPY had higher survival as juveniles, yearlings, and 2-year olds. Although individual multilocus heterozygosity was positively correlated with juvenile survival, there was no correlation with survival of older individuals, which supports the prediction that differences in survival among individuals are maximal early in life (Cohas *et al.* 2009). Extra-pair males were preferentially transient individuals who originated from distant family groups likely to be less genetically similar to the female than the pair male. In general, the alpine marmot mating pattern supports the hypothesis that females choose extra-pair mates to increase the genetic quality of their offspring (Cohas *et al.* 2007b); thus, possibility number two is also supported.

A final question is to what extent does EPP occur in the alpine marmot mating system? In their natal territory, 12–22% of subordinates become dominants and about 50% of the subordinates become dominant in the immediate neighborhood; i.e., take over an adjoining or nearby territory. Where dominant individuals, all offspring and the number of mature male subordinates in the family group were known, 36 of 369 offspring were EPY and 20 of 103 litters contained at least one EPY (Cohas *et al.* 2008). Thus, about 20% of the litters had EPY that represented about 10% of the young. Clearly, the territorial males father most of the offspring and becoming territorial provides the greatest fitness benefits.

The relatively small number of EPY suggests that female choice is limited, as indicated by the lower incidence of EPP when subordinate males were absent and the territorial male could defend against EPP. Female choice is evident in the pattern of EPP. The probability of EPY in a litter is higher for low and high values of genetic similarity between mates. Extra-pair mates were never closely related to the dominant female, but were sometimes related to the dominant male. The dispersal rate of subordinate females increased with their relatedness to the dominant male, but not to that of the dominant female. Subordinate males that obtained EPP were unrelated to the female. Only one mating between mother and son was observed in 20 years. This pattern indicates that inbreeding avoidance is a major factor in female mate choice (Cohas *et al.* 2008). Overall, the mating system is enhanced by the social system in which the territorial male fathers most of the offspring, but in which the territorial female has some opportunity to increase her fitness by avoiding both in- and out-breeding depression.

Summary

In brief, marmot sociality evolved as a consequence of large-bodied individuals living where the growing season is short such that young are retained in their natal group for one or more additional years of growth in order to reach maturity. The social systems are diverse and facultative, and are influenced by demography, habitat conditions, and the mating strategies of males and females.

6 Body-mass variation

Because hibernation is a major component of the annual cycle, marmots must add sufficient mass during their short active seasons to ensure survival through the long period of torpor. The central role of time as a limiting factor in seasonal mass gain and survival will be described.

A question to be explored is how to account for the body-mass diversity among marmot species (Table 2.1). The range in diversity suggests that some factor(s) other than torpor is involved.

Body mass

Among yellow-bellied marmots, body mass varies with age and sex. Mass of male young (range 340–910 g, $\bar{x}$ = 590 g) is significantly (p < 0.001) larger than that of females (range 230–790 g, $\bar{x}$ = 500 g) at weaning (Armitage *et al.* 1976). This difference in body mass is maintained through older age classes (Table 6.1). The pattern of larger mass with increasing age also occurs in members of the restricted and extended family groups (Barash 1973a, 1974b, 1989, Bryant 1998, Arnold 1990a, Blumstein and Arnold 1998).

The difference in body mass among the age groups allows assignment of age to marmots not trapped as young, as long as the marmot is trapped in the first half of the active season. Marmots older than 3 years cannot be reliably distinguished from 3-year olds. In the solitary woodchuck, young are smaller than yearlings who are smaller than adults, but by late summer yearlings cannot be reliably distinguished from adults (Snyder *et al.* 1961).

At all ages yellow-bellied marmot males, on average, are larger than females (Table 6.1). Body mass of adult males is larger than that of females in the Olympic marmot (Barash 1973a), woodchuck (Snyder *et al.* 1961), Vancouver Island marmot (Heard 1977), hoary marmot (Barash 1989), long-tailed marmot (Blumstein and Arnold 1998), and gray and steppe marmots (Arnold 1993a), but not in alpine marmots (Lenti Boero 2001).

Rates of mass gain

Mean rates of mass gain of yellow-bellied marmots vary from 16.8 g/day for reproductive females to 23.8 g/day for male yearlings (Table 6.2). Rates are significantly higher for males than for females (Table 6.3), which accounts for the larger male size at all ages. Rates of mass gain of yearlings is greater than that of adult females and young and the rate of mass gain of young and 2-year-old females exceeds that of older females (Table 6.3)

Table 6.1 Body mass (g) at emergence for different age groups of yellow-bellied marmots. All values rounded to nearest gram. Data for yearlings from Armitage *et al.* (1976). M = male, F = female.

				Statistical Comparisons		
Age (years)	Sex	Mass ($\bar{x} \pm$ SE)	N	Sex	Age	
1	M	1980 ± 49	48	t = 33.6,	1 vs. 2	t = 8.6, p < 0.001
	F	1720 ± 44	56	p < 0.001	1 vs. 2	t = 5.4, p < 0.001
2	M	2445 ± 66	25	t = 40.2	2 vs. 3	t = 4.6, p < 0.001
	F	2035 ± 38	44	p < 0.001	2 vs. 3	t = 6.2, p < 0.001
3 or older	M	2977 ± 94	22	t = 4.6		
	F	2547 ± 72	31	p < 0.001		

Table 6.2 Mean active season mass gain rates (g/day) for yellow-bellied marmots. N = number of individuals, SE = standard error, NR = non-reproductive, R = reproductive. Table modified from Salsbury and Armitage (2003).

		Growth rate	
Sex/Age	N	Mean	SE
Female young	73	19.23	0.86
Male young	71	19.11	0.70
Female yearlings	60	19.72	0.56
Male yearlings	58	23.76	1.14
Female 2-year olds (NR)	34	22.01	3.03
Female 3-years old or older (NR)	39	17.61	1.73
Female adults (R)	31	16.80	1.44
Male adults (R)	58	21.55	1.76

Young, yearlings, and 2-year-old females (whose body mass averages about 76% of adult mass) have significant somatic growth in addition to accumulating fat for hibernation. The body size of yearlings and 2-year-old females at the time of immergence is, in part, a consequence of their having the longest growing season of all the age-sex groups (Table 6.4); they initiate mass gain soon after emergence.

Reproductive females have the lowest rates of mass gain (Table 6.2) which are significantly lower than those of non-reproductive females (Table 6.3). The lower rates are a consequence of the time spent in lactation when marmots gain little or no mass (Armitage *et al.* 1976).

Time is a limiting resource that truncates seasonal growth for both young and reproductive females. On average, reproductive females gain mass for only 42 days after weaning their young. The lack of time is a significant factor related to winter mortality. Of 20 females that weaned litters after mid-July, 9 died as did 59 of their 82 young. The survivorship rate of 0.55 for the adults is considerably lower than the overall population average of 0.71 and the rate of 0.28 for young is much less than the overall average of 0.53 (Oli and Armitage 2004). I also examined survival of young from early (before mid-July) and late (after mid-July) litters of the same female in different years or among females in the same year at

Table 6.3 Analysis of variance and Tukey simultaneous post-hoc tests for the effects of age, sex, location, and year on mass gain rates for yellow-bellied marmots. Only the significant post-hoc comparisons are presented. R = reproductive, NR = not reproductive, d.f. = degrees of freedom. See Salsbury and Armitage (2003) for additional details on the analyses.

General linear model ANOVA

Variable	d.f.	F	p
Age	1	6.3	0.013
Sex	4	18.6	< 0.001
Year	31	4.7	< 0.001
Location	56	3.3	< 0.001

Tukey simultaneous tests

Comparison	T	p
Young < yearlings	4.5	0.0001
Young > 3-year olds (NR)	6.0	< 0.0001
Yearlings > 3-year olds (NR)	8.4	< 0.0001
Yearlings > female adults (R)	3.3	0.0084
2-year-olds (NR) > 3-year olds (NR)	5.1	< 0.0001
3-year-olds (NR) > 3-year olds (R)	5.3	< 0.0001
Males > females	2.5	0.0124

Table 6.4 The number of days of mass gain for yellow-bellied marmots. All values are mean ± standard deviation. Data from 1990 through 1993 when marmots were trapped early and late in the season. (N) = number of individuals. Mass gain begins soon after emergence from hibernation. During lactation, there is typically no mass gain.

Age/sex group	Number of days
Reproductive females	
pre-lactation	39.5 ± 9.9 (15)
post-lactation	42.4 ± 5.7 (10)
Non-reproductive females	
Two-year-old	85.9 ± 9.1 (3)
Three-year-old or older	70.0 ± 18.6 (7)
Adult males	64.3 ± 14.8 (6)
Yearling females	93.1 ± 6.7 (10)
Yearling males	67.8 ± 12.1 (5)
Young females	58.3 ± 14.5 (27)
Young males	61.2 ± 13.6 (30)

the same site. Of 80 young from early litters, 70% survived whereas significantly (G = 18.8, p < 0.001) fewer (26.5% of 68 young) survived from late litters. These results indicate that time, not habitat or female differences, is the critical factor affecting survival.

Growth rates varied among locations and across years (Table 6.3). Relatively low growth rates of location/year combinations frequently occurred in years of low rainfall, but also occurred in years of above normal rainfall (Salsbury and Armitage 2003). High rainfall, associated with cool, windy conditions, decreases above-ground activity, especially of young (Melcher *et al.* 1990), thus reducing foraging time, and as a consequence, reducing growth rates. There is a complex interplay among elevation, habitat quality, and weather patterns that affect growth rates; the relative contributions of these factors are yet to be determined, in part because weather variables are not available for each site. Environmental factors, emergence mass, and year of birth affected post-weaning growth rates of young alpine marmots whose mean growth rate (20.0 ± 0.4 g/day) (Allainé *et al.* 1998) does not differ statistically from that of yellow-bellied marmots. Most likely growth rates of all species of marmots are similarly affected.

Young that are larger at weaning are also larger at hibernation. Further, later weaning dates resulted in lower predicted mass at hibernation, which is associated with lower survival and accounts for the high mortality in late litters. Weaning date and weaning mass were not significantly related at any location for yellow-bellied or alpine marmots (Allainé *et al.* 1998); larger mass at weaning does not compensate for later weaning.

The number of days that yellow-bellied marmots grow during the active season is affected, in part, by the circannual rhythm. Even the provision of additional high-energy and high protein food extends the time of mass gain by only a few days (Woods and Armitage 2003a). Adult males and non-reproductive 3-year-old or older females gain mass, on average, for 64 and 70 days, respectively (Table 6.4). In effect, they cease mass gain up to 35 days before hibernation. This period of mass stasis occurs in all the age-sex groups and the length of stasis varies among years (Armitage 1996a). Yellow-bellied marmots gain mass as early in the season as possible and may complete mass gain (about a 41.6% increase in mass) up to 3–5 weeks before immergence, which reduces the probability that food quality will decline before mass gain is completed.

In general, the peak in plant biomass occurs during the period of marmot mass gain (Fig. 6.1). Plant biomass is low during gestation and lactation, which partially explains why

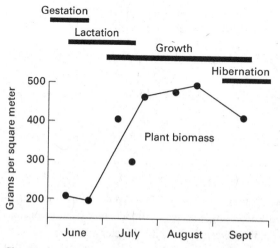

Figure 6.1   Changes in plant biomass in relation to yellow-bellied marmot phenology. Values are means calculated from Kilgore and Armitage (1978) and Frase and Armitagé (1989).

lactating females make little or no mass gain. Plant biomass decline in late summer explains, in part, why late-weaning females and their young experience high mortality. The decline in biomass is likely another reason why marmots tend to complete growth by the time plant biomass decreases. The decline in plant biomass does not indicate the change in the quality of the plants as senescence and the high abundance of plants that marmots do not eat decrease both food choice and food quality (see Chapter 4).

Why do individuals that have completed mass gain delay hibernation, which is expected to follow the peak of body mass (Davis 1976)? Marmots are susceptible to predation and presumably could avoid most predators by plugging their burrows and hibernating. Three explanations for the delay are plausible. First, adult males defend their territories and early hibernation may allow a competitor to achieve residency. Second, reproductive females spend much time in wariness behaviors while their young are foraging; investing time in anti-predator behavior may be a form of parental investment to increase the probability that their young will survive to hibernate. Third, energy use during early hibernation would not be available the following spring when it may be needed if snowmelt is late or the weather is cold and stormy and little or no foraging is possible. The end of winter is unpredictable; energy saved by allocating sufficient time to foraging to maintain body mass in the weeks leading up to hibernation may be essential 8 months later.

Body-mass variation among species

All marmot species gain mass during the active season and the accumulated fat is critical for survival. Can the relationship between body mass and fat accumulation and use explain the variation in body size of marmots? One possible relationship is that body mass or loss of mass during the hibernation period (the difference between immergence mass and emergence mass) is related to the length of the hibernation period. However, there is no correlation among species between measures of mass or mass loss and the length of hibernation (Armitage and Blumstein 2002). Mass loss and immergence mass are highly correlated ($r = 0.91$), which suggests that species are heavier because they require more mass to survive the hibernation period (Armitage 1999).

Marmots can increase mass in two ways: (1) have a larger skeletal frame and (2) increase the amount of mass per frame. Emergence mass and immergence mass are linearly related to body length, but a stronger relationship exists between immergence mass (IM) and body length (BL): IM = 2064.4 g + 14.87X BL (mm); p = 0.008, $R^2 = 0.49$ (Armitage 1999).

Mass change, the difference between immergence and emergence mass, is curvilinearly related to body length (p = 0.026, $R^2 = 0.37$); larger species have a relatively larger change in mass. Thus, both mechanisms of mass increase occur; the skeletal frame is enlarged and more mass per unit of frame is deposited. The latter mechanism has limitations; a rotund marmot would lose both speed and agility. Marmots do not become rotund, body mass did not affect maximum running speed in golden marmots, *M. caudata aurea* (Blumstein 1992). This result suggests that natural selection has acted against developing the ailments of obesity (Young and Sims 1979) in marmots in order to maintain running speed, which doubtless is critical for escaping from predators or aggressive conspecifics, and permits marmots to prepare for hibernation well in advance of any late season environmental factors, e.g., drought, that could negatively impact survival during hibernation.

Why should there be such a wide range of mass loss during the hibernation period? Two factors during torpor can increase mass loss, the time spent in deep torpor (Armitage *et al.* 2000) and low temperature in the hibernaculum. The time spent in deep torpor, when metabolism is at its lowest, varies among marmot species; the less time in deep torpor the greater is the mass loss (Armitage *et al.* 2000). In general, when environmental temperatures decline to approximately 5°C or lower, metabolism increases significantly (Vasiliev 1992, Heldmaier *et al.* 1993a, Armitage *et al.* 2000, 2003). The effect of low temperature on metabolism and mass loss depends on how long the hibernating marmot is exposed to low temperature; this exposure is affected by the amount and timing of snow cover (Arnold 1990b, Arnold *et al.* 1991). Prolonged snow cover increases mortality of alpine marmots (Grimm *et al.* 2003) and patterns of snowfall and length of winter affect young and adult survival, variation in population densities, and variation in net reproductive rate of yellow-bellied marmots (Schwartz and Armitage 2005). These survival and reproductive effects of prolonged snow cover probably explain why marmots prefer to live on south-facing slopes where snow melts earlier than on north-facing slopes (Armitage 2000). Prolonged snow cover probably explains the large body size and mass loss in the Himalayan marmot whose lower elevational distribution is about 3000 m (Nikol'skii and Ulak 2006).

The important relationship between body size and climatic factors is indicated in patterns of variability in intraspecific body size. For *M. bobak*, *M. baibacina*, *M. sibirica*, *M. camtschatica*, *M. menzbieri*, and *M. caudata* body mass and body length in each species increase from south to north and with elevation above sea level (Terekhina and Panteleyev 1994). Similar relationships occur among North American marmots. The most northern subspecies of the hoary marmot, *M. c. caligata* is among the largest, whereas the more southerly *M. c. okanagana* is smaller. High-elevation *M. c. okanagana* may be much larger than low elevation conspecifics. Among subspecies of yellow-bellied marmots, *M. f. nosophora* in southwestern Alberta is large, whereas *M. f. engelhardti* in southwestern Utah is among the smallest. Although latitude and elevation may function as rough estimates of winter conditions, much further study of the relationship between body size and climatic conditions is needed in which actual climatic data are used rather than latitudinal/elevational surrogates.

Factors other than climate, such as reproductive strategies affect body size and mass loss; e.g., several species of marmots lose considerable mass after the termination of torpor because they become euthermic and mate before emergence (Table 3.5).

The discussion thus far indicates that mass loss is related to body size and life-history traits. A factor analysis of 12 traits (6 measures of mass and mass loss, 2 measures of the length of the active/hibernation seasons, and 4 life-history traits) produced 4 interpretable factors that explained 93.2% of the variation. Factor 1, the mass factor, included measures of mass and mass loss. Factor 2, the time factor, included the length of hibernation and age of first reproduction. Factor 3 was characterized by emergence mass and factor 4, a breeding factor, included time of breeding and litter size (Armitage and Blumstein 2002). We concluded that a complex interplay among these traits (and possibly others) accounts for the variation in mass and mass loss among marmots.

Body mass and survival of young

Young is the most vulnerable age group to suffer mortality during hibernation; mean survivorship is less than that of adults and may be as low as about 50% (Barash 1989, Lenti Boero 1994, Bryant 1996b, Blumstein and Arnold 1998, Schwartz *et al.* 1998, Farand *et al.* 2002). Because body mass is a critical factor in young survival and could be affected by time of weaning and snowmelt, I used data from 26 years to explore the possible relationships of time of weaning, time of snowmelt the following spring (measured as the time bare ground appears), and body mass of young on August 30 (Julian day 242), when mass gain typically is completed, to the survival rate of young.

First, all animals were combined into a regression analysis in which each young was recorded as having a mass above or below the group mean and whether the young survived hibernation. Survival was significantly higher in young whose mass was above the mean (G = 32.4, p < 0.001). Next, I grouped the young into 12 mass categories; a linear regression of survivorship against mass was highly significant (Fig. 6.2, p < 0.0001). A young with a mass of 1800 g or larger had an 85% probability of surviving whereas a young weighing <1200 g had only a 35% probability of survival. However, only 70 (13.1%) young weighed <1200 g.

Finally, I analyzed body-mass data for 536 young (259 males, 277 females) in a multiple regression of survivorship as a function of sex and mass. Survivorship was not significantly affected by sex (p = 0.407), but was positively related to body mass (p < 0.0001). The lack of a sex effect was unanticipated because male young are

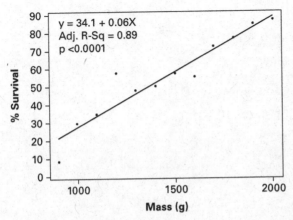

$$y = 34.1 + 0.06X$$
Adj. R-Sq = 0.89
p <0.0001

Figure 6.2 The relationship between survival and mass of young yellow-bellied marmots. Young are grouped into 100 g intervals; e.g., 900–999 g; 1000–1099 g, etc.

significantly larger than female young at hibernation (1628 g vs. 1523 g, p = 0.001). The lack of a sex effect may reflect the contributions of the skeletal frame to body mass. Young not only add fat for hibernation they also increase body length, which may double in gray marmots in 60 days (Bibikow 1996). Adult male yellow-bellied marmots have greater length then adult females (Nee 1969) and this added skeletal frame probably accounts for much of the sex differences in mass.

Up-valley (UV) young are significantly lighter than down-valley (DV) young at hibernation (1313 g vs. 1667 g, t = 5.9, p < 0.001), but survivorship does not differ between the two areas (DV young = 0.577, UV young = 0.571, p > 0.9). The overall results strongly indicate that typically young marmots gain sufficient mass to survive hibernation and that the higher masses of males and DV residents do not confer additional survival benefits.

Mean values of time of weaning and mean time of snowmelt the following spring did not significantly affect survivorship of young in either the UV or DV areas either in regression or rank correlation analyses. Hibernation mass tended to be larger when weaning dates were earlier (UV: $R_S = -0.341$, DV: $R_S = -0.366$, p < 0.1), but apparently only litters that were weaned unusually late suffer high mortality because of low mass gain (Armitage *et al.* 1976).

It is not surprising that there is no consistent pattern of winter mortality. Mortality varies with the local habitat and the range of values that affect mortality are unknown. Weather and snow-cover information collected at a diverse array of sites would greatly assist in untangling the complex relationship between weather and survival. The effects of weather will be discussed in Chapter 8.

7 Hibernation energetics and the circannual rhythm

Marmots have a short active season, the homeothermal period, in which to mate, reproduce, grow, and accumulate fat for the inactive season, the heterothermal period (Fig. 7.1). Sufficient energy must be obtained during the homeothermal period for maintenance, growth, and reproduction. Because marmots do not store food, adequate fat must be accumulated during the active season to provide for maintenance during the heterothermal period, i.e., hibernation. Reproduction can occur only when energy exceeds that needed for maintenance. The widespread occurrence of reproductive skipping (Table 3.4) indicates that marmot reproduction is energy limited. Thus, we expect that marmots should reduce energy expenditures for maintenance in order to allocate more energy to growth (including fattening) and reproduction. In other words, energy conservation should characterize marmot energetics.

Hibernation

The major adaptation for conserving energy is hibernation, which occurs in all species of marmots and evolved in an ancestor of the monophyletic *Marmota* (Armitage 2008). Hibernation is characterized by a series of torpor bouts. Each torpor bout consists of three phases: (1) descent into torpor from euthermy, characterized by decreases in oxygen consumption and body temperature (T_B); (2) torpor, when T_B <30°C; (3) arousal, when T_B and oxygen consumption rapidly increase to euthermy (Fig. 7.2). The length of a torpor bout in the yellow-bellied marmot is shorter early and late in hibernation and longer during mid-hibernation (French 1990). This pattern of torpor appears universal in hibernating Marmotini (Lyman *et al.* 1982, French 1986) and occurs in many hibernating mammals (Geiser and Ruf 1995).

The variation in bout length raises the question of how the timing of arousals is regulated. Lyman *et al.* (1982) suggested that regulation was by a circadian pacemaker. Circadian rhythms of activity were reported for captive alpine marmots (Cochet *et al.* 1992) and for T_B in laboratory yellow-bellied marmots (Florant *et al.* 2000, Woods *et al.* 2002a) and in field marmots prior to and after hibernation, but not during deep hibernation where ambient temperature varied from 5°C to 7°C (Florant *et al.* 2000). By contrast circadian rhythms of T_B were maintained throughout hibernation in the woodchuck (Zervanos *et al.* 2009).

The differences in circadian cycles between the yellow-bellied marmot and the woodchuck may be related to the duration of euthermy during a bout and body temperature during torpor. Euthermic bouts of woodchucks were as long as 51 h; those of

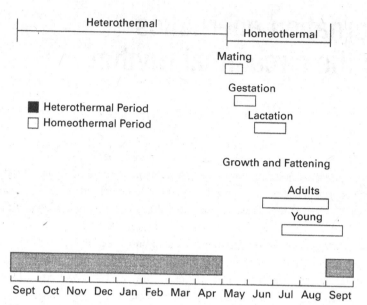

Figure 7.1   The annual activity cycle of the yellow-bellied marmot.

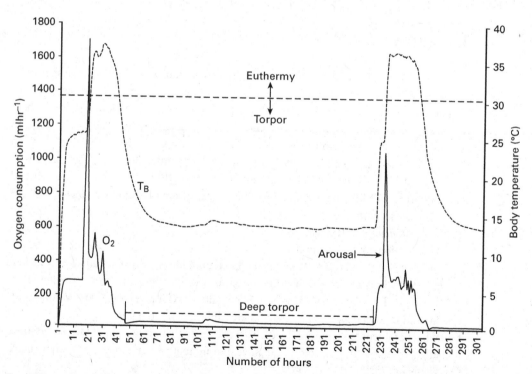

Figure 7.2   A typical torpor bout of a captive young female yellow-bellied marmot maintained at 10°C in constant darkness.

yellow-bellied marmots were too short to detect whether a circadian rhythm was present. Woodchucks maintained a higher mean T_B (females, 10.9°C; males 12.3°C) than yellow-bellied marmots (<10°C). The higher T_B of woodchucks may allow the pacemaker to continue working, or the pacemaker may run during euthermy and be turned off during torpor (Florant *et al.* 2000). More refined T_B measurements of yellow-bellied marmots during torpor in the field may resolve the question of whether a circadian pacemaker functions to regulate the timing of arousal.

The higher body temperature and longer periods of euthermy of male than of female woodchucks during hibernation apparently is related to male reproductive strategy. Some testicular development occurs prior to immergence; spermatogenic activity is greatest at emergence (Christian *et al.* 1972); the testes develop during euthermy (Barnes 1996) and are scrotal at emergence (Concannon *et al.* 1990). Thus, higher T_B and longer periods of euthermy prepare males for reproductive activities upon spring emergence. The longer time spent in euthermy has energetic costs and meeting these costs may be one advantage to larger size in males.

Hibernation can become costly when low burrow temperatures impose metabolic costs on the hibernator. Both the woodchuck and yellow-bellied marmot hibernate at burrow temperatures that remain above the lower critical temperature of about 5–6°C (Armitage *et al.* 2003), but the Alaska marmot may encounter a minimum hibernaculum temperature of −14.97°C, with a 7 month mean of −7.33°C (Lee *et al.* 2009). There are no reported metabolic rates for Alaska marmots, but some insight into how it copes with low hibernation temperatures can be obtained by looking at its body temperature during hibernation in the wild.

The minimal body temperature of *M. broweri* was 1.01°C when defending a 15.5°C thermal gradient. Neither *M. flaviventris* nor *M. monax* permit their body temperatures to decrease below about 5°C in the laboratory when ambient temperature was 2°C (Armitage *et al.* 2000, 2003), a temperature lower than the 5–7°C recorded for field animals. A critical factor involved in energy savings is the length of time spent in deep torpor (Florant *et al.* 2000). *M. broweri* was torpid 87.5% of the time during 224 days of heterothermy. This value is similar to the 92.7% for adult yellow-bellied marmots hibernating in the laboratory at 6°C (Armitage *et al.* 2003) and is much larger than the mean of 52.9% for laboratory woodchucks (Armitage *et al.* 2000). However, during midwinter when mean torpor bout length was 13.9 days, Alaska marmots were torpid 95.2% of the time. Similarly, during the midwinter period, laboratory yellow-bellied marmots at 6°C were torpid 93.3% of the time. In comparison with other marmots measured in the field, Alaska marmots cope with a severe environment by tolerating a depression of body temperature, a long period of hibernation (224 days), and longer torpor bout length (Lee *et al.* 2009).

The circannual rhythm

Hibernation is a phase in a circannual rhythm which drives the annual cycle and the transitions between the heterothermal and homeothermal stages. Under constant photoperiod

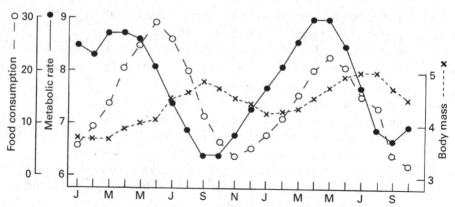

Figure 7.3   The circannual cycle of metabolism (ccO$_2$/g Bm/h), food consumption (g/kg Bm/day), and body mass (kg) in captive yellow-bellied marmots, maintained at 20–22°C at L:D 12:12 photoperiod.

and environmental temperature in the laboratory, circannual rhythms of food consumption, body mass, and metabolism (Fig. 7.3) are expressed in yellow-bellied marmots (Ward and Armitage 1981a). Circannual cycles of food consumption (Davis 1967, Fall 1971), body mass (Davis 1967b), metabolism (Bailey 1965, Armitage 2008), and thyroxine and prolactin levels in males (Concannon *et al.* 1999) occur in woodchucks and probably occur in all species of marmots. The rhythm is expressed in laboratory alpine marmots by a seasonal decline in metabolism and a shift in the thermoneutral zone (TNZ) such that the lower critical temperature shifts from about 10°C in July to about 0°C in September (Ortmann and Heldmaier 1992). Oxygen consumption (ml O$_2$ · h^{-1} and ml O$_2$· kg^{-1} · h^{-1}) of wild-caught marmots also declines linearly (Armitage and Salsbury 1992, 1993) in the 50-day post-molt period prior to hibernation even though body mass increases (Armitage 2004b).

The metabolic cycle apparently drives the food consumption and body mass cycles. When metabolism is high, food consumption increases followed by an increase in body mass. The decline in metabolism is followed by a decline in food consumption and body mass subsequently decreases (Fig. 7.3). The increase in body mass lags behind the increase in food consumption which lags behind the increase in metabolism. Energy balance is negative during hibernation even when food is available (Kortner and Heldmaier 1995). Field and laboratory studies indicate that the circannual rhythm controls food consumption; yellow-bellied marmots ceased gaining body mass for at least 2 weeks before hibernation even though they had access to supplemental food (Woods and Armitage 2003a). Generally, yellow-bellied (Armitage *et al.* 1996) and hoary marmots (Taulman 1990a) spend less time foraging as the time for hibernation approaches.

The circannual rhythm is highly advantageous; it is endogenous; i.e., it is free-running under constant laboratory conditions (Davis 1976). Although the sequence of the phases of the rhythm cannot be altered (see Fig. 7.1), the phases can be shifted. Woodchucks sent from Pennsylvania to Australia shifted the annual phase of maximum and minimum mass by 6 months (Davis and Finnie 1975). The seasonal phenology of marmot activity

Table 7.1 Hibernation characteristics of free-living woodchucks from different latitudes (from Zervanos *et al.* 2010).

Locations	Maine (43°42'N), Pennsylvania (40°22'N), South Carolina (34°40'N)
Length of hibernation season	Directly related to latitude
Interbout arousals	Length decreased with the increase in latitude
Torpor bouts	Length increased with latitude
Arousals	Number increased with latitude
Total time spent euthermic	No differences among the three populations
Total time in torpor	Maine woodchucks spent 68% more time torpid than South Carolina animals
Burrow temperatures	Maine 4.2°C, Pennsylvania 7.3°C, South Carolina 12.4°C

may shift to adjust to seasonal changes in weather patterns associated with differences in elevation, latitude, and aspect (Armitage 2005).

Woodchucks exemplify the phenotypic plasticity in hibernation characteristics of animals at different latitudes and demonstrate the adaptability of the hibernation period (Zervanos *et al.* 2010). Some hibernation characteristics were related to latitude (Table 7.1). Latitude was unrelated to the time spent euthermic but was directly related to the total time spent in torpor which decreased significantly as burrow temperatures increased. Thus woodchucks in each population maximized the time spent euthermic and utilized torpor to the level needed to survive winter hibernation. Despite the differences in time spent in hibernation, there was no significant difference among sites in percent body-mass loss. Woodchucks apparently balance the costs of a depressed metabolism with the benefits of energy conservation (Zervanos *et al.* 2010).

Based on calculated metabolic rates and body-mass loss, woodchucks from the northernmost latitude had significantly higher total energetic costs over the hibernation season, but a significantly lower cost per day than woodchucks from more southern latitudes (Fenn *et al.* 2009). This lower cost per day can be attributed to the greater total time spent torpid when metabolic rate is minimal. Woodchucks brought into the laboratory from the three sites showed similar trends to those found in the field. However, sample sizes were small and some variables, such as torpor bout length, torpor body temperature, percent mass loss, and total time euthermic, did not appear to vary greatly among individuals from the three populations. The combination of some similar trends and the lack of variation among some variables suggest that both genetic (direct gene control) and phenotypic (environmentally induced) factors may be involved in latitudinal differences. Larger sample sizes maintained over a longer time period are required to demonstrate the importance of genetic and phenotypic contributions (Fenn *et al.* 2009).

The shift of the phase of the circannual rhythm in accordance with local environmental conditions suggests that some environmental cue acts as a "zeitgeber" to entrain the rhythm (Pengelley and Asmundson 1974). Free-running circannual reproductive cycles persist in woodchucks exposed to a 12L:12D photoperiod (Concannon *et al.* 1992), but reversal of daily changes in photoperiod can alter metabolic and reproductive cycles in adults within 6–8 months and reverse the cycles by 20–24 months (Concannon *et al.* 1993). When woodchucks were exposed to simulated northern hemisphere (boreal) or southern hemisphere (austral) natural photoperiods, they were entrained and synchronized for 6 years by the boreal photoperiod and re-entrained to the austral photoperiod (Concannon *et al.* 1997). Because photoperiod was the only environmental factor that varied, it seems clear that photoperiod is the natural zeitgeber.

The persistence of the circannual cycle means that hibernation in marmots and other Marmotini is obligate rather than facultative or permissive as in many other mammalian species (Morrison 1960, Pengelley and Asmundson 1974). More importantly, the endogenous rhythm enables marmots to accumulate fat and prepare for immergence for up to 3–5 weeks before immergence (Armitage 1996a). If marmots delay preparation until the onset of cold weather or the first snowfall (Table 8.3), the vegetation will become senescent long before sufficient fat reserves are accumulated (see also Pengelley and Asmundson 1974). Immergence time is facultative and is affected by weather conditions; *M. bobak* immerges earlier in a humid summer and later in a hot summer (Soroka 2000). This variation likely reflects the effects of weather on vegetation and mass gain; e.g., yellow-bellied marmots hibernated later in a drought year when mass gain rates were slower (Lenihan and Van Vuren 1996a).

The timing of spring emergence is also critical in order to have sufficient time to reproduce and prepare for the next hibernation. Most marmot species excavate snow tunnels in order to emerge; some species initiate reproduction before emergence (Table 3.5). It is difficult to imagine any reliable external environmental variable that marmots could detect as a cue for emergence; however, the time of woodchuck emergence was correlated with periods of warm weather (Davis 1977) and April air temperatures were highly predictive of yellow-bellied marmot emergence (Inouye *et al.* 2000).

The timing of emergence is facultative; yellow-bellied marmots were detected earlier when snow melted earlier. The date by which 50% of the group emerged and the date when the first adult female was seen were explained by the date of 50% snow cover (Blumstein 2009). Social structure also affected emergence time. For each additional male in the group, marmots emerged about two days earlier, but the number of females in the group did not affect the date the first marmot was sighted (Blumstein 2009). This male effect occurred in years in which an unusually high number of males was present in a colony and likely reflects the need for males to emerge early in order to compete for mates.

Conditions of snow cover vary widely in the spring, thus yellow-bellied and other marmot species may need fat reserves to provide energy until snowmelt exposes growing vegetation. Marmots with adequate fat reserves are more likely to reproduce successfully (Andersen *et al.* 1976). Fat reserves are conserved because yellow-bellied marmots do not hibernate as soon as they reach hibernation mass (Armitage 1996a), but continue foraging probably to meet current maintenance needs, not to accumulate more fat.

Hibernation: energy savings

Although body temperature is useful for describing torpor bouts, metabolism must be measured to calculate energy savings during hibernation. Average monthly metabolic rates (AMMR) may be calculated for laboratory animals from hourly metabolic rates and the time spent torpid or active (Armitage *et al.* 2000). Energy saved (ES) during hibernation was calculated from an AMMR for which marmots were considered to be euthermic compared to total AMMR (active AMMR + torpid AMMR); i.e., 1 − total AMMR / euthermic AMMR × 100 = %ES. ES should be calculated for euthermic values for both summer and winter. Winter values are available from measurements made during hibernation. Summer euthermic metabolism is based on values available from measurements made during the homeothermic phase. Measurements of oxygen consumption throughout the active season are rare and do not account for the seasonal declines in metabolism. However, summer ES indicates the importance of the circannual cycle in decreasing metabolism and thereby saving considerably more energy than would be saved if euthermic metabolism were as high during hibernation as during the early active season.

Energy savings are considerable. At 5–6°C, winter ES was 43.8% for *M. monax* (Armitage *et al.* 2000), 43.2% for *M. marmota* (Heldmaier *et al.* 1993a), and 83.6% for *M. flaviventris* (Armitage *et al.* 2003). Summer ES was higher than winter ES for all three species, but the results are not comparable because summer euthermic metabolism was chosen differently for each species.

Energy savings depend on the amount of time a marmot is torpid. The length of time spent torpid appears to be driven by the circannual rhythm. In *M. flaviventris* and *M. monax*, the amount of time spent torpid increases as the time in hibernation progresses, then decreases as the time for emergence nears (Armitage *et al.* 2000, 2003, Zervanos and Salsbury 2003). This rhythm was clearly evident in a group of young yellow-bellied marmots; oxygen consumption, time spent torpid, and body temperature followed circannual patterns (Armitage and Woods 2003).

Differences among marmots in ES are reflected in DML, as mg · day^{-1} · g IM. DML values are lower in members of the subgenus *Petromarmota* than in members of the subgenus *Marmota* (Table 7.2A). Most DML values are calculated from field animals that were captured as close to immergence and as soon after emergence as possible (Arnold 1993a), but some uncertain number of days at euthermy contribute to the calculation of DML. I investigated mass loss for yellow-bellied marmots when late season and early season body mass was available (Table 7.2B). DML of adult females is similar to laboratory measurements and those of the young are essentially identical to values measured in the laboratory where DML varied from 1.33 for single young to 1.46 for grouped young (Armitage and Woods 2003).

These results indicate that field measurement of body mass if taken shortly before immergence and soon after emergence yield a reasonable estimate of DML. It is also interesting that young have a higher DML than adults. Although nearly all littermates hibernated in the same hibernaculum (Lenihan and Van Vuren 1996a), group hibernation does not increase survival and there is no effect of litter size on survival (Armitage and Woods 2003). The higher DML of young than of adults may explain, in part, the higher mortality of young than of adults during hibernation. The higher DML may be related to the differences in fur properties of young and adults; juveniles have shorter hair length, a smaller hair diameter, and less fur depth than adults (Melcher 1987); as a consequence of these differences, young are more greatly affected by low ambient temperatures (Melcher *et al.* 1990).

Table 7.2 DML values for marmots housed singly in the laboratory (A) (Armitage 2008) and for field yellow-bellied marmots (B).

A. Laboratory measurements

Species	DML
M. flaviventris	0.94
M. vancouverensis	0.95
M. olympus[a]	1.06
M. monax	1.94
M. marmota	1.95
M. broweri[b]	1.07

B. Field yellow-bellied marmots

Animal	N	Mean number of days between measurements	Mean DML
Adult female	25	258.8	1.19
Young female	18	240.5	1.44
Young male	11	238.6	1.31

[a] Values calculated from field data when time of emergence and date of capture were known (Griffin, pers. com.). A regression analysis calculated mean body mass at emergence; this value was used to calculate DML.
[b] Calculated for a field female from Lee et al. (2009).

DML values suggest that members of the *Petromarmota* clade are energetically more efficient during hibernation than members of the other clade. This difference raises the question of what is the basis for differences in DML. DML for *M. broweri* is similar to the *Petromarmota* values (Table 7.2A). Six animals emerged from the hibernaculum used by the adult female. This pattern of group hibernation suggests that social thermoregulation may account for the low DML (Lee *et al.* 2009). During late summer, all members of a hoary marmot group begin sleeping in the same burrow and presumably hibernate together (Taulman 1990a). Larger social groups have the potential for social thermoregulation that probably is not present in smaller groups (Kyle *et al.* 2007); however, social structure had only a minor influence on survival (Braun *et al.* 2011). Social thermoregulation, which occurs in the extended family group (Table 5.2), will be discussed in the next section.

Because low values of DML are associated with hibernating singly, we would expect the DML of the woodchuck to be similar to the value of the three species of *Petromarmota*. Two studies of free-ranging woodchucks reported DML values only slightly higher than those reported for laboratory animals. In Canada, mean DML for a 5.5-month hibernation period was 2.07 for females (Ferron 1996). Mass loss did not differ for females that hibernated in the same burrow from that of females hibernating singly. This pattern illustrates that because two or more marmots may hibernate in the same burrow does not mean that they engage in social thermoregulation. In Pennsylvania, mean DML was 1.78 for females for a hibernation of about 4 months (Zervanos and Salsbury 2003). As a percentage of body mass, Canadian woodchucks lost an average of about 40%,

Pennsylvania woodchucks, about 22%, and wild-caught yellow-bellied marmots, about 29% (Armitage and Blumstein 2002). Using data from Ferron and from Armitage and Blumstein, and assuming that DML by yellow-bellied marmots was the same as that of woodchucks, I calculated that adult yellow-bellied marmots would lose from 50% to 67% of their IM rather than the 29% measured for field animals. Such high loss could not be sustained as IM is about 42% higher on average than emergence mass. Or to express the relationship in terms of grams, yellow-bellied marmots would lose at least 1715 g during hibernation but gain only 1009 g during the homeothermal period. Thus, the higher energetic efficiency is fundamental to survival of yellow-bellied marmots.

By contrast, woodchucks combine a larger body size with a short hibernation period; large size enables woodchucks to store more fat which provides sufficient energy for the relatively short hibernation. More absolute mass is lost at higher latitudes because of longer hibernation, but DML is only 16% higher than that of woodchucks at lower latitudes; to put it simply, woodchucks with their large body size and short hibernation period can cope with their mass loss without any additional adaptations.

Mechanisms of energy conservation

Energy can be saved at various stages of a torpor bout (Fig. 7.2). Entry into torpor occurs when metabolic rate is depressed which is followed by loss of heat and slowly decreasing body temperature. Energy expenditure is actively down-regulated and controlled at a minimal level despite changes in body temperature (Ortmann and Heldmaier 2000). Metabolism may reach its minimum 20 h before body temperature stabilizes in yellow-bellied marmots (Woods *et al.* 2002b). Similar patterns occur in woodchucks (Armitage *et al.* 2000) and alpine marmots (Heldmaier *et al.* 1993b). Active suppression of metabolic rate saves considerably more energy than would be saved if metabolism decreased as body temperature declined and energy is not wasted dissipating heat in order to reach hypothermia. Active suppression of metabolism to enter torpor is probably a universal process among hibernators (Heldmaier *et al.* 1993b).

The decline in metabolism is followed by a period of deep torpor when metabolism is relatively constant (Fig. 7.2). Metabolic rate during deep torpor is temperature independent over a range (TNZ) from 4°C to 6°C to 10°C in *M. monax* and *M. flaviventris* and to 15°C in *M. marmota* and from −2°C to 2°C in *M. camtschatica* (Armitage 2008). The low TNZ in *M. camtschatica* represents an independent evolutionary event that enables this species to hibernate in permafrost (Solomonov *et al.* 1996). T_B was always below 5°C (Vasiliev and Solomonov 1996).

The mean time it takes to enter deep torpor varies among marmot species; at 6°C, *M. flaviventris* reaches deep torpor in 23.2 h; *M. monax* requires 33.4 h. The time to deep torpor plus the amount of time spent in deep torpor (mean = 90.5 h and 69.3 h for *M. flaviventris* and *M. monax*, respectively) accounts for the lower DML and higher ES in *M. flaviventris* (Armitage 2008).

At burrow temperatures below the TNZ, metabolic rate accelerates; it approximately doubles at 2°C in *M. flaviventris*, *M. monax*, and *M. marmota* (Armitage 2008). Thus the burrow temperature during hibernation can markedly affect energy use. Mean burrow temperature of the yellow-bellied marmot was 9.1°C in early October (Kilgore and Armitage 1978). On average, first snow cover occurs in early November (Table 8.3). Given the time lag between air temperature and burrow temperature, it seems unlikely that burrow temperature cools below 5°C before first snow cover provides insulation. Over a 2-year period in Pennsylvania, mean burrow temperatures of woodchucks ranged from 7.3°C to 7.7°C during hibernation (Zervanos and Salsbury 2003). Thus it is likely that woodchucks rarely encounter stressful burrow temperatures.

Energy savings: social thermoregulation

Temperature in the hibernaculum of the alpine marmot can decrease to almost 0°C by spring and the burrow temperature was <5°C for about two-thirds of the hibernation season (Arnold et al. 1991). Rather than a higher energetic efficiency, the alpine marmot copes with low temperature by social thermoregulation (Arnold 1993a). The essence of social thermoregulation is that adult marmots share heat, especially with young, that has both costs and benefits. One major benefit is that social thermoregulation reduces mortality of young, thus increasing the fitness of the parents, typically the territorial pair (Arnold 1990a, b). Heat is provided mainly by subordinate adults, particularly those that are full sibs of the young. The contribution of the subordinate adults is an example of alloparental care.

Costs usually are expressed in higher DML and mass loss. The DML of the territorial pair hibernating without offspring is 1.60; with offspring, DML increases to 1.82, a value similar to that of laboratory adults hibernating singly (Armitage 2008). When only the territorial pair and young hibernate together, the entire group may die (Arnold 1990b). Winter mortality is significantly reduced when subordinate adults are present. The subordinate adults suffered the cost of additional mass loss when young were present, but lost less mass with increasing group size when young were absent (Arnold 1990b). Thus, there are benefits to group hibernation in this species as the group members lie huddled together (Arnold 1988). Huddling should reduce metabolism; e.g., *M. camtschatica* in a group of three used less oxygen per individual (0.0147 ml $O_2 \cdot g^{-1} \cdot h^{-1}$) than marmots hibernating singly (0.0328 ml $O_2 \cdot g^{-1} \cdot h^{-1}$) (Vasiliev 1992). The lower metabolism reduced body-mass loss to 24–34% of IM compared to 46–50% loss in single animals (Vasiliev and Solomonov 1996). By contrast, oxygen consumption of *M. flaviventris* did not differ significantly between single and group animals (Armitage 2008). This comparison of black-capped and yellow-bellied marmots demonstrates the necessity of quantifying any metabolic benefits of group hibernation rather than assuming that metabolic benefits occur because more than one individual hibernates in a burrow system. Other possible benefits of group hibernation are access to mates and resources.

Social thermoregulation can work only if the torpor bouts are highly synchronized. Temporal patterns of body-temperature changes are highly synchronized in hibernating alpine marmots (Ruf and Arnold 2000). The degree of group synchrony was the most important factor affecting mass loss; the presence of young impaired group synchrony and increased mass loss. Apparently young and yearlings are the last to arouse in a group, which seems to minimize their costs of warming. Because arousal and euthermia account for more than half of the energy expenditure of a hibernating marmot, reducing these costs by utilizing heat produced by adults saves considerable energy for young. This impairment of synchrony in larger groups with young apparently offsets the benefits of improved insulation provided by huddling. The high cost of providing heat to young may explain why no more than a single litter per year is produced in the family group and why dominant females recovering from the energetic cost of reproduction, which includes the cost of young care during hibernation, skip the following reproductive season (Ruf and Arnold 2000).

Kinship is also a factor in social thermoregulation. The DML of subordinate adults with less-related (not offspring or full sibs) young did not differ from the DML when young were absent, but significantly increased when they hibernated with full sibs (Arnold 1993a, b). Yearling DML does not differ when yearlings hibernate without young or with less-related young, but increases when hibernating with full sibs. The increased DML of yearlings hibernating with full sibs is much less than its increase in subordinate adults; the yearlings receive some benefit from heat production of older animals.

Shrunken digestive tract

During hibernation, the gastrointestinal tract, a high-energy-use system, of *M. marmota* (Hume *et al.* 2002) and *M. broweri* (Rausch and Rausch 1971) is greatly shrunken. Reduced size obviously decreases energy use, but this saving has not been quantified. Another benefit of the shrunken digestive tract is that there is a large decrease in the parasite load, especially of helminthes (Callait and Gauthier 2000). *M. marmota* does little or no feeding for the first 2–3 weeks following emergence (Mann *et al.* 1993) and we observed similar patterns in *M. flaviventris*; they do not eat dandelions (*Taraxacum officinale*), a preferred food, when placed on the snow next to their emergence tunnel. If delayed feeding is associated with a shrunken digestive tract, then a reduced tract may characterize marmot hibernation. Because intestinal atrophy occurs in ground squirrels (Carey and Cooke 1991), it may be a general adaptation for hibernating squirrels and be an ancestral trait of marmots.

Summary

To summarize, hibernation is the major mechanism whereby marmots conserve energy during a season of little or no food. Hibernation is part of an annual cycle which is controlled by an endogenous, circannual rhythm. The circannual rhythm prepares marmots for hibernation before immergence is necessary and initiates the onset of the active season. Further energy savings are achieved by reducing the time to reach deep torpor and spending more time in deep torpor.

Part II

Biotic and abiotic environments

8 The environment of the yellow-bellied marmot

The social and population biology of the yellow-bellied marmot is more clearly understood if one can visualize the habitat patches where they live. This chapter describes and illustrates the characteristics of the habitat sites that extend for about 6.4 km along the Upper East River Valley, Colorado, USA. Not all sites were studied each year because some sites did not have marmots every year and some sites were added as the study progressed. In addition to the 11 major sites (Table 8.1), an additional eight minor sites were monitored when they were occupied and we were searching for the location of tagged marmots.

Research sites

Marmots occupy meadow patches in a mosaic of spruce and aspen woodland (Svendsen 1974) at elevations ranging from 2711 to 3043 m (Table 8.1). The meadow vegetation is characteristic of the *Festuca thurberi* community (Langenheim 1955), with several species of *Festuca*, *Poa*, and *Bromus* the dominant grasses. Rock outcrops, boulders, or talus occur in the meadows occupied by marmots (Svendsen 1974). Burrows are located in the talus or under rocks, but also may be dug under the roots of shrubs or trees. Rocks provide some shelter from predators and provide perches where marmots sit or lie (Fig. 8.1) while being wary (Armitage and Chiesura Corona 1994). All major study sites are located on south-, southwest-, or east-facing slopes on the valley floor where sunlight reaches the site only slightly later than on the south-facing slopes.

Marmot study sites vary from 0.15 to 7.24 ha (Table 8.1). The five larger sites, where more than one matriline typically occurs, are designated colonies. Smaller sites, where the presence of more than one adult female is uncommon and mean matriline size was ≤1.34, are called satellites (Svendsen 1974). The Bend/Falls and West Bend/Falls were originally treated as four satellite sites. Because adult females moved between sites from year to year (e.g., between Bend and Falls), I combined each pair of sites into one. Reducing four sites to two increased the areas and the mean number of resident adult females at the two sites. The slope of colony sites ($\bar{x} = 33.0°$) was significantly greater than that ($\bar{x} = 18.4°$) of satellite sites. Colony sites also had significantly more used burrow sites and resident burrows than satellite sites (Svendsen 1974). Area was positively related with the mean number of adult females and negatively with the mean density of adult females, but unrelated with mean matriline size (Table 8.2). The mean number and the mean density

Table 8.1 Major habitat sites of yellow-bellied marmots in the Upper East River Valley, Colorado, USA. The study area extends from 38° 56′ 33.986″ N/ 106° 58′ 59.88″ W to 38° 59′ 12.780″ N/ 107° 0′ 44.892″ W. Satellite sites are indicated with (s). Other sites contain colonies. Location in the valley is indicated as down valley (DV), or up valley (UV). Sites are listed in the order of largest to smallest. (N) = number of years the site was studied. The mean number of adult females is derived from the number of residents in June.

Site	Location	Area (ha)	Elevation (m)	Mean matriline size	Mean number of adult females	Mean density/ha
North Picnic (39)	UV	7.24	2991	1.16	2.46	0.34
Picnic (41)	UV	5.45	3043	1.57	6.35	1.17
River (41)	DV	2.73	2867	1.68	2.97	1.09
Marmot Meadow (41)	UV	2.39	2930	1.4	1.63	0.69
Cliff (27)	UV	1.88	3006	1.9	2.11	1.12
Bend/Falls (s) (20)	DV	1.75	2769	1.2	1.7	0.97
West Bend/Falls (s) (20)	DV	1.75	2834	1.0	1.43	0.82
RMBL (36)	DV	1.6	2881	1.43	3.47	2.17
River Annex (s) (18)	DV	0.5	2711	1.0	1.25	2.5
Boulder (s) (40)	UV	0.2	2977	1.34	1.84	9.2
Beaver Talus (s) (23)	UV	0.15	2901	1.14	1.6	10.7

Figure 8.1   Yellow-bellied marmot lying on rock near its burrow at Marmot Meadow Colony.

Table 8.2 Rank correlation analyses.

Comparisons	r_s	p
Area vs. mean number of adult females	0.548	0.05
Area vs. mean density of adult females	−0.629	0.05
Area vs. mean matriline size	0.235	ns
Mean number vs. mean density of adult females	−0.1	ns

of adult females were unrelated (Table 8.2). Clearly, area explains a small proportion of the variation in social or population structure at yellow-bellied marmot sites.

Some of the variation in mean population numbers and social structure is apparent in the distribution of the major resources, i.e., burrow sites and foraging areas, in the marmot habitats. River (Fig. 8.2a), the southernmost colony, is located adjacent to the East River. Most burrow sites occur on the cliff face above the river or on the flat area between the river and the gully (Fig. 8.2b) that separates the burrow area from extensive meadows. Marmots also colonize the rocky area where the river forms a large bend at a tree-lined mound and the cabin area (Bench) to the north (Fig. 8.2a). Numerous burrow sites and open meadows to the east, southeast, and north of the main burrow area support a relatively large number of adult females at a moderate density. Density is limited, in part, because potential foraging areas that are distant from burrows are not used.

Movement into meadows is risky because of the danger of predation, which is evident in the marmots' behavior. When moving away from the burrow to a foraging area, a marmot may be vigilant for several minutes before initiating foraging (Fig. 8.3). The danger of predation apparently restricts the distance marmots will move from a burrow. Marmots reduce the distance to a burrow by digging new flight or escape burrows (Armitage 2003d) to which they retreat when alarmed by a predator. But these burrows are not circumscribed by rocks, a marmot in such a burrow is vulnerable to a digging predator, such as a badger, *Taxidea taxus* (Andersen and Johns 1977).

Three satellite sites are located along the East River south of River Colony. River Annex (Fig. 8.4) has the smallest mean number of adult females (Table 8.1). Bend/Falls is located down river from River Annex; the meadow is continuous between the two sites. Burrows are located in trees along the river with some scattered burrows under rocks. West Bend/Falls, across the river from Bend/Falls, contains numerous burrow sites in the steep bank with sparse vegetation above the river (Fig. 8.5). Marmots must either travel far to or use burrows close to a meadow area to the south. In effect, marmots are restricted to few burrow sites and foraging areas and the mean number of adult females is the second lowest (Table 8.1). The more extensive meadow area at Bend/Falls supports a slightly larger mean number of adult females than West Bend/Falls (Table 8.1).

Beaver Talus consists of a small patch of talus and meadow nearly enclosed by aspen and spruce trees (Fig. 8.6). Cliff occurs north of Beaver Talus on the west side of the East River. Although the area is relatively large, much of it consists of patches of trees and shrubs (Fig. 8.7). Marmot activity was restricted by avoidance of vegetation over 90–100 cm high (Travis and Armitage 1972).

Figure 8.2   (a) River Colony. An overview of the River/Bench area photographed from the slopes of Snodgrass Mountain to the west. Marmots dig burrows under the cabins (Bench area on the left) and forage in the meadows at the River area. Marmots utilize meadows on the far side (east side) of the East River; the major burrow area is in the steep banks on the right. Because marmots shift residency between Bench and River areas, the two sites are combined into one and recorded as River. (b) View from the east side of the River area looking toward the flat areas separated from meadow foraging areas (foreground) by a gully. Fence posts are used as sitting sites.

Marmot Meadow lies northeast of Cliff on the east side of the river. The site has two major burrow areas (Fig. 8.8), the rocky outcrop is the main burrow and the aspen site is occupied only during times of population increase. Foraging areas are extensive, but the lack of more quality burrow sites apparently limits the number of residents.

Picnic Colony, the second largest, extends over an elevational gradient of more than 100 m. Burrows are abundant in the talus and in other rock outcrops on the slope (Fig. 8.9). The talus area has an estimated 78 burrows in various stages of use in about 0.85 ha (Svendsen 1974). The distribution of burrow sites and the extensive meadows allow marmots to spread across the slope and support the highest mean number of adult females (Table 8.1).

Boulder lies north of Picnic in a small meadow enclosed by trees (Fig. 8.10). Boulder has the largest mean number of resident adult females of the satellite sites (Table 8.1).

Figure 8.3 A vigilant marmot at River Colony at a foraging area about 30 m distant from the main burrow area.

Figure 8.4 River Annex. The burrows are located at the north end in the trees downslope from the large meadow where the marmots forage.

North Picnic is the largest and most northerly of the sites. Despite its large area, it supported the lowest density of adult females of all sites (Table 8.1). The low density apparently is a consequence of the extensive distribution of trees and shrubs and few burrow sites (Fig. 8.11). Although there is an extensive meadow area on this steep slope, the vegetation is relatively sparse and there is considerable bare ground. Predation may be a factor causing the low density; the high vegetation provides cover for coyotes and predation was observed here (Armitage 1982b).

To summarize, I conclude that a large area is a necessary but insufficient requirement for the development of a large yellow-bellied marmot population. The distribution of burrow sites with respect to foraging in nearby meadows is of critical importance. This relationship is well illustrated at the RMBL site, where cabins and other buildings are distributed through a meadow. Cabins serve as burrow sites and each site is adjacent to

Figure 8.5   West Bend/Falls area with marmots on the steep slope with sparse vegetation above the East River.

Figure 8.6   Beaver Talus. Marmots occupy the talus and forage in the adjacent meadow. Burrow sites and foraging areas are limited and the mean number of adult females is low.

Figure 8.7   Cliff Colony. The major burrow area is located in the rocky outcrop at the right center, other burrow sites are located in the open areas in the middle and left centers.

Figure 8.8   Marmot Meadow. The major burrow area (Main Talus) is the rocky area on the left side of the meadow and a second area (Aspen Burrow) is in the group of rocks at the edge of a grove of aspen on the right side of the meadow where rocks and boulders extend into the meadow foraging area toward Main Talus.

foraging areas. Hence, RMBL, the smallest colonial site, supported the second highest mean number of adult females (Table 8.1).

Effects of weather

Because these habitats occur over a south–north distance of 6.4 km with an increase of 332 m elevation, weather factors could affect population numbers at the various sites. Long-term weather data are available only from the weather station in Crested Butte,

Figure 8.9   Picnic Colony. The talus slope is the major area of activity, but other rocky areas above the aspen are also used. The area from the aspen downslope is called Lower Picnic; the area near the spruce trees above and to the left of the aspen is Middle Picnic; the area above the aspen is Upper Picnic.

Figure 8.10   Boulder. The large boulder is the site of the major home burrow area; marmots forage in the surrounding meadow.

Figure 8.11   North Picnic. The major area of occupancy is the steep slope below the willows and spruces in the center. Other burrow sites are scattered throughout the site, such as at the cliff at the top of the site, but none have consistently housed marmots.

about 10 km distant, but some snowfall data and snow-cover data are available from RMBL.

The major factors affecting marmot biology are temperature and precipitation. The amount and kind (rain or snow) of precipitation is critically important, especially its timing, such as when snowfall and snow cover first appear in the fall and when the last snowfall and snow cover occur in the spring. From snowfall and snow-cover data, the length of winter and length of the growing or active season can be estimated.

On average the first snowfall occurs about mid-October and the first permanent snow cover (of 2.5 cm or more), about mid-November (Table 8.3). These snow events occur after marmots have hibernated as most individuals enter torpor by mid-September (Armitage 1998). The period between hibernation and snow cover can be important. Mean air temperature during this period is about one-fourth of that during the summer (Table 8.3). If the temperature of the hibernaculum falls below 5°C, marmots increase metabolic rate and this increase persists throughout hibernation, which causes an increase in energy expenditure, which in turn can cause mortality or decrease the likelihood of reproduction in the subsequent active season (Arnold *et al.* 1991). As described in Chapter 3, plugging the burrow excludes cold air from penetrating to the hibernaculum; thus soil temperature at the level of the hibernaculum must decrease to 5°C or less. Although a multiple regression model attributed significant effects of weather on density ($R^2 = 0.41$), there was no significant effect of the timing of fall snowfall or snow cover (Schwartz and Armitage 2005). At RMBL, over a 21-year period, on average, over 200 cm of snow falls by the end of November (Barr, pers. com.). Thus, I conclude that typically snow cover occurs before low temperatures reach the hibernaculum, marmots are well insulated, and fall and winter temperatures do not significantly affect mortality. However, the timing and length of snow cover affect survival. Temporal variation over

Table 8.3 Weather variables. Data were obtained for Crested Butte, Colorado, which is about 10 km from the study area (Anonymous 1963–97).

	Mean ± SD
Precipitation: snow (Julian date)	
First snowfall	287.3 ± 17.2
First snow cover	315.9 ± 15.7
Last snowfall	127.4 ± 14.3
Last snow cover	115.2 ± 14.6
Precipitation (cm)	
Winter (September–May)	40.0 ± 12.5
Summer (June–August)	15.0 ± 4.0
Temperature (°C)	
September–November	2.9 ± 0.19
June–August	11.7 ± 1.9
Growing season (number of days)	
Last vernal snow cover until first autumnal snow cover	200.0 ± 20.6
Length of winter (number of days)	
First fall snowfall until last vernal snowfall	199.2 ± 39.0
First fall snow cover until last vernal snow cover	165.2 ± 22.0

28 years in juvenile survival rates, but not in adult survival rates, was negatively affected by the duration of permanent snow cover (Ozgul *et al.* 2006b).

Weather data from a single source does not adequately describe the extent of snow cover at our study sites. Snow melts later at the UV sites, such that the time of 50% snow cover on average is about 20 days later at Picnic, Cliff, Boulder, and North Picnic than at River. The time of 50% snow cover affects several life-history traits; frequency of reproduction (measured as the number of litters per adult female), mean litter size, and mass of young in August decrease with prolonged snow cover (Van Vuren and Armitage 1991). Site-specific variation in juvenile survival was negatively influenced by elevation (Ozgul *et al.* 2006b). Snow cover lasts longer at higher elevations; thus the effect of elevation on juvenile survival is likely caused by prolonged snow cover and not elevation per se.

Site-specific variation in juvenile survival rates was positively influenced by aspect (Ozgul *et al.* 2006b). Survival rates were higher on southwest-facing slopes than on northeast-facing slopes. Snow cover melts earlier on those sites with greater exposure to sunlight; thus, the significance of aspect of site on juvenile survival rates can also be attributed to the duration of snow cover (Ozgul *et al.* 2006b).

Other demographic variables were affected by weather variables (Schwartz and Armitage 2005). The timing of the first snowfall explained 18% of the variation in net reproductive rate. A positive relationship with temperature and total precipitation and a negative relationship with the number of frost free days accounted for 34% of the variation in variances of litter size, but no weather variable significantly affected mean litter size. Temperature, total precipitation, time of last snowfall, and length of growing season explained 38% of the variation in the percentage of adult females that reproduced.

Temperature effects were minor; the other major variables are related to snowfall and snow cover.

Clearly, weather affects the survival and reproduction of yellow-bellied marmots. Weather effects are expressed mainly by the timing of snow cover. Early snow cover in the autumn insulates burrows against the effects of low environmental temperatures. The importance of snow cover as insulation was apparent when all hibernacula that were located in May, 1972, were situated in places where snow cover persisted and emerging marmots tunneled up through the snow (Svendsen 1974). Prolonged snow cover in the spring may also increase mortality, especially of overwintering young, because food does not become readily available until snowmelt occurs (Svendsen 1974, Inouye *et al.* 2000).

The importance of rainfall for growth and reproduction was demonstrated in another analysis using weather data from 1962 to 1986. Juvenile (young and yearlings) survival was positively correlated with mean precipitation in the current year (partial correlation coefficient = 0.62) and the percentage of females reproducing was positively related to the mean precipitation of the previous year (partial r = 0.65). Precipitation is associated with good conditions for plant growth and quality and thus for mass gain by marmots which results in better survival of juveniles and better conditions for reproduction by adult females (Schwartz and Armitage 2002).

It is not surprising that there is no consistent pattern of weather effects on winter mortality. Marmots have undergone selection for thousands of years that should favor behaviors and physiology that enhance survival. However, marmots may not always settle in optimal habitat and litters weaned in suboptimal habitat may be more vulnerable to weather effects. This problem is difficult to analyze because suboptimal habitat is identified as a burrow location where marmots do not persist or suffer high mortality. Thus, circular reasoning is a problem since habitat quality is defined by survival. Weather and snow-cover information collected at a diverse array of sites would greatly assist in untangling the complex relationship between weather and survival.

There is also a lack of certainty as to which weather variables are critical. The different effects of weather variables on survival and reproduction reported in the three major analyses (Schwartz and Armitage 2002, 2005, Ozgul *et al.* 2006b) probably are, in part, a consequence of the weather variables used and the time periods included in the studies. For example, when and where precipitation occurs are more important than total precipitation, which was used in the major analyses. In conclusion, normal weather patterns have little effect on the survival of adults and yearlings, but do affect survival of young. Extreme or unusual weather patterns significantly impact demography and social structure (see Chapter 17) but probably occur with insufficient frequency to significantly affect the analyses of long-term data sets.

Environmental physiology

In Chapter 7 mechanisms that reduce energy expenditure during hibernation were described. Because fat accumulation is critical for marmot survival and reproduction, maintenance metabolism during the active season should be minimized. Also, marmots must shift from coping with low temperatures of hibernation to the higher temperatures of summer (Table 8.3), especially during daytime when temperature can exceed 20°C. Marmots must also adjust to the problem of gas exchange in their burrows and the reduced partial pressure of oxygen at high elevations. In addition, marmots must meet other requirements, such as water needs.

Resting and field metabolic rates

Basal or resting metabolism of yellow-bellied marmots is about 70–80% of that predicted from body-mass:metabolic-rate equations (Armitage and Salsbury 1992). Resting metabolic rate (RMR) of both adult males and females declines after midsummer (Armitage 2004b). The lower than predicted metabolic rate is probably a consequence of the seasonal decline in metabolism, which is controlled by the circannual rhythm (Fig. 7.3). This seasonal decline is enhanced by the summer molt; post-molt metabolic rate is 63% of pre-molt metabolic rate. Body temperature does not differ between pre-molt and post-molt adult females, but conductance is significantly lower in post-molt females. Although body mass increased from pre-molt to post-molt, both mass-specific and total, whole organism metabolism decreased. Thus, the decrease in metabolism, measured in the same individuals, was attributed to decreased conductance (Armitage and Salsbury 1993).

If low conductance is important for conserving energy, why is conductance relatively high during most of the active season? Marmots lose hair and increase conductance after emergence from hibernation. This loss of insulation appears to be essential for coping with the thermal load of summer (to be discussed below). Molt increases insulation in preparation for hibernation; lower conductance reduces heat loss at the low temperatures of the hibernaculum.

Drought is associated with reduced growth rates, especially of young (Armitage 1994, Lenihan and Van Vuren 1996a). One likely factor that could cause decreased growth is a higher maintenance metabolism. On average, RMR is 16% higher during drought than in normal years (Table 9.1). Adult females were less affected by drought probably because non-reproductive females had completed mass gain by the time drought occurred (Armitage 1994).

Table 9.1 The effect of drought on resting metabolic rates of young and adult yellow-bellied marmots. $VO_2 = mlO_2 \cdot kg^{-1} \cdot h^{-1}$. Values are mean ± standard error. Because of small sample sizes, values for adult females were combined.

	Drought year	Pre-drought year	Post-drought year	P
Young	322.7 ± 9.4	265.4 ± 11.1	286.4 ± 12.4	0.004
Adult females	197.8 ± 11.8		170.7 ± 10.9	0.102

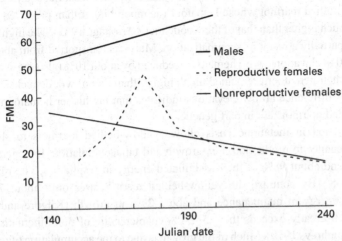

Figure 9.1   Field metabolic rates ($L\ CO_2 \cdot kg^{-1} \cdot d^{-1}$) of reproductive and non-reproductive adult females and of adult males.

Field metabolic rates (FMRs) measure total metabolism, which includes all activities. In reproductive females, FMR increased to a peak during lactation and declined thereafter (Fig. 9.1) (Armitage 2004b). For adult males, FMR significantly increased to a peak in early summer; because males are difficult to trap after midsummer, data are inadequate to determine if FMR declined. However, a value of 16.7 for a male on day 221 indicated a decline, which accords with the seasonal decline in RMR for non-reproductive adult females whose FMR was maximal in the month following emergence, then declined (Fig. 9.1). Except during lactation, mean measured FMR was significantly lower than mean predicted FMR (predicted FMR is derived from metabolism:body-mass equations). The lower than predicted FMR is consistent with the lower than predicted RMR and confirms that marmots conserve energy by reducing metabolic costs.

Although yellow-bellied marmots have lower than predicted metabolic rates, they must be capable of delivering oxygen to the tissues, especially when active, in a high-elevation environment. Marmots had a consistently lower respiratory frequency and higher oxygen extraction than expected based on a generalized allometry for mammalian resting ventilation (Chappell 1992). The reduced ventilatory response, also evident in woodchucks, is ubiquitous among burrowing mammals. The reduced response probably conserves energy as increased ventilation would have little utility in the high CO_2 environment of the burrow (Boggs *et al.* 1984). Oxygen uptake is enhanced by the high affinity of hemoglobin for oxygen; the O_2 dissociation curve lies to the left of the human curve and shifts further to the left during hibernation. There is also a considerably greater Bohr affect (Harkness *et al.* 1974), which would enhance O_2 uptake in the burrow. Yellow-bellied marmots and woodchucks prepare for hibernation by increasing the number of red blood cells and the concentration of hemoglobin and hematocrit in late summer compared to early summer (McBirnie *et al.* 1953, Hock 1967, Harkness *et al.* 1974). Hematocrit is inversely related to body mass in semi-fossorial squirrels except for the yellow-bellied marmot whose hematocrit is much higher than predicted from body mass and much higher than that of the woodchuck (Armitage 1983). The high hematocrit values are partially a result of acclimatization. Marmots transported from about 2900 m to about 300 m elevation have a hematocrit reduced by about 10% (also see Winders *et al.* 1974), but their hematocrit remained much higher than that of woodchucks. The persistence of this difference at low elevations indicates that the higher hematocrit values of yellow-bellied marmots are in part genetic.

The reduction in metabolic costs allows yellow-bellied marmots to shift energy from maintenance to production; i.e., growth and fat accumulation. In general, homeotherms expend about 98% of their assimilated energy in respiration; i.e., maintenance (Golley 1968). By contrast, the yellow-bellied marmot uses only about 77% of its assimilated energy in maintenance and about 23% in growth (Kilgore and Armitage 1978), which greatly exceeds the 1.5–3.1% characteristic of non-hibernating homeotherms (Humphreys 1979). Much of the growth is due to the accumulation of fat, which is used during the hibernation period and when food is unavailable in the early post-emergence season.

The ability of yellow-bellied marmots to conserve energy was investigated by quantifying energy intake and expenditure in free-ranging animals (Melcher *et al.* 1989). Energy intake ranged from 666 KJ d^{-1} for a yearling female in September to 3283 KJ d^{-1} for two lactating females in July. The yearling consumed 75% less food in September than in June, which is consistent with the seasonal changes in metabolism. Daily energy expenditures (DEE) ranged from 539 KJ d^{-1} for a female yearling in June to 1017 KJ d^{-1} for a lactating female. The highest proportion of DEE expenditures occurred during time in the burrow and the lowest, during thermoregulation (Table 9.2). These results verified that energy costs of activity are low and that low DEE chiefly reflects a reduced RMR.

The low cost of temperature regulation suggests that marmots have a wide range of temperature independence. Because marmots hibernate for many months at relatively low environmental temperatures, we predicted that the lower critical temperature (the temperature below the TNZ where metabolism increases) should be relatively low. We tested two populations, one from our mesic study area in Colorado and one from a semi-arid area of eastern Washington at an elevation of 393 m. For both populations, there was a narrow TNZ between 15 and 20°C, oxygen consumption increased significantly at 10 and 5°C, and declined at 25°C (Armitage *et al.* 1990).

Table 9.2 Distribution of daily energy expenditures (DEE) for yellow-bellied marmots (from Melcher *et al.* 1989).

Activity	Percent of daily DEE	
	Mean	Range
Time in the burrow	49.7	41–60
Foraging	32.3	11–51
Sitting	13.7	1–28
Thermoregulation	4.3	1–6

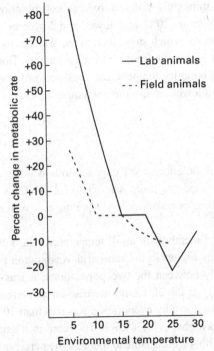

Figure 9.2   The relationship between environmental temperature (°C) and metabolic rates as percent change from TNZ value (= 0) for field and laboratory adult yellow-bellied marmots.

Temperature acclimation

The increase in oxygen consumption of 35.6% from 15 to 10°C was inconsistent with mean summer burrow temperatures centering on 10°C. Because yellow-bellied marmots spend nearly 18 h each day in their burrows, they would suffer a considerable metabolic cost. Therefore, we measured oxygen consumption of wild-caught marmots over a temperature range of 5–25° C (Fig. 9.2). In these field-acclimated marmots, the TNZ was shifted 5°C lower (to the left) in both adults and yearlings. Metabolism at 5°C was significantly greater than at any other temperature. For adults, metabolism decreased at 20 and 25°C (Armitage 2004b). It seems most likely that the laboratory marmots, maintained at 20–22°C, acclimated to the higher temperatures such that 20°C was not a stress. However, yellow-bellied marmots in the field spend most of their time at environmental temperatures of 15°C or less; surface activity markedly declines when air temperature reaches 20°C (Armitage 1962, Travis and Armitage 1972, Melcher *et al*. 1990) and they are acclimated to a lower temperature regimen. This comparison indicates that yellow-bellied marmots can occupy wide-ranging environments because they can acclimate to prevailing temperature conditions. The ability to acclimate to high temperatures is limited as is evident in the depressed metabolism at 25°C followed by an increase at 30°C in the laboratory animals (Armitage *et al*. 1990). In this context, it is interesting to compare the Colorado and Washington marmots. When accounting for body-size differences (Washington marmots were significantly smaller), oxygen consumption was significantly higher at all temperatures from 20°C and lower and lower at the higher temperatures in the Washington marmots, which suggests that the Washington marmots were genetically adapted to living at higher environmental temperatures. Both populations were maintained at the same laboratory temperature; thus it is not known if the Washington marmots can acclimate to lower environmental temperatures.

Water budgets

Both the Colorado and Washington populations were exposed to two water regimens, one *ad libitum* and the other restricted. A three-way factorial ANOVA revealed a significant second order interaction: population × temperature × water regimen (Armitage *et al*. 1990).

A restricted water regimen reduced metabolism at all temperatures in both populations. Total conductance was generally higher in the semi-arid Washington population but overall did not differ significantly between the two populations; it was relatively stable from 5–15°C, increased slightly at 20–25°C and increased exponentially above 25°C. The percentage of heat production lost by evaporation was less than 10% in both populations at 20°C and lower and was higher in the semi-arid population at temperatures greater than 10°C. The Colorado marmots were unable to increase the percentage of heat lost by evaporation at temperatures above 20°C whereas the semi-arid marmots increased that percentage. Finally, both populations produced sufficient metabolic water at 5–15°C to meet evaporative water losses. Because metabolic water meets water budget requirements at low ambient temperatures, marmots do not require water from external sources during hibernation and excrete small amounts of urine.

Water budgets of the two populations differ (Ward and Armitage 1981b). The smaller semi-arid marmot needs only half as much water per day as the larger montane marmot. The semi-arid marmots lose more water by evaporation and have more concentrated urine (2591 vs. 1686 milliosmoles/l) than the montane marmots that lost more water via urine.

Both populations conserve energy through low rates of conductance; in montane marmots conductance is 82.5%, and in semi-arid marmots, 88.8% of that predicted from body-size regressions (Armitage *et al*. 1990). Low conductance saves energy not only by reducing dry heat loss but also by low evaporative water loss. A trade-off is that the capacity to increase conductance at high temperatures is greatly reduced. The semi-arid marmots cope with their thermal environment through smaller size, which reduces heat production, by conserving water by forming a highly concentrated urine, by reducing the slope of increase in metabolism when body mass increases, and by the ability to evaporate water at high environmental temperatures. The inability of the Colorado marmots to increase water evaporation at high temperatures is not because of limited water. Mass-specific water influx of free-ranging marmots was up to five times greater than the use by *ad libitum* marmots in the laboratory (Melcher *et al*. 1989).

Daily activity cycle

Marmots avoid heat stress by adjusting when they are active on the surface. Early and late in the season when temperatures are relatively cool, the daily patterns of activity of yellow-bellied marmots and woodchucks are unimodal; either there is no difference in the number of animals active at any time during the day or there is an increase in activity during the warmer midday period (Armitage 1962, 1965, Bronson 1962). As seasonal temperatures increase, the activity pattern becomes bimodal (Fig. 9.3). Similar patterns occur in alpine yellow-bellied marmots (Pattie 1967) and in the alpine (Perrin *et al*. 1992, Sala *et al*. 1992, Semenov *et al*. 2000), Vancouver Island (Heard 1977), hoary (Barash 1973b, Taulman 1990a) gray, Menzbier's, and black-capped marmots (Semenov *et al*. 2001a). The bimodal cycle of activity is associated with a circadian cycle of body temperature in the woodchuck; peak temperature occurs in late afternoon (Hayes 1976). In the yellow-bellied marmot, body temperature ranged from a nighttime average low of 36.6°C to an average daytime high of 39.8°C (Melcher *et al*. 1990).

It is unlikely that a body-temperature cycle synchronizes the daily activity cycle. Although body temperature of yellow-bellied marmots generally rose before they emerged in the morning, it sometimes decreased. Emergence and immergence times shifted seasonally with the times of sunrise and sunset (Melcher *et al*. 1990). The activity

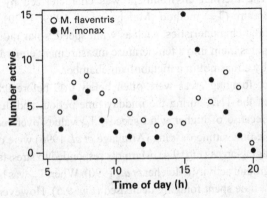

Figure 9.3   The daily activity cycle of the woodchuck (*M. monax*) and yellow-bellied marmot (*M. flaviventris*).

cycle is more likely synchronized by the light–dark cycle (Golombek and Rosenstein 2010). However, black-capped marmots of Yakutia, Russia at 71° 56′ N and 127° 19′ E (north of the Polar Circle) live under conditions of continuous light; the sun was permanently close to or above the horizon. The activity cycle was synchronized with solar altitude: when mean daily temperatures were <5°C, activity started when the sun was 35° above the horizon and ended when it dropped below 28°. When mean daily temperatures were >5°C, the onset of activity began when solar altitude was about 17–18° and ended when it dropped below 10°. Thus temperature acted on the response to solar altitude so that activity began later during the cold season and earlier during the warm season when the characteristic midday avoidance of thermal stress occurred (Semenov *et al.* 2001a).

The bimodal activity cycle is apparent when the amount of time spent can be compared with the amount of time available for above-ground activity. The day was divided into three periods; (1) morning, from the time sunshine first strikes the habitat until 10:00; (2) midday, from 10:00 to 16:00, and (3) afternoon, from 16:00 until all marmots entered their burrows. On average, across three study sites (River, Marmot Meadow, Picnic) from mid-June to mid-August marmots were active for 75% of the available time in the morning, for 43% during midday, and for 64% in the afternoon (Armitage *et al.* 1996). Above-ground activity varied seasonally; it was lowest during gestation, increased during lactation, remained high during the first 20 days of post-lactation, and declined thereafter. The decrease in activity in early August coincides with the seasonal decline in FMR and the time when many marmots have completed mass gain prior to hibernation.

Standard operative temperature

The previous discussion indicates that yellow-bellied marmot physiology has two major characteristics; energy conservation that is associated with a high growth efficiency and heat intolerance that is associated with adaptations for coping with low environmental temperatures. Heat intolerance is not only a function of ambient temperature but also is strongly affected by thermal energy exchange between marmots and their environment. The thermal environment was characterized by using the standard operative temperature (T_{es}) method (Melcher *et al.* 1989, 1990). In effect, T_{es} incorporates microhabitat characteristics, such as wind speed, solar radiation, and air temperature, and integrates them into a temperature measurement equivalent to the temperature experienced by a marmot in a metabolism chamber.

T_{es} values calculated for foraging areas were often below 0°C before 08:00, often exceeded the upper limit of the TNZ during the midday time period, and decreased in the afternoon (Fig. 9.4). Because of higher wind speeds, T_{es} values over rocks where marmots spend considerable time sitting or lying (Armitage *et al.* 1996) were often 10°C lower than those in the foraging areas (Fig. 9.4). Marmots responded to stressful high T_{es} values by reducing above-ground activity (Melcher *et al.* 1990). When T_{es} was higher than the upper limit of the TNZ, time spent foraging decreased (Fig. 9.5). However, foraging time of an adult did not decrease at low temperatures; in fact, temperatures in or below the

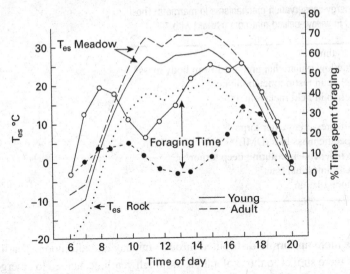

Figure 9.4   General pattern of T_{es} and foraging time for young and adult yellow-bellied marmots. Note the considerable difference in T_{es} on rocks and in the meadow.

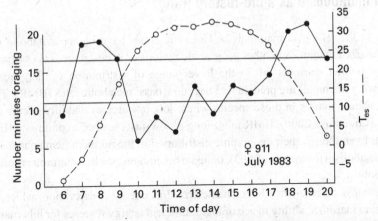

Figure 9.5   Foraging pattern in relation to T_{es} for an adult female throughout the day. The two horizontal lines represent the range of T_{es} that is of low or no stress. Note that T_{es} is most often above or below that range and that foraging is most frequent at low T_{es}.

TNZ were preferred. While marmots were foraging, values for body temperature (T_B) usually rose at rates of 0.114°C min^{-1} for young and 0.075°C min^{-1} for adults. Marmots tolerated short-term increases in T_B but when T_B approached 40°C, marmots either entered their burrows or perched on rocks, where T_{es} is much lower than in the meadow (Fig. 9.4).

Young were more strongly affected by T_{es} than adults. Young responded to stressfully low T_{es} by modifying the daily activity cycle to avoid stress. Low T_{es} in the morning and evening was avoided by emerging later in the morning and immerging earlier in the evening (Melcher *et al.* 1990). Because young are more affected by convection, foraging

Table 9.3 Energy conservation mechanisms in marmots. Those demonstrated in yellow-bellied marmots marked with *.

Circannual rhythm*
Metabolic rate lower than that predicted from body mass*
Seasonal decline in resting metabolic rate*
Seasonal decline in field metabolic rate*
Hibernation*
Metabolism depressed to enter torpor*
Low daily specific mass loss (DML)*
Temperature independence during deep torpor*
Shrunken gastrointestinal tract
Social thermoregulation

activity was more uniformly distributed through midday and afternoon. Yearling foraging patterns were similar to those of adults, although yearlings tended to emerge slightly later in the morning and immerge somewhat earlier in the evening (Melcher *et al.* 1990).

Slow metabolism as a life-history trait

Large (>358 g) mesic mammals in Holarctic zones have higher metabolic rates than similar-sized mammals in other geographic regions or in desert areas (Lovegrove 2000). Yellow-bellied marmots fall in the lower range of distribution of "largeness" where burrowing mammals are predicted to have low basal metabolic rates (BMR). The lowest BMR should evolve in those species where low resource availability occurs in time or space (Lovegrove 2000). BMR falls along a slow-fast metabolic continuum (Lovegrove 2003). Even though their geographic distribution in the northern hemisphere suggests a fast metabolism (Lovegrove 2003), obligate hibernators, such as marmots, express a slow metabolism.

This slow metabolism I have characterized as energy conservation, which is fundamental to marmots' ability to accumulate sufficient energy reserves for hibernation while also reproducing and growing. The lower than predicted high metabolic rates are achieved by reducing conductance (Armitage and Salsbury 1993). Because marmots spend most of their lives at relatively low temperatures, their physiology is focused on coping with cold environments. Their capacity to cope with hot environments is limited and behavioral mechanisms are used to avoid heat stress. Mechanisms of energy conservation are summarized in Table 9.3.

Part III

Social structure and behavior of the yellow-bellied marmot

10 The role of kinship: resource sharing

In Chapter 5, I argued that marmot sociality evolved as a consequence of the retention of offspring in their natal site through at least the first hibernation. The retention of kin to form social groups of varying complexity suggests that kin selection operates in the structure and function of marmot societies. Inclusive fitness theory states that a gene can be transmitted to the next generation by favoring individuals who carry the gene identical by decent. The gene may promote the reproductive success of the individual bearer or that of relatives (Gardner *et al.* 2010). The theory of inclusive fitness may be expressed as "Hamilton's Rule" which states that natural selection favors a behavior when

$$rb - c > 0$$

(Hamilton 1964, Dugatkin 1997, Bourke 2011), where b = the benefit associated with the trait, c = the cost of expressing the trait, and r = the coefficient of relatedness. The contribution to inclusive fitness by the individual bearer is the direct fitness component and the contribution of relatives (collateral kin) is the indirect component (Brown 1980).

Hamilton's rule is deceptively simple; in reality, it is very difficult to quantify the benefits (b) and costs (c) in natural populations over the lifetime of the involved individuals. For example, if marmot A chases marmot B from its home range, what are the costs and benefits of the agonistic interaction. Presumably marmot B has a cost because it is denied access to a resource. Marmot A gains a benefit by maintaining exclusive use of the resource and accrued a cost, the metabolic cost of the social interaction. Does the energy expended in the chase exceed the possible energy gain from foraging in that part of its home range? Did the time expended in the agonistic interaction impose a cost by reducing foraging time? Would a single interaction have any fitness costs over the lifetime of the two individuals?

The great difficulty or virtual impossibility of quantifying the costs and benefits of behaviors does not preclude assessing the importance of kinship in social and population dynamics of animals. The relative contributions of direct and indirect fitness to inclusive fitness (Oli 2003) can be estimated by assessing the way individuals behave toward and share resources with kin of various degrees of relatedness.

The relative importance of direct and indirect components is an on-going problem and controversy (Beckerman *et al.* 2011). One question is does an individual direct benefits to its neighbors according to the degree of relatedness appropriate to the situation (Hamilton 1964)? The benefits may be distributed in proportion to the degree of relatedness (Weigel 1981, Bourke 2011) or preferentially directed only toward close relatives (Altmann

1979). Rubenstein and Wrangham (1980) argued that altruism toward kin should be rare because if an individual invests in kin (e.g., siblings), asymmetries in investment or the uncertainty in return may lead to a loss of inclusive fitness; thus the individual should invest in offspring.

Briefly, there are two models; (1) an individual should distribute benefits according to the degree of relatedness, and (2) benefits should be limited to close kin. In this and the next chapter, these models will be tested by examining the relationship between kinship and resource sharing and social behavior. In addition, I will also explore whether the relationship depends on who the relatives are; e.g., aunt–niece or grandmother–grandoffspring, each group related by $r = 0.25$. Because the critical resources of burrows and foraging areas are spatially distributed, space overlap serves as the metric for resource sharing.

Asocial

Kinship affects settlement patterns of *M. monax* in Maine (Maher 2009a). Male and female adults and yearlings lived near kin and overlapped spatially to a greater extent with closer relatives. For both sexes, the amount of home-range overlap increased with increasing relatedness. However, overlap was small; females shared only 5.2% of the 95% distribution of locations. Opportunities for successful dispersal in this high-density population appear to be limited. Direct fitness may increase by accepting offspring as neighbors who may occupy part of their natal home range. As a consequence of the small degree of home range overlap, resource sharing is probably minimal.

Settling near closely related kin forms kin clusters (Michener 1983), which provides an opportunity for kin-biased benefits. In the Columbian ground squirrel (*Urocitellus columbianus*), adult females with nearby close kin had significantly greater direct and inclusive fitness than females with no co-surviving close kin (Dobson *et al.* 2013). Littermate sisters maintained burrows that were significantly closer than burrows of non-littermate sisters which suggest that the social system favors the association of close kin with whom an individual has associated from birth.

Restricted family group

I noted in Chapter 4 that home range area is shared with all members of restricted and extended family groups. Thus, the question becomes, to what extent do these families consist of closely related kin? Are these kin groups a product of direct or of indirect fitness behaviors or some combination of these? Data are limited, but are available for each of the two family groups.

In *M. caligata* most non-reproductive adults consisted of offspring from within the group and mate selection occurs within the social group (Kyle *et al.* 2007). Thus, the potential exists for resource sharing to be based on kinship, but relatedness values for members of social groups are not available. Because some individuals move between social groups, average r, the coefficient of relatedness, is likely to be much less than 0.5.

Extended family group

Families of *M. marmota* consist of the territorial, dominant pair and their offspring of subordinate rank of various ages (Arnold 1993a). There is little overlap (about 10%) between social groups (Allainé *et al*. 1994). Social group structure and pattern of space use suggests that resource sharing is kin-biased and forms primarily through direct fitness behaviors that incorporate offspring into the family; average $r = 0.33$ between dominants and subordinates (Allainé 2000). This relatively low r is a consequence of demographic processes. Only 15% in Vanoise, France, and 18.3% in Berchtesgaden, Germany, of the philopatric females were those who inherited their natal territory. Thus, 80 to 85% of the females disperse; on average 17.5% (France) to 21.5% (Germany) of these females become resident in directly neighboring territories. Intruder males may displace the resident territorial male and chase out the subordinate adult males (Sala *et al*. 1993, Perrin *et al*. 1996). Mate change in this species is considered to be "forced divorce," when one member of the pair (either sex) remains (stayer) and the other member is forced to disperse. Some of the evicted formed new territories and some re-paired on existing territories, but most disappeared or were found dead. Reproductive success of the stayers did not improve (Lardy *et al*. 2011). These forced evictions appear to be non-adaptive for the stayer, but the evictor probably gains direct fitness by gaining reproductive opportunities, as most of the evictors are subordinates from other territories. This process also illustrates that male and female reproductive strategies may conflict.

New immigrants from outside a colony reduce r (Lenti Boero 1994). For example, a pregnant immigrant female may displace the resident female in a family; thereby reducing r (Goossens *et al*. 1996). In the Grand Nominen colony, no female became reproductive in her natal family and yearlings were present only when their parents were still present (Lenti Boero 1999). As a consequence of the movements into and out of families, these extended families may include juveniles, yearlings, and subordinate adults that are not descendents of the territorial pair (Arnold 1993a). When the territorial pair is unrelated to the other members of the social group, the younger marmots apparently disperse. Thus resource sharing among the territorials and subordinates of all ages is a direct fitness strategy; keeping descendents at home until they can compete for residency, either in the natal family or in an adjacent family in the colony, increases the probability that a marmot will produce reproductive offspring (Armitage 1996c).

The ascendency of some individuals to territorial dominance and the movements of individuals among families within a colony can result in partners of dominant pairs ($r = 0.214$) being more closely related than that of all possible pairs ($r = 0.009$). There is no evidence that mating is based on kinship; extra-pair mating (mating between a territorial female and a male other than the territorial male) enhanced offspring heterozygosity and decreased mother–offspring and full-sibs genetic similarity (Cohas *et al*. 2007a). Furthermore in extra-pair matings between the territorial female and subordinate males, the subordinates were unrelated to the dominant male or female (Cohas *et al*. 2006). To summarize, resource sharing in the alpine marmot is directed toward enhancing direct fitness.

Female kin group: matrilines

Resource sharing among yellow-bellied marmots varies with kinship (Armitage 1996c). The relationship between resource sharing and kinship is more complex than in other marmot species because matrilines may persist for several generations and several matrilines may occur simultaneously in a colony (Armitage 1991a). The relationship between kinship and resource sharing was extensively analyzed for this chapter. Differences in the amount of space overlap were tested by the Mann-Whitney non-parametric test. All coefficients of relatedness (r) were determined from pedigrees.

Percent overlap is directly and significantly related to kinship (r) in each of the four colonies (Fig. 10.1, all $p < 0.001$ except North Picnic $p = 0.002$). Adjusted R^2 varied from 27.6% at North Picnic to 40.5% at Picnic. Considerable variation in overlap is not explained by r; e.g., overlap of unrelated marmots may be higher than overlap of individuals related by $r = 0.5$ (Fig. 10.1). What requires explanation is the widespread

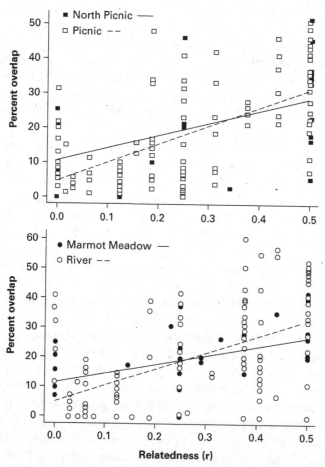

Figure 10.1 . The relationship between percent space overlap and relatedness (r). Upper, North Picnic and Picnic Colonies; lower, Marmot Meadow and River Colonies.

Table 10.1 Mean space overlap (%) of adult, female yellow-bellied marmots at Picnic and River Colonies. A lower case letter indicates that space overlap of the group where the letter first appears in the table does not differ significantly from overlap of those groups lower in the table with the same letter (p > 0.05). When litters intermingled and the mother could not be identified, each young was assigned the average r for the group; e.g., young from two mothers would be assigned a r of 0.375 (0.5 for offspring, 0.25 for nieces and nephews).

Kin group	Overlap	r	N						
Picnic Colony									
Half-sisters/maternal cousins	48.4	0.375	8	a					
Littermate sisters	44.5	0.5	21	a	b				
Mother:daughter	41.6	0.5	24	a	b	c			
Granddaughter (adults and yearlings)	31.7	0.25	6		b	c			
Unrelated	16.9	0	13	d					
Cousins or half-nieces	12.7	0.156	12	d	e				
Non-littermate sisters	11.5	0.5	33	d	e	f			
Aunt:niece	8.8	0.25	17	d	e	f	g		
Half-niece or grandniece	5.3	0.06–0.125	24			f	g	h	
Cousins	4.2	0.03–0.125	10	i				h	
Paternal half-sisters	3.3	0.25	48	i					
River Colony									
Littermate sisters	52.4	0.5	28	a					
Mother:daughter	45.1	0.5	28		b				
Non-littermate sisters	41.0	0.5	3	a	b	c			
Daughter or maternal half-sisters	28.8	0.375	4			c	d		
Mother or aunt or grandmother	27.6	0.375	37				d		
Daughter or niece or maternal half-sister	16.9	0.375	22	e			d		
Cousins	15.0	0.0625	14	e	f				
Aunt:grandniece	11.3	0.125	3	e	f	g	d		
Sibs or paternal half-sibs/maternal cousins or aunt:half-niece	7.1	0.4375	31	h	f				
Cousins	4.7	0.125	20	h	g				
Aunt:niece	1.2	0.25	14						

variation in overlap for any value of r. The variation appears to be related to the history of genealogies and differences in the habitat area and resource distribution of the four colonies. The differences among colonies emphasize the necessity of long-term studies in diverse habitats. I first examine the nature of this variation for adult females at the four major colonies.

At Picnic, overlap varies from 3.3% among paternal half-sisters ($r = 0.25$) to 48.4% among half-sister/maternal cousins ($r = 0.375$; Table 10.1). This difference is a consequence of matriline organization. Half-sister/maternal cousins were born in the same matriline; they have the same father and their mothers are sisters. Paternal half-sisters have the same father, whose territory includes two different matrilines, but their mothers are unrelated. Non-littermate sisters ($r = 0.5$) are in the same matriline and their overlap is significantly greater than that of paternal half-sisters and significantly less that that of littermate sisters (Table 10.1). Although littermate and non-littermate sisters are in the same matriline, the littermate sisters associate from birth and are recruited together into the

Table 10.2 Changes in the relatedness of adult female marmots at Picnic Colony from 1974 to 1980.

Year	Participants	r	Process
1974	Mother (683):daughter (976) matriline. Sibs 1177, 1194 granddaughters ($r = 0.25$) of 683 and grandnieces ($r = 0.125$) of 976	0.5 Average r of the four females = 0.29	The granddaughters forming a matriline independent of the mother:daughter matriline
1978	Two distinct matrilines that are related as aunt:niece (0.25) or aunt:grandniece (0.125)	0.5 within each matriline. Average r of the four females = 0.23	Daughters (301, 349) of 683 form matriline with little overlap with 1194 and her daughter 920
1980	Two distinct matrilines mother:daughter, aunt:niece	Average r for three females = 0.25, 0.5, 0.25	Female 920 budding off an independent matriline, overlap much lower than average for mother:daughter (Table 10.1)

matriline. Non-littermate sisters are members of different year-classes, do not associate together as juveniles, and are more likely to be recruited when a matriline is spreading over a large part of the site and dividing into separate matrilines (see Fig. 17.3).

Overlap among unrelated females is unexpectedly high; it does not differ significantly from non-littermate sisters ($r = 0.5$) or from adults from mixed litters or those that are half-sister/maternal cousins ($r = 0.375$). There are two reasons for this similarity between unrelated females and females of high r. First, females in adjoining matrilines that were in conflict had high overlap because females invaded adjoining home ranges (Armitage 1992). Second, only those females that actually had some space overlap were included in the analysis. If all females distributed across the habitat are included, unrelated females have the lowest overlap (x = 2.6%, N = 84).

The pattern of overlap and relatedness varies with the demographic and genealogical history of the resident adults. Two examples illustrate this variation. When a matriline consisting of two sisters was separating from a matriline consisting of their grandmother and an aunt (Table 10.2), overlap among the four adults was relatively low (Fig. 10.2). As matrilines became more distinct, average overlap decreased (Table 10.2; Fig. 10.3). In the seventh year, the matrilines had changed home range areas, there was some change in matriline membership (Table 10.2), and one of the two matrilines was undergoing fission, which reduced mother:daughter overlap (Fig. 10.4) The fission of the mother:daughter matriline illustrates the process whereby an individual gains some independent space (possibly escaping costs of sharing space) while at the same time allowing the matriline to utilize a larger share of the habitat.

Demographic history may result in high space overlap of a group of paternal half-sister/maternal cousins (Fig. 10.1). Two 3-year-old sisters weaned young who intermingled; thus average r among the young was 0.375 (Table 10.3). When the young were 2 years old, they formed a matriline of five (Fig. 17.3); space overlap was high among the 2-year olds and did not differ statistically from that of littermate sisters (Tables 10.1, 10.3). Over the next several years, two matrilines developed and space overlap between

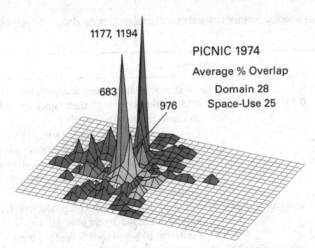

Figure 10.2　Pattern of space overlap among members of a related group of adult females at Picnic Colony in 1974; 683 is the grandmother of sisters 1177 and 1194 and mother of 976. Domain refers to the two-dimensional area that comprises the home range and space use indicates the frequency with which a particular grid was occupied and is represented by the height of the peaks.

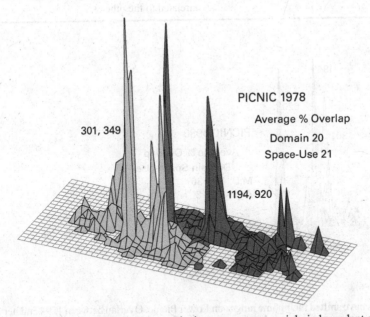

Figure 10.3　Two distinct matrilines at Lower Picnic occupy space mainly independent of each other, 301 and 349 are sisters; 1194 is the mother of 920 and the niece of the two sisters.

the older and younger females declined while remaining high between two surviving younger females.

In River Colony, low and high values of kin-group overlap differed from those at Picnic. Overlap varied from 1.2% for the aunt:niece kin group ($r = 0.25$) to 52.4% for

Table 10.3 The demographic process whereby paternal half-sisters/maternal cousins develop a relatively high space overlap. Location at Upper Picnic.

Year	Participants	r	Process
1980	Two 3-year-old sisters, young	0.5 0.375	Each weaned a litter, young intermingled, related by 0.25 (half-sibs) through their father and 0.125 (cousins) through their mother
1982	Matriline of five, one 4-year old, four 2-year olds	0.375	The 4-year old was the mother or aunt of the 2-year olds. Two-year olds associated together as young and yearlings and the mean space overlap among group members was 50.3% and 36% with the mother/aunt
1983	Mother/aunt forming a separate matriline	Overall r = 0.375	Two 3-year olds present, non-reproductive. Overlap was 43% between the 3-year olds, but only 25% between the 3-year olds and their 5-year old mother/aunt
1984	Mother/aunt 6-year-old. The two 4-year olds reproductive	Overall r = 0.375	Mean space overlap between the mother/aunt and the 4-year olds declined to 15%; space overlap between the 4-year olds remained at 43%
1985	Only a 5-year old present		She had no overlap with two other adult females on Lower Picnic with whom she never associated. She was related by r = 0.094 to one of the females and unrelated to the other.

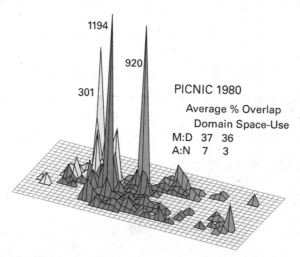

1194
920
301

PICNIC 1980

Average % Overlap
Domain Space-Use
M:D 37 36
A:N   7   3

Figure 10.4   Matrilines have shifted their home ranges on Lower Picnic. Overlap between 1194 and her daughter 920 is lower than the average for mother:adult daughter overlap because 920 is in the process of budding off an independent matriline. M:D = mother:daughter; A:D = aunt:niece.

littermate sisters ($r = 0.5$; Table 10.1). In contrast to Picnic Colony, overlap among non-littermate sisters did not differ significantly from that of littermate sisters. The matriline with both groups of sisters consisted of the mother and daughters from two litters: four 2-year-old littermates and the yearling non-littermate. Behavior among the females was amicable; overlap between the yearling and her 2-year-old sisters averaged 52.5%. In

Table 10.4 Space overlap and relatedness dynamics for adult females at River Colony.

Participants	r	Process
Mother (7-years old) and two littermate daughters (4-years old)	0.5	All females reproduced; litters intermingled. High space overlap among adults and young.
15 young		Kinship of the young was either sibs, paternal half-sibs/
Relatedness to adults	0.375	maternal cousins, or aunt/half-niece.
Relatedness among young	0.4375	
Mother/aunt (6-years old) and five 2-year old daughters or nieces	0.375	The daughters or nieces were chased by the mother/aunt. They spread into unoccupied space from the mound
Relatedness among younger female cohort	0.4375	with the spruce tree (Fig. 8.2b) up-river to the big bend (Fig. 8.2a) to form three new matrilines.

Table 10.5 Mean space overlap (%) of adult female yellow-bellied marmots at Marmot Meadow and North Picnic Colonies. Space overlap of kin groups with the same letter does not differ significantly. The lack of significance between littermate sisters and aunt:niece overlaps probably is a consequence of small sample size.

Kin group	Overlap	r	N	
Marmot Meadow Colony				
Littermate sisters	35.0	0.5	8	A
Mother:daughter	20.4	0.5	5	A
Aunt:niece	20.3	0.25	3	A
Unrelated	7.5	0	12	
North Picnic Colony				
Sister or daughter	23.0	0.5	3	
Unrelated	5.1	0	21	

contrast to the demographic pattern at Picnic, where the matriline increased the space it occupied, the matriline at River in 1991 was constrained in the amount of space available by a contiguous matriline, which consisted of related females (Fig. 16.1). All females were non-reproductive (no adult male was present) perhaps reducing competition. The next year the non-littermate female wandered extensively and had high overlap with the dominant female (a non-littermate sister) of the adjoining matriline. The extensive wandering may account for the high overlap as she dispersed by mid-June.

A mother/aunt group (Table 10.1) occurred when three females weaned litters that intermingled (Table 10.4). The relatively high space overlap between young and adults ($r = 0.375$; Fig. 10.1) reflects their cohesive relationships (Rayor and Armitage 1991). Age structure changed when the old female and one of the 4-year olds died. High overlap persisted because an older, reproductive female frequently entered the home ranges of the younger females and chased them. Subsequently, dispersion of the younger female cohort (Table 10.4) resulted in one of the lowest space overlaps (7.1%; Table 10.1).

The number of kin groups at Marmot Meadow and North Picnic are few compared to Picnic and River Colonies (Table 10.5; Fig. 10.1). This difference is related to habitat characteristics. The higher number of major burrow sites at Picnic (6) and River (5) than at Marmot Meadow (2) and North Picnic (2) provides more opportunity for diverse kin groups to develop.

Demography explains the similarity of mother:daughter and aunt:niece space overlap at Marmot Meadow. A young female was the daughter of one of two littermate sisters and the niece of the other. All three shared the same burrow system after young were weaned, but usually foraged in different areas. Because the daughter/niece tended to avoid the older females, low, but similar overlap patterns developed between the two kin groups.

The only kin associated with an adult female at North Picnic was either a sister or daughter. Although North Picnic is the largest site, its habitat structure is similar to an assemblage of satellite sites. It has a high rate of immigration; 65.6% of adult residents at North Picnic Colony were immigrants compared to only 18% at Picnic Colony. As a consequence, North Picnic has the largest number of overlaps among unrelated females (Tables 10.1, 10.5). The few overlaps among females related by $r = 0.5$ indicates that matriline structure is poorly developed at this site. Poor matriline structure also character-izes satellite sites; mean matriline size at North Picnic is similar to that of most satellite sites and much smaller than that of other colonies (Table 8.1).

Unrelated females occur in colonies when immigrants colonize unoccupied space (Armitage 2003i). Overlap among unrelated adult females is low as no immigrant ever joined an existing matriline and overlap occurs primarily at the edges of home ranges. Note that unrelated adult females did not co-occur at River Colony (Table 10.1) as immigration occurred only when all matrilines died out in the same year.

Several common features occur among the kin-group characteristics of the colonial adult females. Littermate sisters and mother:daughter kin groups are always among the three groups with the highest overlap. Cousin and aunt:niece kin groups are always among the groups with the lowest overlaps. Kin groups are formed among adult females by individuals that have associated together, usually from birth. This process indicates that social bonds are formed by familiarity rather than by kin recognition (e.g., Holmes 1984c). However, kin recognition is not precluded, but has not been demonstrated in marmots. These space overlap patterns do not support the hypothesis that benefits are distributed in proportion to r and strongly support preferential distribution to close relatives and that direct fitness drives the formation of kin groups.

Space overlap when $r = 0.5$

The relationship between relatedness and overlap can be examined by describing space overlap among kin groups related by $r = 0.5$. In contrast to the previous discussion when only adult females were considered, all age/sex groups are included. Space overlap in these groups varied widely (Fig. 10.1).

Space overlap varied from 11.5% for non-littermate sisters at Picnic Colony (Table 10.6) to 52.4% for adult littermate sisters at River Colony (Table 10.7). In each colony, the top three kin groups did not differ significantly. Young sibs were in the top three kin groups in all four colonies (Tables 10.6–10.8) as were yearling sibs (Table 10.8), yearling sisters (Table 10.7), or yearling brothers (Table 10.6). Adult

Table 10.6 Mean space overlap (%) of yellow-bellied marmots related by r = 0.5 at Picnic Colony. A lower case letter indicates that space overlap of the group where the letter first appears in the table does not differ significantly from that of those groups lower in the table with the same letter (p > 0.05).

Kin group	Overlap	N						
Young sibs	51.2	127	a					
Yearling brothers	48.4	16	a	b				
Adult littermate sisters	44.5	21	a	b	c			
Mother:adult daughter	41.6	24		b	c	d		
Yearling:young sibs	41.0	44	e		c	d		
Yearling sisters	37.0	20	e	f	c	d		
Mother:yearling daughters	36.3	27	e	f	g	d		
Mother:yearling sons	34.6	40		f	g	d	h	
Mother:young	33.7	81	i	f		d	h	
Adult female:yearling sisters	31.4	21	i	f	g	j	h	
Adult male:yearling daughter	28.2	30	i	k	g	j	h	
Adult female:full sib young	25.2	19	l	k		j		
Adult female:yearling brother	24.3	13	l	k	m	j		
Adult male:adult daughter	22.8	16	l	k	m	j		n
Adult male:yearling son	21.8	43	l	o	m			n
Adult male:young offspring	19.2	109		o				n
Non-littermate adult sisters	11.5	33						

Table 10.7 Mean space overlap (%) of yellow-bellied marmots related by r = 0.5 at River Colony. A lower case letter indicates that space overlap of the group where the letter first appears in the table does not differ significantly from that of those groups lower in the table with the same letter (p > 0.05).

Kin group	Overlap	N						
Adult littermate sisters	52.4	28	a					
Young sibs	51.3	93	a	b				
Yearling sisters	49.3	18	a	b	c			
Adult female:yearlings sibs	48.3	10		b	c	d		
Yearling female:young sibs	46.9	8		b	c	d	e	
Adult mother:adult daughter	45.9	28			c	d	e	f
Non-littermate adult sisters	41.0	3		g	c	d	e	f
Yearling brothers	38.7	20	h	g	c		e	f
Adult female:young daughter	35.5	31	h	g		i		
Adult male:yearling daughter	33.3	15	h	g	j	i		
Adult female:yearling offspring	32.4	24	h	g	j	i		
Adult female:young sons	27.9	24		g	j	i	k	
Adult male:young offspring	27.8	99			J		k	
Adult male:adult daughter	22.6	10	l					
Adult male:yearling son	20.0	14	l					

Table 10.8 Mean space overlap (%) of yellow-bellied marmots related by $r = 0.5$ at Marmot Meadow and North Picnic Colonies. A lower case letter indicates that space overlap of the group where the letter first appears in the table does not differ significantly from that of those groups lower in the table with the same letter (p > 0.05).

Kin group	Overlap	N							
Marmot Meadow									
Yearling sibs	41.9	80	a						
Young sibs	39.8	154	a	b					
Adult littermate sisters	35.0	8	a	b	c				
Adult female:young sibs	28.8	5		b	c	d			
Yearling female:young brothers	27.3	13			c	d	e		
Mother:offspring	26.3	122	f		c	d	e		
Adult male:young offspring	25.4	43	f	g	c	d	e		
Adult male:yearling offspring	21.7	21	f	g		d	e	h	
Adult female:yearling sibs	21.3	20	f	g		d	e	h	
Adult female:yearling offspring	21.1	34		g		d	e	h	i
Adult mother:adult daughter	20.4	5	f	g	j	d	e	h	i
Yearling and young sisters	14.6	5			j			h	i
North Picnic									
Young sibs	51.9	21	a						
Yearling sibs	45.7	27	a	b					
Adult brothers	37.0	2	a	b	c				
Mother:young offspring	36.1	12			c	d			
Mother:yearling offspring	34.7	17	e	b	c	d			
Adult male:young offspring	29.1	10	e		c	d	f		
Adult sisters or mother:daughter	23.0	3	e	g	c		f		
Adult male:yearling offspring	16.3	4	e	g	c	h	f		
Yearling and young sisters	5.7	3		g	c	h			

littermate sisters was the third member of the top group in three colonies and adult brothers was the third member at North Picnic Colony (Table 10.8) where the adult littermate kin group did not occur. Overall, high space overlap is maintained by littermates from young through adulthood. This pattern supports the importance of familiarity in establishing kin groups. A study using social network analysis, but excluding young, verified the importance of kinship in establishing social networks (Wey and Blumstein 2010), which probably coincide with matrilines.

In contrast, space overlap between yearlings and young is significantly lower than that among young sibs or yearling sibs in three of the four colonies. The one exception is River Colony (Table 10.7); overlap was relatively high among four yearling females and two young sibs (Table 10.9A). Low overlap between yearling and young sisters occurred when a yearling female moved to a different burrow system (Table 10.9B). This pattern is consistent with the interpretation that individuals attempt to maximize direct fitness.

Space overlap between adult males and young or yearling offspring varies among the four colonies, is generally low, and may not differ between adult males and young or yearlings (Table 10.8). In North Picnic, yearlings were rarely present with their fathers.

Table 10.9 Social structure patterns affecting space overlap among kin related by *r* = 0.5.

Colony	Social structure	Contributing factor
A. River	Father, mother, 4 yearling females, two young	An adjoining matriline limited the space available to matriline members
B. Marmot Meadow	Adult female, mother of a yearling female, and three young females	Yearling moved to a different burrow system and established a new matriline as a 2-year-old
C. North Picnic	Adult and two yearling males	Yearling brothers philopatric when a new male became resident in early summer. The 2-year-old adult brothers, had high space overlap (Table 10.8) prior to dispersing

At Picnic (Table 10.6) and River (Table 10.7), adult males had significantly higher overlap with yearling daughters than with yearling sons. The low overlap with yearling sons is a consequence primarily of avoidance of adult males by yearling males who are subjected to agonistic behavior from adult males (Armitage 1974).

Overlap of adult males and yearling daughters may (Table 10.7) or may not (Table 10.6) differ significantly from that of adult males and young offspring. Low overlap between adult males and their offspring (Table 10.6) occurs when one territorial male occupies an entire colony. His daily movements do not include all of the home ranges of the young; thus average overlap is low. One unexpected result was the high overlap of adult brothers at North Picnic, the only instance where this kin group occurred (Table 10.9C).

Mother:offspring space overlap was always significantly lower than that among young or yearling sibs. This difference is surprising as a close relationship between mother and young is expected. Because young have significantly smaller home ranges than adults, home range of young overlaps with only a portion of the adult's home range. If space overlap between young and mother for the entire summer is compared with that of the post-lactation period when young are above ground, space overlap between young and mother is significantly higher during the post-lactation period.

Another rare kin relationship is the adult male with his adult daughter. This kin group occurred at Picnic (Table 10.6) and River (Table 10.7) and overlap was among the lowest of all groups. This kin group is driven by demography. For it to occur, the male must be resident for at least 3 years and his daughters must be recruited into the population and survive to at least 2 years. This situation is rare; father:daughter matings represent about 10% of potential matings (Armitage 2004c). Although rare, inbreeding does occur and its consequences will be discussed in Chapter 15.

To summarize, there are important similarities in kinship patterns when overlaps of adult females are compared with overlaps of all age/sex groups related by *r* = 0.5. Familiarity, which develops within litters and among matriline members, is associated with high home-range overlap. Some kin groups occur as a result of demographic processes

Table 10.10 Mean space overlap (%) of yellow-bellied marmots related by $r = 0.25$ at Picnic Colony. A lower case letter indicates that space overlap of the group where the letter first appears in the table does not differ significantly from that of those groups lower in the table with the same letter ($p > 0.05$).

Kin groups	Overlap	N							
Maternal half-sisters	46.5	4	a						
Yearling male:maternal half-sib young	41.3	4	a	b					
Adult female:adult or yearling granddaughters	31.7	6	a	b	c				
Yearling:young maternal half-sibs or niece or nephew	28.1	10	a	b	c	d			
Yearling female:yearling half-sibs	23.2	10	a	b	c	d	e		
Adult male:young paternal half-sibs	15.3	4	a		c	d	e	f	
Adult female:young granddaughter	10.6	11					e	f	g
Adult female:yearling niece or nephew	10.5	111	h				e	f	g
Adult aunt:adult niece	8.8	17	h	i			e	f	g
Yearling female:young niece or nephew	8.8	20		i	j		e	f	g
Adult female:young niece or nephew	8.0	240	h	i	j	k	e		g
Adult female:yearling male half-sibs	5.9	11	h	i	j	k	e	l	g
Yearling half-sibs	3.4	21	m	i	j	k	e	l	g
Adult female:paternal half-sisters	3.3	48	m		n	k	e		
Yearling paternal half-brothers	3.0	11	m	i	n	k	e	l	o
Adult female:young half-sibs or niece or nephew	1.7	6	m	i	n	k	e	p	o
Adult female:yearling paternal half-sisters	0.6	10	m		n			p	o

occurring over several years. Finally, differences occur among colonies. These differences are related to the distribution and availability of resources in the colonies.

Space overlap when $r = 0.25$

To test whether these generalizations apply to other degrees of relatedness, I examined overlap patterns for kin groups related by $r = 0.25$ (Tables 10.10, 10.11). At Picnic Colony there is the same number of kin groups related by $r = 0.25$ or $r = 0.5$ (Fig. 10.1, upper). At the other three colonies, the number of kin groups related by $r = 0.25$ averages less than half the number of $r = 0.5$ kin groups (Fig. 10.1, lower).

Picnic Colony is a large area with multiple burrow sites and foraging areas. Among the five kin groups with the highest overlaps (Table 10.10), one consists of adult females with granddaughters and the other four have half-sibs as one of the groups (Table 10.12A). Overlap between young paternal half-sibs and an adult male occurred once and is another example of demography producing a short-lived relationship (Table 10.12B). The lowest values (<8.0%) of overlap (bottom 6, Table 10.10) include mostly adult females or yearlings who lived in different matrilines (Table 10.12C). The patterns of overlap among these kin groups is much more strongly related to matriline membership than to $r = 0.25$ per se.

At River Colony, all kinship groups had an adult component (Table 10.11). The lowest overlaps were adult females with adult or yearling nieces or nephews who lived in

Table 10.11 Mean space overlap (%) of yellow-bellied marmots related by $r = 0.25$ at River, Marmot Meadow, and North Picnic Colonies. A lower case letter indicates that space overlap of the group where the letter first appears in the table does not differ significantly from that of those groups lower in the table with the same letter (p > 0.05).

Kin group	Overlap	N				
River Colony						
Adult female:young niece or nephew	33.6	40	a			
Adult female:yearling granddaughter	31.3	3	a	b		
Adult female:young maternal half-brothers	26.2	21		b	c	
Adult female:young grandoffspring	22.0	28		b	c	d
Adult female:young maternal half-sisters	18.5	13		b	e	
Adult female:yearling maternal half-sibs	17.9	14	f	b	e	d
Adult aunt:yearling niece or nephew	14.9	17	f	b	e	
Adult aunt:adult niece	1.2	14				
Marmot Meadow						
Adult female:young nephew or niece	23.4	7	a			
Adult aunt:adult niece	20.3	3	a	b		
Adult female:yearling nephew or niece	15.1	7	a	b	c	
Adult female:maternal half-sib young	9.6	9		b	c	d
Yearlings:maternal half-sib young	0.6	16				d
North Picnic						
Adult female:yearling maternal half-sisters	24.5	4	a			
Adult female:grandoffspring	20.5	2	a	b		
Yearling paternal half-sibs	6.4	12		b		

different matrilines. The high values of overlap of adult females with young nieces or nephews occurred in one year when four reproductive females formed an expanded matriline (Table 10.12D). Although the females weaned young in different burrows, the young freely intermingled in the matrilineal home range. Each of the three sisters was both a mother and an aunt and did not discriminate among the young; there was no significant difference (p = 0.146) in overlap between mother and young (29.8%, N = 20) and aunt and niece/nephew overlap (32.4%, N = 26). Once again we note that overlap is related more strongly to matriline structure and familiarity than to relatedness per se or the composition of the kin group.

The relatively high overlap between the adult female and yearling granddaughters occurred within a matriline (Table 10.12E). In the absence of their mother the high overlap was not sustainable within the matriline, there was no unoccupied space available where the yearlings could establish an independent existence, and the granddaughters dispersed. Analysis of the causes of dispersal revealed that either the presence of the mother or unoccupied space was required for a yearling female to become philopatric (Armitage *et al.* 2011).

Relatively high overlap occurred between adult females and maternal half-brothers or half-sisters (Table 10.11). This grouping was common in 3 years, each year a single matriline occupied the site (Table 10.12F). In all years, overlap among the adults ranged

Table 10.12 Space overlap and relatedness dynamics for marmots related by $r = 0.25$.

Colony	Kinship pattern	Process
A. Picnic	Adults or yearlings with half-sibs	All half-sibs had same mother, thus these groups developed within their natal matrilines.
B. Picnic	Two-year-old male and four young	Adult and young had same father. Overlap occurred in foraging area between the burrow sites of two unrelated matrilines.
C. Picnic	Yearling half-sibs or adult female with half-sibs or nieces and nephews	All half-sibs lived in different matrilines.
D. River	Adult females and young nieces or nephews. Mother with average overlap of 25% with adult daughters; which was lower than that among the three sisters (46.3%, p = 0.08), and significantly lower than for offspring (p < 0.001) or for aunt:nieces/nephews (p < 0.001)	Mother and three littermate daughters reproductive in a matriline. Because mother somewhat spatially separated from daughters, she had little overlap (15.9%, N = 13) with grandoffspring.
E. River	Mother, adult daughter, 3 yearlings. Another matriline present	Mother and four adult daughters reduced to mother and adult daughter. Daughter who weaned the yearlings absent.
F. River	1985. Adult female and three 2-year-old daughters 1988. Older female and her daughter 1993. 10-year-old female and her four 4-year-old daughters	Adult's offspring half-sibs of the daughters. Older female's offspring half-sibs of her daughter. Only 10-year old reproduced, young half-sibs of the 4-year olds.

from 43% to 61%, but the young limited their activity in the home ranges of the adult daughters. Again, birth and residence within the matriline are the critical factors rather than degree of relatedness.

Most kin groups at Marmot Meadow had an adult female component. The overlap between adult aunt and adult niece was unexpectedly high (Table 10.11). This relationship occurred in 2 years when two sisters formed a matriline and the daughter of one was recruited and all three adults occupied the same burrow system. Within this matriline higher than expected overlap also occurred between an adult female and her young or yearling nieces/nephews (Tables 10.11). All young and yearlings were either a niece or nephew of one of the two sisters.

Overlap was strongly affected by the social dynamics of the two sisters. In the first year, space overlap of young did not differ between the two adults (Table 10.13). However, in the second year mother:young overlap was much greater than aunt:young niece/nephew overlap. The aunt disappeared from the meadow for much of the summer but returned to maintain the matriline (Frase and Armitage 1984). During the first half of the summer when both females and yearlings were present, mother:yearling overlap averaged 39%, about 50% greater then the overlap between aunt:niece/nephew yearlings (25%). In the third year, overlap with the yearlings was low, they were treated agonistically by all three adults, and the yearlings disappeared (presumed dispersers) by mid-June.

Table 10.13 The affect of social dynamics on space overlap in kin groups at Marmot Meadow and North Picnic Colonies.

Kin group	Process
Marmot Meadow	
1st year, 2-year-old 911 reproduces	Mother:young overlap (56%) differs little from adult aunt:young niece/nephew overlap (48%).
2nd year, 3-year-old 918 reproduces	Mother:young overlap (27%) much greater than aunt: young niece/nephew overlap (0%).
3rd year, 4-year-old sisters 911 and 918, and 2-year-old daughter (179) of 911 reproduce	Overlap with yearlings (mother:offspring, 10%, aunt: niece/nephew 8%). Yearlings were offspring of 918.
North Picnic	
Adult female and yearlings	Occupied the same area that they occupied with their mother the previous years. Mother did not return and yearlings presumably dispersed after mid-July.

At Marmot Meadow, as at Picnic and River Colonies, demography within a matriline produces kinship groups of $r = 0.25$ such that matriline structure and function are more important in determining the amount of overlap than relatedness per se. The importance of matrilineal organization is apparent in the low overlap with maternal half-sib young (Table 10.11) which occurs because the half-sib young and adult female or yearlings are in different matrilines.

Matrilineal structure accounts for the high values of overlap between adult females and their yearling maternal half-sisters at North Picnic (Table 10.11). The adult female and yearlings had the same mother and were active in the same home range until the yearlings dispersed (Table 10.13). Low overlap among paternal half-sibs occurred because they were in different matrilines.

Summary

A consistent pattern runs through all analyses of overlap between kin groups. The matriline is the fundamental social structure within which kin groups form. Resources are shared primarily with descendent kin or siblings; resource sharing with collateral kin ($r \leq 0.25$) occurs when demographic processes produce such kin groups, some of which are temporary, within the matriline. Groups that have the same degree of relatedness that form in different matrilines do not share resources. The way that social structure and demography affect space overlap among kin is illustrated in Figure 10.5. Young have high overlap when their mothers do. Within matrilines of two females, overlap values between littermates and non-littermates are similar and much higher than overlap with non-littermates when $r = 0.375$.

The process that leads to a particular kinship configuration may occur over several years. Thus, measuring relatedness in a single year or colony may lead to an erroneous

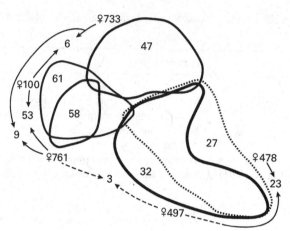

Figure 10.5 Space overlap among young of five adult sisters that formed three matrilines. Overlap among adult females is indicated by the space within overlapping boundaries. The identification number of each female is placed adjacent to her home range. Numbers within the home range boundaries are the mean domain overlap (%) of the littermate young. Space overlap between young of different litters is indicated by arrows connecting the overlap values and adult females. There was no overlap of young of 733 with young of 478 and 497 or of young of 761 with young of 478. The average *r* of young between litters was 0.375 (paternal half-sibs/maternal cousins) and 0.5 for littermates.

interpretation of the role and importance of the levels of kinship to inclusive fitness. Space overlap (resource sharing) does not support the deduction from inclusive fitness theory that benefits are distributed in proportion to *r* but supports the argument that benefits are directed to close kin. Evidence, therefore, indicates that individuals attempt to maximize direct fitness benefits by sharing resources with descendent kin. Indirect fitness benefits, when they occur, are primarily the result of demographic processes leading to sharing space as a result of direct fitness strategies.

11 The role of kinship: social behavior and matriline dynamics

Social dynamics of ground-dwelling sciurids have cooperative and competitive behaviors (Armitage 1987a). Based on the relationships between social structure, kinship, and resource sharing, I predict that amicable behavior will predominate within matrilines or families and between closely related kin and agonistic behavior will characterize social interactions between social groups and among more distantly related kin. Although group membership may affect whether a social interaction between two individuals is amicable or agonistic, group membership may be less important than other factors that determine rates of social interactions.

Seasonal change

In all marmot species, social interactions typically are high in the spring and decline thereafter to low values in late summer (Bronson 1964, Barash 1974a, b, c, 1989, Heard 1977, Sala *et al.* 1992, Perrin *et al.* 1993b, Armitage 2003a). This seasonal pattern suggests that the rate of social interactions is strongly associated with the reproductive season. In a colony of yellow-bellied marmots, agonistic behavior among adult females peaked about midway through gestation and then declined as females moved away from a center of activity to other areas in the colony, which reduced overlap in home ranges (Armitage 1965). In general, little information is available on social behavior during mating and gestation because snow cover and weather conditions make field observations very difficult. Also, some species of marmots with the family social structure terminate hibernation and mate in the burrow before any individuals emerge (Armitage and Blumstein 2002). Under these conditions it obviously is not possible to observe how social behavior affects which female in the family reproduces.

For an activity that is considered to have a critical role in fitness, social behavior utilizes a surprisingly small part of the daily activity of marmots. In the alpine marmot social behavior is considered a minor activity which accounts for less than 3% of the above-ground time budget (Sala *et al.* 1992). In woodchucks interactions averaged about one per 3 h (Bronson 1964). Woodchucks spent relatively less time interacting with more closely related individuals; this pattern persisted for females but not for males when sexes were analyzed separately. The percentage of time spent interacting usually was <5% (Maher 2009b). For adult female Olympic marmots, social behavior utilized less than 5%

of the activity budget (my estimate from figures in Barash 1973a). Hoary marmots also expended little time in social behavior (Holmes 1984a).

Among yellow-bellied marmots, social behavior among animal cohorts averaged over the summer varied from 1.5–5.2 min per day or 0.43–1.7% of the activity budget. Social behavior varied among colonies; e.g., rates of social behavior at Picnic Colony were higher during the midday time period than at the other colonies and the effect of season was most pronounced at Marmot Meadow where an adult female and her young were highly interactive. In general, variation among colonies was attributable to differences in their age/sex cohorts (Armitage et al. 1996).

Behavior and relatedness

We now explore the distribution of amicable and agonistic behaviors between individuals and whether these behaviors are associated with degrees of relatedness. In the relatively asocial woodchuck, rates of amicable behavior increased with increasing relatedness but rates of agonistic behavior were not correlated with relatedness. Among littermate siblings of different ages, rates of agonistic behavior did not vary with age but rates of amicable behavior among adults were lower than those between juveniles and yearlings (Maher 2009b). Rates of amicable interactions between mother:offspring and littermate:sibling dyads were higher than expected, but fewer amicable interactions occurred between non-littermate siblings and more distant kin. The decrease in amicable behavior among non-littermate siblings suggests that familiarity may play a role in the distribution of social behaviors. Interestingly, mothers behaved more amicably towards younger offspring and more agonistically towards older offspring, especially females. This pattern suggests that both cooperation and competition occur among close kin (Maher 2009b) and that competition increases with reproductive maturity. Patterns of social behavior among yellow-bellied marmots were also characterized as having cooperative and competitive elements (Armitage 1989). Relatedness introduces variation into social interactions; amicable behavior predominates between individuals related by $r = 0.5$ and agonistic behavior predominates between individuals related by $r < 0.5$. Why both amicable and agonistic behavior occurs among both closely related and more distantly related yellow-bellied marmots will be discussed later in this chapter.

Among marmot species living in family groups, amicable behavior characterizes social interactions within families and agonistic behavior, between families (e.g., Perrin et al. 1996). In alpine marmots, agonistic behavior occurs primarily during the reproductive period; male aggression toward other males characterizes this behavior (Sala et al. 1992, Mann et al. 1993). Cohesive interactions increased just before immergence when family members come together at the hibernaculum (Perrin et al. 1993b). Because relatedness in alpine marmot families is much less than 0.5, social behaviors must be strongly influence by familiarity. However, as will be discussed further in the chapters on reproductive success, both competition and cooperation occur in these families and kinship has an important role at that time.

Among Vancouver Island marmots, greeting and play-fighting accounted for 65% of all behaviors and behaviors were similar among the age/sex classes. There was more aggression between adult and yearling females than for other dyads and a high rate of greetings between adult and 2-year-old females. There appeared to be a dominance hierarchy: adult male > adult female > 2-year-old female > yearling male > yearling female (Heard 1977). Dominant individuals performed play-fight invitations, social grooming, and mounting.

Among monogamous hoary marmots, adult males responded to the intrusion of adult males and to 2-year olds whereas adult females responded to the intrusions of adult females; most responses were agonistic (Holmes 1984a). In polygynous colonies where all animals shared the colony site, greetings were common and chases were associated with play bouts; adult males were the most frequent chasers and 2-year-olds and yearlings were the most frequent chasees (Barash 1974b). Chases also occurred when an adult male chased a peripheral male.

Social behavior in an Olympic marmot family is primarily greeting, which commonly occurred whenever two marmots met after a period of separation and was most common following the morning and afternoon emergence (Barash 1973a). Greeting frequencies varied among age/sex groups daily and seasonally; relatedness among members of a colony was unknown. Patterns of age/sex demography suggest that some female offspring are recruited into their natal population which would produce relatively high r values. Recent studies that report little dispersal and virtually no immigration in Olympic marmot colonies undergoing population decline (Griffin *et al.* 2008) also suggest that recruitment of females in their natal families produces groups of high r values. More long-term studies of marmots that live in family groups are needed to determine the significance of relatedness in the formation and functions of these groups.

The most extensive studies of relatedness and social behavior are of yellow-bellied marmots. Early studies revealed that social behavior of burrowmates was predominantly amicable and that of non-burrowmates was chiefly agonistic (Armitage and Johns 1982). These analyses did not explain the basis of why both amicable and agonistic behavior occurs within a social group nor was relatedness a part of the analysis.

Amicable behavior

Rates of amicable behavior among females are greater at all r values within matrilines than between matrilines (Table 11.1). Amicable behavior was never observed between adult females ($r = 0.5$) living in different matrilines, was rarely observed between adults and yearlings living in different matrilines, and was commonly observed within matrilines (Table 11.1). Rates of amicable behavior among adults were greater at $r = 0.5$ than at $r \leq 0.375$ (Table 11.2A); whereas they were statistically identical among adult females and between adults and yearlings ($r = 0.5$), and between adults and yearlings related by $r = 0.5$ or $r < 0.5$ (Table 11.2B). Amicable behavior between adult and yearling females living in different matrilines was rare with low rates regardless of r (Table 11.1).

Table 11.1 Rates of amicable behavior (number/animal/1000 h) and relatedness for females in four colonies of yellow-bellied marmots. (N) = number of years the colony was sampled or the number of times the social interactions between the relatedness groups occurred. All values are mean ± SE.

Social groups and relatedness	Within matrilines			
	Picnic (28)	River (15)	Marmot Meadow (18)	North Picnic (10)
Among adults				
0.5	43.9 ± 6.2 (29)	63.9 ± 12.8 (15)	66.1 ± 20.6 (11)	41.0 ± 9.0 (2)
0.4375		48.4 ± 11.4 (5)		
0.375	60.5 ± 14.5 (2)	7.0 (1)	7.0 (1)	
0.25	22.3 ± 7.2 (3)	41.0 (2)		
0.125	22.0 (1)			
Between adults and yearlings				
0.5	55.3 ± 14.3 (12)	58.8 ± 19.5 (6)	48.7 ± 12.2 (6)	61.0 (1)
0.406		18.0 (1)		
0.375	53.0 ± 25.2 (5)	8.0 (1)	39.5 ± 25.5 (2)	
0.25	50.8 ± 24.8 (4)	10.5 ± 2.5 (2)		
0.1875	18.0 ± 13.0 (2)			
0.0625	44.0 (1)			
Among yearlings				
0.5	53.0 ± 10.4 (9)	106.2 (47.3) (4)	94.8 ± 41.0 (6)	115.0 (1)
0.4375		198.0 (1)	72.0 (1)	
0.375	24.5 ± 4.5 (2)			
0.3125			22.0 (1)	
0.25	29.0 (1)			
	Between matrilines			
Among adults				
0.5	4.5 ± 0.5 (2)			
0.375		7.0 (1)		
0.25	15.0 (1)			
0.125	16.0 (1)			
Between adults and yearlings				
0.5			4.0 (1)	
0.33		4.0 (1)		
0.067	6.0 (1)			
Among yearlings				
0.5			5.0 (1)	

High mean rates of amicable behavior among yearling females within the matriline were unrelated to r and did not differ with mean rates between female yearlings and adults or among adults (Table 11.2C). Only one instance of amicable behavior among yearling females ($r = 0.5$) living in different matrilines was observed (Table 11.1).

Agonistic behavior

Agonistic behavior occurs among all relatedness groups both within and between matrilines. Unrelated females occur only in different matrilines and their social

Table 11.2 Statistical analyses of rates of social behavior among female yellow-bellied marmots. When sample sizes were small, relatedness groups were combined and recorded as $r < 0.5$.

Comparison	
A. Adult females within matrilines across all colonies	Rate of amicable behavior significantly greater ($t = 2.52$, $0.02 > p > 0.01$) at $r = 0.5$ than at $r \leq 0.375$
B. Among adult females and between adult and yearling females, $r = 0.5$	Amicable behavior, $t = 0.09$, $p > 0.9$
Adult and yearling females amicable behavior	$r = 0.5$ vs. $r < 0.5$, $t = 1.37$, $p > 0.1$
C. Female yearlings within a matriline	Amicable behavior unrelated to r, $t = 0.54$, $p > 0.5$
Amicable behavior among yearlings	
vs. among adults	$t = 1.49$, $p > 0.1$
vs. between yearlings and adults	$t = 1.35$, $p > 0.1$
D. Rates of agonistic behavior among adult females ($r = 0.5$) or	Significantly less than rates of amicable behavior, $t = 35.3$, $p < 0.001$
adult females related by < 0.5	Rates of amicable and agonistic behavior do not differ, $t = 0.79$, $p > 0.4$
adult females where $r = 0.5$ compared with adult females where $r < 0.5$	No difference in rates of agonistic behavior between the two groups, $t = 0.25$, $p > 0.5$
E. Adult females living in different matrilines where $r < 0.5$	No difference between rates of amicable and agonistic behavior, $t = 1.45$, $p > 0.1$
Adult and yearling females living in different matrilines where $r < 0.5$	Rate of agonistic behavior greater than that of amicable behavior, $t = 2.76$, $0.02 > p > 0.01$
F. Rates of behavior between yearling and adult females in the same matriline:	No difference when $r = 0.5$ or < 0.5
amicable	$t = 0.51$, $p > 0.5$
agonistic	$t = 1.37$, $p > 0.1$

interactions are always agonistic (Table 11.3). Agonistic interactions between yearling females were rare regardless of r and whether the yearlings lived in the same or in different matrilines (Table 11.3).

Within matrilines, rates of agonistic behavior among adult females were lower than rates of amicable behavior. Agonistic behavior among adult females within matrilines is not determined by relatedness; rates of amicable and agonistic behavior do not differ where $r < 0.5$ (Table 11.2D).

Where $r < 0.5$ among adults or between adults and yearlings living in different matrilines, rates of amicable and agonistic behavior did not differ among adult females, but the rate of agonistic behavior was greater in adult:yearling social interactions (Table 11.2E). This pattern suggests that adults may be more agonistic to unfamiliar yearlings from a different matriline even though distantly related. Such discrimination does not occur when adults and yearlings live in the same matriline; rates of amicable and agonistic behavior are unrelated to kinship (Table 11.2F).

Table 11.3 Rates of agonistic behavior (number/animal/1000 h) and relatedness for females in four colonies of yellow-bellied marmots. All values are mean ± SE. (N) = number of times the social interaction occurred.

Social groups and relatedness	Within matrilines			
	Picnic	River	Marmot Meadow	North Picnic
Among adults				
0.5	13.3 ± 2.4 (13)	16.5 ± 4.4 (11)	29.5 ± 7.7 (8)	
0.4375		10.0 ± 0.6 (3)		
0.375	7.0 ± 4.0 (2)			
0.25	12.5 ± 7.5 (2)		119.0 (1)	
0.125	12.5 ± 9.5 (2)			
Between adults and yearlings				
0.5	15.3 ± 7.0 (7)	33.3 ± 15.9 (4)	104.4 ± 30.4 (10)	24.5 ± 9.5 (2)
0.375	11.7 ± 2.9 (3)	77.0 (1)	69.0 ± 37.0 (2)	
0.33		34.0 (1)		
0.25	15.0 ± 11.0 (2)	3.0 (1)	15.0 (1)	
0.1875	10.0 (1)			
Among yearlings				
0.5			8.0 (1)	
0.4375			3.0 (1)	

	Between matrilines			
Among adults				
Unrelated	26.6 ± 12.4 (7)		5.0 ± 3.0 (2)	15.0 (1)
0.5	2.5 ± 0.5 (2)		25.7 ± 10.3 (4)	
0.4375		6.0 ± 0 (2)		
0.375	24.0 (1)	22.0 ± 7.2 (3)		
0.25	10.7 ± 5.8 (3)	61.0 (1)	7.0 (1)	
0.125	29.3 ± 10.2 (3)	1.0 (1)		
0.0625	20.3 ± 11.7 (3)	8.0 (1)		
0.03125	4.0 (1)			
Between adults and yearlings				
Unrelated	27.3 ± 22.6 (4)		21.5 ± 13.5 (2)	
0.375		16.0 ± 7.0 (2)		
0.33		9.0 (1)		
0.25	7.0 (1)	35.0 (1)	19.0 (1)	
0.133	3.0 (1)			
0.1875	16.0 (1)	37.0 (1)		
0.125	85.0 (1)	2.0 (1)		
0.067	9.0 (1)			
0.0625	10.8 ± 4.9 (4)	35.0 (1)		
0.046	3.0 (1)			
0.03125	4.5 ± 0.5 (2)			
0.0156	3.0 (1)			
Among yearlings				
0.375			19.0 (1)	
0.0625	7.0 (1)			

Summary of behavior and relatedness

To summarize, relatedness is less important than living in the same or different matrilines in determining rates of social behavior among females. This pattern further indicates that familiarity through group living is more important than relatedness in forming matrilines. Because mother:daughter:sister associations characterize matrilines (Armitage 1984, 1998), relatedness of $r = 0.5$ predominates. In effect, familiarity among group members produces social groups of high relatedness; the average relatedness of yearling females with adult females in matrilines is 0.421 (Table 18.3). Two important points require emphasis. First, not all behavior within matrilines is amicable or cohesive; agonistic behavior also occurs (Table 11.2). Second, individuals related by $r < 0.5$ also occur within matrilines. The dynamics of matriline formation and the consequence of agonistic behavior or competition may require several years for the process to unfold. Several examples of this process follow.

Matriline dynamics: formation of groups when *r* < 0.5

The first example illustrates the way that individuals related by $r < 0.5$ form a matriline. A new matriline formed when two 2-year-old females, either full sibs or paternal half-sibs/ maternal cousins moved away from their mother or aunt (see Fig. 19.1) in their natal matriline (Table 11.4). Social interactions between the 2-year olds were only amicable, but were entirely agonistic with non-matriline adult females, including their putative mother/aunt.

Table 11.4 Process of matriline formation when individuals related by $r < 0.5$ at Picnic Colony from 1978 to 1989.

Year	Social structure and matriline activity
1978	Two females (167, 174) born into intermingled litter of mean $r = 0.4375$. 167 and 174 together as yearlings, then moved to form new matriline.
1980	Each reproduced, young intermingled. Only amicable interactions between adults, among young, and between adults and young.
1981	Two adults (3-years old) associated with seven yearlings related on average by 0.4375, average r between adults and yearlings was 0.375. Adults, yearlings, and five weaned young intermingled freely.
1982	Matriline consisted of 4-year-old 174, and 2-year-olds 553, 556, 483, 571. No young weaned. Three yearling females related by $r = 0.5$.
1983	Five-year-old 174 moved to form a separate matriline, she weaned a litter. Three-year-olds 556 and 571 remain in original matrilineal home range, a litter of four was weaned; none survived.
1984	One litter weaned, the 4-year-old mother had agonistic interactions with her older mother or aunt living in her own matriline. Interactions with unrelated adults and yearling were agonistic.
1985	Five-year-old 556 and yearling female 326 present; 556 weaned a litter.
1986	Two-year-old 326 and yearling 381 present, no reproduction. Both females had the same father and distantly related through their mother.
1988	Four-year-old 326, two of her yearling daughters, and one yearling niece present.
1989	Five-year-old 326, three 2-year-olds; one 2-year-old killed by coyote.

In 1981 the matriline consisted of the two adults and seven yearling females who were full sibs or paternal half-sibs/maternal cousins and were either offspring or nieces of the adults. All social interactions within the matriline were amicable or play; amicable behavior between the yearlings and adults was three times more frequent than agonistic. No social interactions were observed with members of other matrilines. Mean relatedness of all matriline members was $r = 0.413$. Three yearlings disappeared during the summer.

In 1982 (Table 11.4), all social interactions among the yearlings were amicable or play, but agonistic interactions were 15 times more frequent than amicable between adults and yearlings. Average relatedness between yearlings and adults was $r = 0.464$, and among all members of the matriline increased to $r = 0.435$. Despite the relatively high relatedness between adults and yearlings, agonistic behavior characterized their social interactions and all yearlings dispersed.

In 1983, the original matriline divided to form one matriline with the old mother/aunt and one with two 3-year olds, with an average r of 0.4375 (Table 11.4). Social interactions between the two 3-year olds were equally divided between amicable and agonistic, they had no social interactions with their mother/aunt and social interactions with two females from two other matrilines were all agonistic. One of the females from one of the other matrilines was unrelated to the 3-year olds and the other female was related by $r = 0.156$. Thus, in 1984 relatedness did not change, all social behaviors between the females were amicable, both reproduced, but only one litter was successfully weaned. All social interactions between the two females and with the young were amicable, but otherwise were agonistic (Table 11.4).

In 1985, one yearling and one adult female, related by $r = 0.218$, were present. All social interactions involving the adult, yearling, and young were amicable. This stable relationship with low r continued in 1986 when the two residents were related by $r = 0.359$ (Table 11.4). Social interactions were rare and mainly amicable. In 1987, both females weaned a litter. All social interactions between the adult females were amicable and both behaved cohesively with young from both litters.

In 1988, the average r between the resident adult and three yearling females was 0.39. Only amicable behavior occurred between the adult and yearlings and among the yearlings. All yearlings were recruited and returned to the same matriline with the older female the next year. The death of the 2-year-old niece restored average r to 0.5.

To summarize, this matriline began with two adults who were either sisters ($r = 0.5$) or paternal half-sisters/maternal cousins ($r = 0.4375$). Changes in average r were driven by patterns of recruitment and mortality (Armitage 1996b). Regardless of relatedness, amicable behavior prevailed within the matriline where individuals were familiar with each other from their natal year except in 1982 when adults were highly agonistic to yearlings of uncertain parentage that came from intermingled litters. Recruitment is less likely when parentage is uncertain; marmots apparently opt to seek future reproductive success rather than recruiting a possible non-descendent. In effect, it is better to enhance the probability of one's own reproduction rather than assisting relatives to reproduce (Oli and Armitage 2008). Some of the benefits of living in a matriline of two or more adult females will be discussed in a later chapter.

Matriline dynamics: fission when *r* = 0.5

An immigrant female settled in unoccupied space and in her fourth year of residency gained effective control of the entire lower slope of Picnic Colony. She recruited a daughter in her fourth year; space overlap was high among all matriline members (Fig. 11.1) and all behavior was amicable (Table 11.5). In the fifth year, when only 855 weaned a litter, space overlap was high among all members of the matriline (Fig. 11.1).

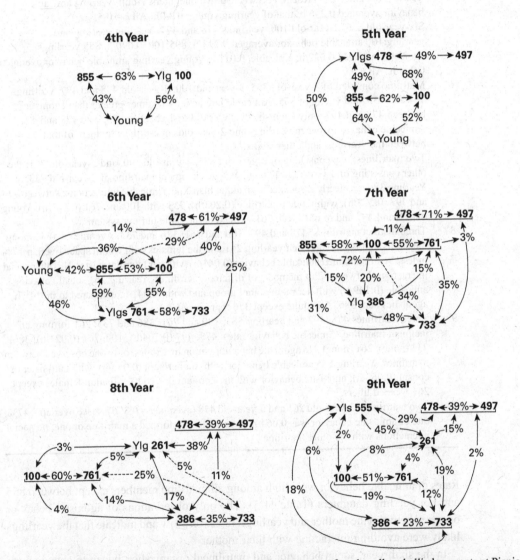

Figure 11.1 Matriline structures and patterns of space overlap among female yellow-bellied marmots at Picnic Colony. Matrilines are underlined. Note that 733 is forming an incipient matriline in the seventh year, but maintained higher overlap with the three-female matriline. Yearling 261 was born in the 478/497 matriline. Numbers on the lines connecting marmots are the percent overlaps of space use. Individuals are identified in bold.

Table 11.5 Process of matriline fission when all adult females are closely related at Picnic Colony.

Year	Social structure and matriline activity
1	Immigrant female 855 occupies open space; behaves agnostically with an adult and two yearling females in two other matrilines. Subsequently causes 573 to move to new location (Armitage 1992).
3	Amicable behavior with female young.
4	Yearling daughter 100 present; 855 weaned two daughters. Adult: yearling amicable behavior averaged 0.296/h; that of yearling:young, 0.34/h. All $r = 0.5$.
5	Six-year-old 855, 2-year-old 100, yearlings 478 and 497. Yearling:yearling and yearling:100 amicable behavior averaged 0.213/h, 895:100, 0.090/h. 855:yearlings: agonistic averaged 0.022/h; amicable, 0.011/h, young:yearling amicable behavior averaged 0.244/h. All $r = 0.5$.
6	Matriline consisted of 7-year-old 855, 3-year-old 100, 2-year-old 478 and 497, yearlings 761 and 733, and young of 855. Amicable interactions 10 times greater than agonistic between 478 and 497, only agonistic behavior (0.065/h) between 855 and 478 and 497. Amicable behavior between yearlings and 2-year olds one-fifth as frequent of that between the yearlings and other adults.
7	Two matrilines, one consisting of 8-year-old 855, 4-year-old 100, and 2-year-old 761; the other consisting of 3-year-olds 478 and 497. An incipient matriline of 2-year-old 733. Yearling 386 not clearly associated with either matriline. Amicable interactions: between 478 and 497 (0.373/h); within other matriline (0.203/h); 733 with 100 and 761 (0.147/h). Young of 478 and 497 and of 855, 100, 761 weaned in the same burrow systems.
8	Three distinct matrilines: 478 and 497; 100 and 761; 386 and 733. Average $r = 0.5$ among the adults, but is 0.25 between yearling 261 and the females with whom space overlap 17% or less. 761 maintained amicable behavior (0.041/h) with her littermate sister. No other social interactions between adult members of different matrilines. Yearling 261 social behaviors amicable (0.149/h) with her mother/aunt group and with (0.041/h, compared to 0.0149/h agonistic) aunt 386. All adults except 386 reproduced.
9	Three matrilines 478, 497 and yearling 555; 100 and 761; 386 and 733. 261 forming an incipient matriline. Amicable behavior rates: 478, 497 (0.104/h); 100, 761 (0.021/h); 386, 733 (none); 261 (none). No agonistic behavior within matrilines, only one observed between matrilines. Yearling 555 amicable behavior with mother/aunt (0.052/h), with half-sister or cousin (0.063/h); agonistic behavior with mother/aunt (0.010/h). All adult females except 261 weaned litters.
10	Two matrilines: 3-year-old 261 and 6-year-old 478 (average $r = 0.375$); space overlap = 47%, rate of amicable behavior was 0.064/h; 4-year-old 386 formed a matriline of one, no social interactions with the other matriline.

Rates of amicable behavior were high among all matriline members except between 855 and her yearling daughters (Table 11.5). The only observations of agonistic behavior occurred between the mother and yearlings; rates were low and indicate that the yearlings likely were avoiding interacting with their mother.

In the sixth year, social behavior and matrilineal organization began to change even though all marmots were related by $r = 0.5$. The two 2-year-old females began to bud off an independent matriline; space overlap among the 2-year olds was high, but markedly reduced with the other matriline members who maintained the original core matriline with high space overlap with each other (Fig. 11.1). Amicable behavior predominated within each of the two groups and between the yearlings and the 2-year-old group, but only agonistic behavior occurred between the 2-year-olds and their mother (Table 11.5).

Although all adults were related by 0.5, two distinct matrilines were present in the seventh year (Fig. 11.1). Social affiliations shifted; 761 joined her mother and older sister in a matriline of three while her sister 733 formed an incipient, independent matriline. The yearling had greater space overlap with 733; this association was a precursor of a matriline the following year. Social interactions were primarily amicable within matrilines; agonistic interactions were one-twelfth as common as amicable. Social interactions between the matrilines were rare and were both amicable and agonistic. A major demographic change occurred when badger predation killed 855, two adult males, and several young in late July (Armitage 2004a).

Three distinct matrilines were evident in the eighth year. Space overlap was relatively high within matrilines and relatively low between members of different matrilines (Fig. 11.1). The highest space overlap between members of different matrilines was between littermate sisters (Fig. 11.1), 761 and 733, who interacted amicably (Table 11.5). The space overlap and behavioral pattern indicates that familiarity from birth may maintain a cohesive relationship even when living in different matrilines. The yearling had relatively high overlap with her mother/aunt matriline (mean $r = 0.375$) and relatively low overlap with all other females who were her aunts ($r = 0.25$). Her social interactions were amicable, including with an aunt (Table 11.5). The space overlap and behavior between aunt and niece illustrate that demographic processes over several years can produce kinship of low r within a larger group of high r.

The same matrilines of two females each returned in the ninth year. The relatively low space overlap of 261 with the three matrilines indicated that she was in the process of budding off an independent matriline (Fig. 11.1). The yearling female had high space overlap with her maternal matriline even though average r was 0.1875. Her amicable interactions greatly exceeded her agonistic interactions (Table 11.5). Overall social interactions among residents were few and more a function of matrilineal relationships than kinship (Table 11.5).

A late summer drought in the ninth year reduced population numbers such that matrilineal reorganization characterized the tenth year (Table 11.5). One matriline of two females was characterized by high space overlap and amicable behavior and had no association with the only member of the other matriline (Table 11.5). Two-year-old female 555 dispersed to a nearby site; her space overlap decreased from high values the previous year (Fig. 10.1) to less than 3% and her only social behavior was agonistic.

It is unclear why 555 failed to establish social bonds given her high degree of sociality the previous year (Fig. 11.1, ninth year) and in her yearling year (Table 11.5, ninth year). Her low degree of relatedness (average $r = 0.29$) with the three older adults may have been a factor, but individuals of low r are incorporated into their natal matriline. The low overlap with the other adults indicates that she was unable to settle in good habitat and dispersed to increase the probability of reproducing, but she failed to reproduce in her new habitat. She may have been a masculinized female as her littermates were 71% males; masculinized females from male-biased litters are more likely to disperse and less likely to reproduce (Monclus and Blumstein 2012). Thus, maternal effects seem the more likely cause of reproductive failure.

Summary of major points

I have presented this 10-year account of the fate of a group of female descendents from the same mother to emphasize several points. First, an understanding of the relative importance of the relationship of kinship and familiarity to social behavior and matriline organization and function usually requires several years of studying the same individuals in the same habitat. Second, either kinship or familiarity or both may be critical in the year-to-year dynamics of matrilineal structure and function. Third, a high degree of relatedness among females over several years does not result in continued growth of a matriline. Rather than remaining in one group in which all individuals are related by $r = 0.5$, individuals spread over the habitat to bud off daughter matrilines. Despite the high degree of relatedness, social behavior and space overlap decreased among individuals forming independent matrilines.

This process of matriline fission raises the question of why mothers and daughters and full sisters separate to occupy different matrilines. This fission of matrilines when $r = 0.5$ seems puzzling given that social behavior within and between matrilines was primarily amicable. The answer lies in reproductive competition that is not apparent in social behavior during the field season. Reproductive competition is expressed as reproductive suppression of younger adult females by older females (Armitage 2003f). In the fifth and sixth years, only 855 reproduced, reproductively mature 2- and 3-year olds did not reproduce. When the females budded off independent matrilines and reduced or eliminated reproductive suppression, they weaned litters. Females attempt to maximize inclusive fitness by maximizing direct fitness and do not compensate for loss of direct fitness through indirect fitness; e.g., by assisting kin (Oli and Armitage 2008). Reproductive competition will be discussed extensively in the chapter on female reproductive success.

Matriline dynamics: growth and fission in a single lineage

This lineage began in 1983 at River Colony when 3-year-old immigrant 915 settled in the South Mound area and weaned three daughters; all were recruited. No other females were present; thus this matriline could spread over the entire area. All females reproduced in 1985, space overlap was high and social interactions were primarily amicable (Table 11.6). The matriline was unstable; over several years two daughters moved to the River Mound/ Bench area to form matrilines of one; neither female recruited daughters.

In 1987, the three adults reproduced, young intermingled and their space overlap with the adults was relatively low. Space overlap between the adults and their offspring (now yearlings) remained low the following year; social interactions between adults and yearlings were predominantly agonistic. However, the five yearlings had adequate space, especially at Spruce Mound and to the north, to avoid the adults and all became residents (Table 11.6).

Over the next 2 years the remaining older adult occupied South Mound and three 5-year-old recruits spread over the remaining area and began to form new matrilines. The 8-year-old 959 recruited four daughters; space overlap within this group was high and

Table 11.6 Matriline dynamics at River Colony. SO = space overlap (%), Am = amicable and Ag = agonistic behavior (#/animal/h), M = mother, D = daughter, Ylg(s) = yearling(s).

Year	Matriline composition	Matriline characteristics
1985	Five-year-old and three 2-year-old daughters All r = 0.5. All weaned litters.	SO. M:D = 23, Among D = 46.3. Am: M:D = 0.009, Among D = 0.169. Ag: M:D = 0.039. Among D = 0.004.
1986	Six-year-old and two 3-year olds; one 3-year old moved to and reproduced at Bench area; yearlings dispersed.	SO among adults = 39.3, adults:ylg = 16, among adults and young = 34.8. Among adults Am = 0.146, Ag = 0.035. Average r between adults and ylgs = 0.29.
1987	All three adults reproduced, young intermingled. No ylg residents.	SO = 45 among adults, 26.5 between adults and young; 4-year-old 960 had lower SO with M 915. Ag 3 × Am between 915 and 960; Am 4 × Ag between 915 and 4-year-old 959.
1988	Eight-year-old 915 and 5-year-old 959 reproduced; 960 moved to River Mound/Bench area. r among ylgs (sisters or paternal half-sisters/maternal cousins averaged 0.406, adults:ylgs (daughters, nieces, granddaughters, maternal half-sisters) = 0.375. Five yearling residents.	Among adults, SO = 43; 915:ylgs (daughters or granddaughters) = 23.6; 959:ylgs (daughters, nieces, or maternal half-sisters) = 28.4; among ylgs = 43.3. Only Am behavior among ylgs (0.195). Adult:ylg mainly Ag (915: ylgs = 0.106), (959:ylgs = 0.019). Behavior between adults rare, Ag twice as frequent as Am.
1989	Six-year-old 959 reproductive, five 2-year-old recruits. 959 related to 2-year olds by 0.375 (daughters, nieces, or maternal half-sisters).	Between 959 and 2-year olds, SO = 24; among 2-year olds, SO = 21.6; 959 and young, SO = 40.3; 2-year olds and young SO = 23.4. Behavior between 959:2-year olds agonistic (0.083). Among 2-year olds Ag = 0.006; Am = 0.025.
1990	Seven-year-old 959 reproductive. Three-year-olds: 38 at south end of South Mound, 431 and 801 centered on Spruce Mound and 43 and 818 centered on River Mound. Latter two groups initiating new matrilines. Four ylg daughters of 959 recruited.	SO 959:38 (28), 959:431 and 801 at Spruce Mound (28.5), 43 and 818 at River Mound (0). SO of ylgs:959 (51.3), with 38 (27.8), with Spruce Mound females (19.5), with River Mound females (1.3). SO among ylgs = 59, among 3-year olds = 9.7. SO with young: 959 = 41, ylgs = 47.8, 3-year olds = 15.6. All social behavior among ylgs AM (0.213), between 959 and ylgs = 0.096. Between 959 and 3-year olds, Ag 5 × Am; among 3-year olds, all amicable 3-year olds, Ag 5 × Am; among 3-year olds, all amicable limited to within new matrilines.

Table 11.6 (cont.)

Year	Matriline composition	Matriline characteristics
1992	Population decline, one 2-year old and one 4-year old died. Late male, 5-year-old 431 weaned seven daughters. Five matrilines: 9-year-old 959 and three 3-year olds at South Mound, 431 at Spruce Mound; 38 south of South Mound; 818 and 43 in River Mound area. 801 and 960 moved independently between Bench and River Mound.	SO: 959:D, 55; among D, 56.3; 959:5-year olds 9; 3-year olds:5-year olds; 12 ("high" because of contiguous solitary matrilines). No amicable behaviors among 5-year olds; between 959:3-year olds = 0.084, among 3-year olds Am = 0.097. Ag rare, only between matrilines.
1993	Four matrilines: (1) South: 10-year-old 959 (litter) and 4-year-old daughters 279, 281 (litter), 284. (2) Spruce: 6-year-old 431 and three resident ylg daughters. (3) River Mound area 6-year-old 43 (litter). (4) River Mound area 6-year-old 818 (litter).	SO: 959:4-year olds 53, among 4-year olds 62.3, South: Spruce matrilines 14, Spruce:River Mound females 15, between River Mound females 38; South group:ylgs 34.8; Spruce female ylgs: 31.7; River Mound females:ylgs 4.5; among ylgs 51.3. Am among South group 0.075, among ylgs 0.053; no other Am behavior. Ag: South group:ylgs 0.024, within South group 0.007, Spruce:South adults 0.001, ylg daughters 0.003.
1994	Four matrilines: (1) South 11-year-old 959, 5-year-olds 279 (litter), 281 (litter) and 284. (2) Spruce 7-year-old 431, 2-year-old 1322, 1399, and 1409 (litter). (3) and (4) River Mound area 7-year-old 818, 11-year-old 960.	SO. Among South group 57, among Spruce group 52.3, South:Spruce 14.1, Spruce:River Mound (0). Social behavior: Am among South group 0.023, none between groups; among Spruce 0.087. Ag among South group 0.013, among Spruce group 0.011, between groups 0.008.
1995	Four matrilines: (1) South 6-year-old 279, 281 and 284 (disappeared). (2) Spruce 8-year-old 431 (litter), 3-year-old 1399. (3) and (4) River Mound/Bench area 8-year-old 818; 12-year-old 960.	SO. Among South group 38.7, among Spruce group 35, between South and Spruce groups 5.3, no SO with females at River Bend/Bench area. Social behavior rare, no Ag, Am only within matriline $r = 0.5$ within each matriline, 0.243 between matrilines (possible daughter, niece, maternal half-sisters, cousin, paternal half-sister/maternal cousin, half-aunt relationships).
1996	Four matrilines. (1) South 7-year-old 279 and 281 (both lost litters before weaning). (2) Spruce 9-year-old 431 (litter and 4-year-old 1399 (litter). (3) and (4) at River Mound/Bench area 13-year-old 960 (litter), 9-year-old 818 (litter).	SO. Within South group 54, within Spruce group 47, between groups 10.3, none with River Mound/Bench area females. Am within groups, South 0.049, Spruce 0.039. One Ag between groups.

social behavior was primarily amicable. Space overlap with the resident 4-year olds and among the 4-year olds was low. The South Mound matriline maintained high space overlap for the next several years, in part because marmots at Spruce Mound and to the north limited their movements. A major matriline developed at Spruce Mound when 431 recruited three daughters (Table 11.6).

For several years four matrilines were present, but the two in the River Mound/Bench area never recruited additional members. Overall amicable behavior prevailed within groups and only agonistic behavior occurred between matrilines. All residents were related; within matrilines $r = 0.5$; between matrilines $r = 0.243$ (Table 11.6). The two multi-female matrilines persisted until 1999, but the South Mound group failed to recruit new members, whereas the Spruce Mound matriline gained a net of five females and utilized the area from Spruce Mound to River Mound. After 2000, the South Mound matriline went extinct and the Spruce Mound matriline divided into two groups. Nine-year-old 1399 moved to South Mound and three 4-year olds remained at Spruce Mound. The South Mound female was either the mother or aunt of the Spruce Mound females (average $r = 0.375$); the Spruce Mound group consisted of either full sisters or paternal half-sisters/maternal cousins (average $r = 0.4375$).

Although matriline dynamics is characterized by the formation of new matrilines and the waning of others, from the perspective of female 915, the progenitor of this lineage, her reproductive success was extraordinary. All residents were her descendents (see Fig. 16.1) and the lineage persisted until at least 2004. Kinship was usually relatively high, but the movements of females from kin groups to become independent and the recruitment of daughters from litters that were not intermingled indicate that females are highly individualistic and strive to maximize direct fitness. This direct fitness strategy is evident in matriline formation; 90% of matrilines are mother:daughter groups (Armitage 2002).

Matriline organization and littermate sister competition

Although social behavior within matrilines is primarily amicable, conflict occurs and is reduced when one or more members of the matriline shift their home range to form a new matriline. In this section, we examine a situation in which littermate sisters remain in a matriline and their relationship is characterized by conflict.

Two female yearlings, resident with their mother at Main Talus at Marmot Meadow, inherited the burrow area as 2-year olds when their mother failed to return. Social structure and social dynamics were highly cohesive (Table 11.7) except when reproductive 911 behaved agonistically with her non-littermate sister ($r = 0.5$). When 918 became reproductive the next year, social structure and behavior were dispersive (Table 11.7). The yearling remained in the meadow at a secondary burrow site, whereas her mother (911) was absent from the meadow for much of the summer; this movement occurred apparently to avoid her agonistic littermate sister. She could not move to Aspen Burrow (see Fig. 8.8) because it was occupied by an unrelated, reproductive immigrant.

In the next year (year 4), the sisters occupied Main Burrow, their daughter/niece used a nearby secondary burrow, and their 3-year-old sister lived in a peripheral burrow. The

Table 11.7 The effects of littermate sister competition on social structure and dynamics when resources restricted.

Year	Social structure	Social and spatial dynamics
1	Six-year-old female, yearling daughters 911 and 918, and young daughter 84	Amicable behavior between yearlings (0.152/h) and between yearlings and young (0.3/h) was high.
2	Three sisters: two-year-old 911 and 918, yearling 84. Young 179 daughter of 911	Reproductive 911 agonistic to 84 (0.159/h) who dispersed. Amicable behavior between the adults was high (0.248/h) and agonistic behavior was low (0.018/h). Their space overlap was 57%. 179 with high space overlap with mother (44%) and aunt (41%); highly amicable with adults, 0.227/h for each.
3	Three-year-old 911 and 918, and yearling 179	Reproductive 918 agonistic to 911 (0.101/h) and 179 (0.218/h) Overlap low between sisters (16%). Overlap of 179 with mother (36%) higher than with aunt (27%). 179 amicable (0.05/h) with mother. High rate (0.364/h) of amicable behavior between mother and two young daughters.
4	Four-year-old 911 and 918, 2-year-old 179, 3-year-old peripheral 84, 2-yearling daughters, 3 litters	Space overlap: 911:918 (46%), 4-year olds:179 (28%), 179:84 (12%), 84:4-year olds (1%). No social behavior between 911:179. Agonistic behavior: 84:three adult kin (0.058/h), 4-year olds:yearling (0.143/h). Two sisters: agonistic (0.131/h), amicable (0.102/h); Aunt:niece: amicable and agonistic both 0.015/h.
5	Five-year-old 911 and 918, 3-year-old 179, 7 yearling females, 3 became recruits	Space overlap: 911:918 (32%), 911:179 (19%), 918:179 (21%), yearling recruits:adults (29%), yearling dispersers:adults (25%). Amicable behavior: 911:179 (0.038/h), adults:yearlings (0.375/h), 911:918 (0.05/h), 918:179 (0). Agonistic behavior: 911:179 (0.013/h), 911:918 (0.075/h), 918:179 (0.225/h), adults:yearlings (0.25/h).

social environment was agonistic (Table 11.7); the yearling females dispersed and the peripheral-living sister was chased whenever she ventured into the meadow. A few days prior to weaning, 179 moved her litter to Main Talus; subsequently three intermingled litters emerged. At Aspen Burrow, three yearling daughters of the immigrant female established residency. Their presence probably prevented individuals from Main Talus from moving to this site to escape agonistic behavior.

In the last year, the overall social environment was similar to the previous year. Again, 179 gave birth in a secondary burrow and moved her young to Main Talus a few days before weaning. Rates of agonistic behavior between aunt (918) and niece (179) were high (Table 11.7). Apparently, the niece challenged her aunt by lying crouched, facing her aunt, and emitting low chirps. In virtually all encounters, the aunt chased her niece, who fled, but always returned. Although 918 was clearly aggressive, she could not exclude niece 179 or sister 911 from Main Talus. Her aggressive behavior may not have had fitness effects by reducing reproduction as all three females weaned litters, but her aggressiveness probably decreased recruitment as yearlings, including her daughters, usually dispersed. The three recruits had much higher levels of amicable than agonistic behavior with the adults. The high level of competition within this matriline accounts for much of the high rate of agonistic behavior between adults and yearlings related by $r = 0.375$ in this colony (Table 11.2).

Badger predation after midsummer killed all marmots at both burrow sites except for 911 and the three yearling recruits. Although average r between 911 and the now 2-year olds was 0.33, they formed a matriline. Eventually, these survivors of badger predation died without recruiting descendents. This matriline became extinct despite its early reproductive output.

In conclusion, littermate sisters may engage in competition, high relatedness does not guarantee cooperation. In this example, neither adult achieved dominance. Only one female was recruited as agonistic behavior between adults and yearlings was usually high. Why was 918 agonistic to her yearling daughters? One possible interpretation is that space (= resources) was limited, in part because the matriline could not spread to the other occupied burrow site. Presumably, 918 would gain more fitness by producing more daughters ($r = 0.5$) then granddaughters ($r = 0.25$). She was unable to exclude the other adults despite treating them aggressively; in other words, she could not achieve social or reproductive dominance. There was no evidence of overt cooperation. However, net reproductive rate increases with matriline size and all females may have benefitted by this relationship (Armitage and Schwartz 2000).

Littermate sister competition: effect of resource availability

In the previous example, I argued that high levels of agonistic behavior occurred among sisters and their offspring and that recruitment failed because the space that could have permitted the matriline to spread to additional burrow sites and foraging areas was occupied by an unrelated matriline. We now examine what happens when space is available for a matriline to spread over the colony area. At the same time, we demonstrate how the fitness of a female is affected by the behavior of her descendents.

At Marmot Meadow, three male and three female yearling littermates occupied Main Talus. The males dispersed soon after an immigrant adult male appeared. All behavior was amicable or play. Female 2007 was less socially engaged then 2009 and 2019, suggesting that 2007 was subordinate to the other two.

In year 3, each 2-year old weaned a litter. Early in the season, 2019 and 2007 spent time at Aspen Burrow, but all were at Main Talus when the young emerged to form one large, social group. All social behavior was amicable. Scent-marking behavior provided infor-mation on the social status of the females. Female 2009 scent-marked more frequently than her sisters (Table 11.8); e.g., she marked 47 times, whereas her sisters marked 33 and 12 times (Brady and Armitage 1999). These results coupled with her low rate of initiating

Table 11.8 The pattern of social dynamics when littermate sisters compete where more resources are available.

Year	Social structure	Social and spatial dynamics
1	Three-year-old 1540, 6 young	Only litter of 1540; she and resident male died over winter with heavy snowfall
2	Three female and three male yearlings	All females engaged in amicable behavior or play, 2007 played less and engaged in fewer amicable behaviors than 2009 and 2019. Space overlap identical among these three
3	Three reproductive 2-year olds, $r = 0.5$	Social behavior amicable: 2019 initiated twice as many greetings as the others, 2009 initiated the fewest. 2009 scent-marked twice as many rocks as 2019 and 2007 and marked 100% and 75% of rocks marked by her sisters
4	Two matrilines, 2009 at Main Talus; 2007 and 2019 at Aspen Burrow, $r = 0.5$	Agonistic behavior between the groups (0.133/ h), between 2007:2019 (0.067/h). Amicable behavior: 2007:2019 (0.167/h)
5	Two matrilines as in year 4, adults 4-years old. Yearling 98 from Aspen Burrow settled nearby	Amicable behavior: 2019 and 2007 (0.032/h) Agonistic behavior: 2009 and 3 yearling daughters averaged 0.194/h
6	Two-year-old 98 formed new matriline near Aspen Burrow	Female 2009 agonistic to yearling daughter (0.4/h) who weaned a litter at age 3 at Cliff Colony. 2009 related to 98 ($r = 0.25$) and her young ($r = 0.125$)
7	All 6-year-old females reproduced; 19 young weaned	Agonistic behavior characterized social interactions between adults and yearlings (3 at Main Talus, 5 at Aspen Burrow)
8	Two 7-year-olds at Aspen Burrow, two yearling daughters of 2009 at Main Talus	Amicable behavior between yearlings at Main Talus and agonistic with their aunts ($r = 0.25$) at Aspen Burrow. Amicable behavior between adults at Aspen Burrow

greetings and that 2019 and 2007 spent some time at a different burrow system where they avoided 2009 suggest that 2009 was dominant.

In year 4, all females weaned litters in two matrilines (Table 11.8). Although the females were highly related, all social interactions between the two groups were agonistic whereas social behavior between 2007 and 2019 was primarily amicable. All social interactions between adults and yearlings (average r was 0.375) were agonistic; all yearlings disappeared by mid-July.

The matrilineal organization was maintained in the next and subsequent years. Space overlap between the two matrilines did not occur and no agonistic behavior was observed among any of the adults. Amicable behavior between 2019 and 2007 was infrequent. This low rate is not unexpected as rates of social behavior frequently decrease between individuals that live together for two or more years (Armitage 1977). All behaviors with yearling daughters were agonistic; rates between 2009 and her yearling daughters were especially high (Table 11.8). All yearlings dispersed, but one settled nearby.

All females reproduced in year 6. Female 2009 was agonistic toward her yearling daughter and attacked the young of her niece. Agonistic behavior between adults and yearlings predominated in year 7 (Table 11.8); all yearlings dispersed and one reproduced at Cliff Colony 2 years later.

In the eighth year (Table 11.8), the dominant 2009 failed to return and neither of her two yearling daughters returned as adults; thus the dominant 2009 failed to recruit any offspring to inherit her territory. Cohesive behavior prevailed at Aspen Burrow; the three yearlings returned the next year as 2-year olds. Only 2007 survived to be an 8-year old. Two of the 2-year olds formed a matriline of three with her at Aspen Burrow; the other 2-year old established a matriline of one at Main Talus. Only the 8-year old weaned a litter.

The main points can be summarized as follows. The availability of the unoccupied Aspen Burrow area allowed two subordinate sisters to form a matriline independent of that occupied by the dominant sister. This separation reduced overt conflict among the females. The dispersion of the subordinate sisters maintained the overall kin group in the meadow and all three females weaned offspring. Average relatedness remained high; the young of one matriline were nieces/nephews of the adults of the other matriline ($r = 0.25$). In any one year, the young of both matrilines had the same father and were first cousins through the mothers of the young of the other matriline, average $r = 0.375$. Despite the relatively high relatedness between members of the two matrilines, only agonistic behavior occurred; in effect, competition between the matrilines continued. When considered from the viewpoint of successful recruitment of descendents who would inherit the site, the dominant female had no fitness because she failed to recruit any daughters. The two subordinate females recruited three daughters who occupied the entire colony. Thus, dominance does not necessarily increase fitness. Although, one daughter of the dominant female reproduced at Cliff Colony, none of the grandoffspring survived. Because dispersal and immigration are less successful than recruitment (Armitage 2003i), forced dispersal by a dominant female is not a successful reproductive strategy.

When fitness is considered from the viewpoints of the progenitor (1540) of the matrilines and of her three daughters, clear differences are apparent. The fitness of the mother was considerable as she had 46 granddaughter and two great granddaughter descendents from her one litter. By contrast, one daughter produced no granddaughter recruits because she failed to recruit any daughters from her five litters, whereas the two daughters at Aspen Burrow, who recruited daughters, did have granddaughters in subsequent years. The importance of considering fitness from the viewpoints of mother and daughter and the significance of recruitment will be discussed at greater length in later chapters.

Summary

The discussion in this and the previous chapter lead to six major conclusions. One, resource sharing and social behavior are not proportional to the degree of relatedness.

Two, cooperative activities (amicable behavior and resource sharing) are centered on the matriline and competitive activities, such as agonistic behavior, characterize inter-matrilineal interactions. Three, the degree of relatedness within a matriline derives from recruitment of descendents and demographic processes, usually over many years. Four, familiarity rather than relatedness per se determines matrilineal composition. Familiarity most frequently involves closely related kin, thus it produces social groups of high r. Five, cooperation and competition characterize matrilineal processes and their dynamics and outcomes are influenced by whether neighbors are present. Six, prefer-ential recruitment of known daughters plus fission of large matrilines to form smaller social groups indicate that yellow-bellied marmots are highly individualistic and form a matriline when the added members increase the reproductive success (and direct fitness) of its progenitor.

12 Social behavior: play and individuality

The presence of individual differences in animal behavior is well established (Clark and Ehlinger 1987); e.g., anti-snake behavior of California ground squirrels, *Otospermophilus beecheyi* (Coss and Biardi 1997), and responses of yellow-bellied marmots to mirror-image stimulation (MIS) (Svendsen and Armitage 1973). Our early studies identified three major factors or axes which were characterized by the behaviors that had high factor scores on each axis. These were designated as social, avoider, and aggressive or approach. For example, among four 2-year-old females, amicable behavior was inversely related to their rank order on MIS axis I, the aggressive/approach axis (Armitage 1986a). The sample size of adult females was too small for statistical analysis and this situation prevails in marmot social groups where most social interactions occur. Because social play by young and yearlings provides adequate sample sizes for analysis, it illustrates the role of individual behavioral phenotypes. First, I discuss behavioral phenotypes, then describe social play, and finally relate play to behavioral phenotypes.

Individual behavioral phenotypes

Behavioral phenotype was determined by placing a marmot in an arena with a covered mirror (Fig. 12.1); after a short period of acclimation, the mirror was uncovered and 19 behaviors were recorded for 10 minutes (Armitage and Van Vuren 2003). Linear typal analysis was used to group individuals based on the similarities of the MIS scores and principal components analysis was used to evaluate patterns of variation in behavior. The analyses were used to produce both two and three behavioral axes; we chose the two axes results for further analyses because the two axes were behaviorally interpretable (Table 12.1). Details on the procedure and analysis of MIS and the behaviors are reported in Armitage and Van Vuren (2003).

Analysis produced two major axes, "shy" and "bold." Shy animals approached the mirror late in the run, spent time in the back half of the arena, and frequently chirped at the image (Table 12.1). Overall, shy animals had little or no contact with the image. By contrast, bold animals approached the image early, made more contact with the image, and spent considerable time investigating the arena. Overall, bold animals focused on making contact with the image. This pattern suggests that the bold marmots were more social. Once they made contact with the unfamiliar marmot (the image), they acted as if they were confident and that they were socially accepted. It is critical to consider that in MIS, the marmot controls the signal sent to the image. If the subject is sociable and

Table 12.1 The major behaviors measured during mirror-image stimulation that determined two major axes from linear typal analyses (LTA) (after Armitage and Van Vuren 2003).

Axis I (Shy)	Axis II (Bold)
Chirping at image	Eating food placed in the arena
Time of nose contact with mirror	Investigation (explores the arena)
Time spent in back half of arena	Nose contact with the mirror image
Tooth-chatter	Pawing/nuzzling of the mirror image
	Time spent in the front half of the arena
	Tail-wag (waving the tail from side-to-side)

Figure 12.1   Arena and observation tower for MIS.

presents an amicable or sociable signal, it will receive that signal from the image. Hence, nose contact, much like the amicable greeting behavior, follows.

The shy animals often lay and stared at the image in much the same posture that was observed in agonistic encounters between adult marmots in the field. Tooth-chatter, a threat behavior, often occurred. Overall, the behaviors represent aggressive elements with some submissive or avoidance elements, such as spending time in the back half of the arena, incorporated into the shy axis. When the shy marmot encounters a stranger (the image) its response is agonistic. Unlike the bold marmot, rather than being affiliative and possibly submissive, the shy marmot is more likely to be dominant or attempt to achieve dominance (Armitage and Van Vuren 2003). In general, bold marmots are more, and shy marmots are less, affiliative. Each individual has a score on each axis and the individuals can be arranged linearly from high to low scores. Early results using the three-axis analyses identified a "sociability" axis (Svendsen 1974). Individuals from satellite sites ranked low on this axis; only colonial animals ranked high on the axis. These results suggest a possible relationship between habitat and behavioral phenotype.

This relationship was investigated by examining the behavioral phenotypes of six satellite and 24 colonial young (Rains 1979). The young were ranked on the axis on which they had their highest score. There was no significant difference in the distribution of young·from satellite and colonial sites that ranked high on the social or avoider axes ($G = 2.0$, $p > 0.1$). There were no sex differences in the rankings on the social and avoider axes (Mann-Whitney tests). There was some evidence for a social effect. Of the 14 young whose highest scores were on the social axis, those from single females ranked significantly lower than those from a group of two females ($U = 5$, $p = 0.01$). However, possible group effects are likely only one factor that determines an individual's behavioral phenotype as behavioral phenotypes change with age.

The behavior of young is primarily bold; only 32.8% of female young were classified as shy. The proportion shy significantly increases to 51.0% in yearlings and to 75% in adults. These shifts in the frequency of shy phenotypes indicate a strong influence of experience in their development; there is very weak evidence for inheritance of behavioral phenotypes (Armitage and Van Vuren 2003). Play could be the critical behavioral experience in the shift from bold to shy phenotypes.

Social play

Play, sometimes referred to as play-fighting, occurs in all species of marmots (Bibikow 1996, Armitage 2003a). In woodchucks, play develops shortly after emergence from the nest at about day 42 (Ferron and Ouellet 1991) and continues for only a few days (Barash 1989). However, a few instances of play were observed between a yearling female and an adult male or female (Meier 1992). In yellow-bellied marmots, play occurs among young and among yearlings but rarely occurs between young and yearlings, does not occur among adults (Nowicki and Armitage 1979), and occurs rarely between mothers and their yearling offspring (Armitage et al. 1996). Play occurs more widely in marmots with a family social structure. Play is common among infant and yearling Olympic marmots and occurs in 2-year olds and adults (Barash 1973a). Young, yearling, and 2-year-old hoary marmots play most often; adults play at low rates (Barash 1974b). Among Vancouver Island marmots, adult males play with adults, 2-year olds, and yearlings; adult females play with 2-year-old females and yearlings, but not with adult females or infants. Two-year-old females play with yearlings, and infants play with other infants (Heard 1977).

Play (play-fighting) includes a variety of motor patterns, including mount, mouth-spar, chase/flee, hide/seek, and grapple (Fig. 12.2). Play motor patterns are associated with transition motor patterns; e.g., escaping, and play-associated motor patterns, which

Figure 12.2   Two young yellow-bellied marmots grappling during play. Drawing by Sara Taliaferro from author's photograph.

include approach, greetings, allogroom, autogroom, social investigation, and withdraw (Jamieson and Armitage 1987). Play may escalate into aggressiveness, especially in those species in which adults play. Thus, 39% of the upright play bouts between adult hoary marmots were followed by a chase (Barash 1974b) and play bouts between Olympic marmot adults were more often characteristic of fights (Barash 1973a).

The relationship between the age/sex groups involved in play and social organization and the escalation of play into agonistic behavior suggest that play has some future benefit (Burghardt 1998). But play is not easily defined functionally because it is not obvious that play serves any particular function either during play or at some future time. Social play may have different functions in different species, ages, and sexes (Bekoff and Allen 1998). One popular suggestion is that play is a means of practicing fighting skills. But the correlations between play and agonistic encounters appear to be crude and do not account for those behaviors that are not observed in agonistic behavior. However, the possibility that some aspects of play may be used in adult encounters cannot be excluded (Pellis and Pellis 1998). Play may be a mechanism for managing development; the individual receives immediate feedback on its competence at manipulative, locomotor, or social skills (Thompson 1998).

What can we infer about the function of play in yellow-bellied marmots? It is interesting to note that play occurs among young and yearlings, but agonistic behavior does not occur among young and is rare among yearlings. It is as if play substitutes for agonistic behavior in these age groups. Perhaps play is a means of developing competitive or social skills in a way that reduces the likelihood of injury. Play may allow the age group to both form social bonds and establish social rank. That is, play could be essentially cohesive whereas agonistic behavior is usually dispersive. Before hypothesizing about the function of play, we need to examine some characteristics of yellow-bellied marmot play.

Both yearlings and young play with the same sex at rates higher than expected and males initiate play with females more than the reverse. Young males engage in mouth-spar, grapple, wrestle, and allogroom significantly more than females (Nowicki and Armitage 1979). Role reversal (flips) occurs whenever marmots change top and bottom positions in wrestling. Role reversal occurs more often when two males are involved and much less often in male–female or two-female pairs. Female young with larger ano-genital (AG) distance index tend to initiate more play and play with more partners than female young with shorter AG distances (Monclus *et al.* 2012). These masculinized females behave more like males which suggests that the benefits of play may be more important for males.

Yearling females are more likely to be in the bottom or flee position and males in the top or chase position. Females are more likely to terminate play. Pouncing occurs as a transition between bouts; males are responsible for about 70% of the pounces (Jamieson and Armitage 1987). These patterns are consistent with adult male dominance. Mount, which has the second longest mean duration (8.2 s) during dyadic play, is very similar to the sex-grapple of adult females by adult males (Armitage 1974, Jamieson and Armitage 1987). Sometimes females rebuff the males with an open mouth threat and mouth-sparring, pre-grapple, and grappling usually follow. The male asserts dominance; the female subsequently escapes without harm. Perhaps play establishes male dominance thereby reducing conflict between adult males and females. Reduced conflict should facilitate social bonding required for reproduction. Adult male:female relationships will be discussed in Chapter 15.

Yearling females that play are less likely to disperse, which suggests that play by the females enhances social cohesion (Armitage *et al.* 2011). If the function of play by female yearlings is to enhance social cohesion, female yearlings would be expected to initiate play more with other females than with males. I examined 11 data sets from three colonies where the number of females and males was sufficient so that choice of playmate was possible. I calculated a simple relationship: did females initiate play with other females more than with males in proportion to the frequency of each sex in the population? Females initiated play more frequently than expected in 6 years and less than expected in 5 years. Clearly, there is no consistent trend for female preference to initiate play with females. However, individual variability adds an additional complexity to the analysis. The frequency of play is affected by the behavioral phenotype of the yearling (Armitage and Van Vuren 2003).

Play and behavioral phenotypes

Seventeen yearling females were ranked from high to low values on the shy axis and by their rates of amicable, agonistic, and play behavior. The rates of social behavior were significantly higher the higher a female ranked on the shy axis, but the correlation was much stronger with play. In other words, yearlings that were less dominant (low ranking on shy axis) would be expected to play less regardless of sex of the play partner.

The higher rates of play by shy females suggest that one function of play is to establish social rank. Relative dominance rank determined from the directional outcomes of play were correlated with later rank relationships calculated from agonistic interactions (Blumstein *et al.* 2013). However, this relationship attenuated over time and potential fitness consequences were not determined. Individual identity (described below as individual behavioral phenotype) was important above and beyond the effect of play and explained 51.6% of the variation in relative dominance rank.

Potential fitness consequences of social rank are evident in the relationship between play, individuality, and recruitment. Shy yearlings are far more likely than bold yearlings to be recruited into their natal colony regardless of the behavioral phenotypes of the adult females and shy yearlings are more likely than bold yearlings to be recruited when the mother is absent (Armitage and Van Vuren 2003). Because mortality is lower in yearling recruits than yearling dispersers (Van Vuren and Armitage 1994a), the recruits gain, at the minimum, a survival advantage.

Play was more frequent among 27 male yearlings the higher they ranked on the shy axis, but there was no significant relationship with amicable or agonistic behavior. For both sexes, play is the behavior most strongly related to behavioral phenotype, which supports the interpretation that play is a major experience shaping behavioral phenotypes. The fitness advantage for males is undetermined as all males dispersed and their subsequent reproductive success is unknown.

Two earlier studies related social dynamics to behavioral phenotype. In an experimental population of eight young, play was not related to kinship but to one of three phenotypic categories. Aggressive young participated in far more play bouts than young classified as avoider or sociable (Armitage 1982c). Among six yearlings, those that had the highest levels of avoidance of their image during MIS had the highest rates of social interactions in the field (Armitage 1986b). These results are consistent with the higher rates of play by the shy individuals. Sample sizes were too small to determine fitness benefits.

Whatever the functional significance of play, it (similar to social behavior in general) is a minor activity: the time allocated to play by juvenile males is $0.012 \pm 0.021\%$ of the daily time budget and is $0.004 \pm 0.012\%$, for females. A play bout averaged 12.4 ± 21.5 s for males and 4.3 ± 9.4 s for females. More play occurred in the morning than in the afternoon or evening and during the second 10-day time period following weaning. Longer bouts rather than more bouts occurred in the morning and during the late summer peak. Yearlings played more in the early season, both as percentage of time and the total number of seconds; play rarely occurred after mid-July (Armitage *et al.* 1996). Yearlings also played more in the morning and males played more than females: 3.5% for males versus 0.8% of time for females, and 137.8 s for males versus 12.8 s for females.

The frequency of play is only one indication of social rank because frequency does not inform the observer about the intensity of play. When an animal quits a play bout, does it do so because it is losing or because it has won and need not play further? Even the initiation of play is ambiguous. Many play bouts seem to start as a consequence of one animal rebuffing the other and the interaction continues as a play bout.

Although play is consistently related to behavioral phenotype, its functions most likely differ between the sexes. In males, establishing dominance/subordinate relationships seems to be the major function, which is evident in the almost universal dominance of adult males over adult females. In females, play may have two functions. One, the formation of social bonds enhances the probability that a yearling female is recruited in its natal colony. Two, dominance, evident in high ranking on the shy axis, increases the likelihood of recruitment when the social environment is less favorable; e.g., when mother is absent. At the least, play illustrates widespread individuality in yellow-bellied marmots.

13 Social behavior: communication

Communication provides a means of obtaining and sending information concerning the functions and intentions of both conspecifics and other species. Chemical communication (scent) apparently is only intraspecific, whereas vocal communication occurs in both intraspecific and interspecific interactions. In this chapter the characteristics of and the contexts in which communication occurs are described and illustrated.

Scent communication

Sudoriferous facial and anal glands have been described for most species of marmots (Table 13.1); it can be safely assumed that scent glands occur in all marmots. The function of anal glands is little studied, in part because their use is difficult to observe. Anal glands are not used in scent-marking but possibly function in defense or fear (Koenig 1957). Many yellow-bellied marmots of all ages and sexes extrude their anal glands during handling. The same response was reported in other nearctic marmots (Rausch and Bridgens 1989). I observed that when two adult male yellow-bellied marmots are in conflict, one will raise its tail and move away and the other will follow; both males tail-flag (tail raised in an arc and waved from side-to-side). Then the males change roles with the former follower now moving away and the former leader now following. I had the impression that each male was exposing his anal glands to the other male. Tail-waving in a figure eight or horizontal direction also occurs in alpine marmots during aggressive, behavior (Koenig 1957). The question is: do males test relative dominance by the amount or quality of the scent from the anal gland?

In a recent study anal gland secretions were collected on cotton swabs and subsequently the swabs were placed on tiles near the burrow of a female yellow-bellied marmot. One swab was a blank control, one contained the secretion from a neighbor, and the third contained scent from a stranger. Marmots spent significantly more time sniffing the anal gland secretions from neighbors and strangers than the control, but the time spent sniffing the secretion from a neighbor did not differ from that of a stranger. Kinship did not influence the response of residents to strangers or neighbors (Cross *et al.* 2013). Thus anal glands apparently are not used in territorial marking and their function requires much further study.

There are at least two sets of head glands in marmots. Facial glands are situated bilaterally between the eye and ear (Koenig 1957). Their structure in the arctic marmot is described

Table 13.1 Marmot species for whom facial and anal glands have been described.

Species	Reference
M. marmota	Koenig (1957)
M. monax	Walro *et al.* (1983)
M. caudata	Shubin and Spivakova (1993)
M. bobak	Shubin and Spivakova (1993)
M. baibacina	Shubin and Spivakova (1993)
M. broweri	Rausch and Bridgens (1989)
M. caligata	Rausch and Bridgens (1989)
M. vancouverensis	Heard (1977)
M. olympus	Barash (1973a)
M. flaviventris	Armitage (1976)
M. sibirica	Bibikow (1996)
M. menzbieri	Bibikow (1996)

and figured in Rausch and Bridgens (1989). The second set of glands is associated with the oral angle and is called the perioral glands. In the woodchuck the glands were active in all ages and sexes examined; young, yearlings, and adults scent-mark by rubbing their muzzles on objects (Walro *et al.* 1983).

Plantar glands were described in *M. bobak*, *M. caudata*, and *M. baibacina*; their function is unknown (Shubin and Spivakova 1993). They may be used in territorial marking; patrolling adult territorial *M. marmota* paw-rubbed on vertical surfaces (Lenti Boero 1995). Marmots may also have chin glands. Yellow-bellied marmots of all ages and sexes chin-rub, which occurs in the same context as cheek-rubbing (Table 13.2A).

Usually scent-marking is reported as cheek-rubbing (e.g., Koenig 1957, Rausch and Bridgens 1989, Bel *et al.* 1995), but is also reported as rubbing the muzzle on objects (Walro *et al.* 1983). The oral gland secretions were successfully used in experimental studies of the function of scent-marking (Meier 1991, Brady and Armitage 1999). Thus, it is likely that scent-marking involves both the cheek glands and the perioral glands. These glands may also function in individual recognition. When greeting occurs, the marmot's nose is directed toward the perioral angle or cheek of the other animal.

All age/sex groups scent-mark in *M. monax* (Walro *et al.* 1983) and *M. flaviventris* where adult males mark significantly more often than expected (Table 13.3). Likewise, adult males accounted for 68.3% and 65.8% of the scent-marking in *M. bobak* and *M. caudata*, respectively (Shubin and Spivakova 1993) and the probability of cheek-rubbing was 0.63 for the resident adult male, 0.35 for the 2-year olds, and 0.16 for yearling *M. marmota* (Bel *et al.* 1995). Young *M. marmota* were not observed to scent-mark (Bel *et al.* 1995, Lenti Boero 1995). Older animals are the primary scent-markers: mainly *M. broweri* older than two (Rausch and Bridgens 1989); adult *M. caudata aurea* scent-marked more than sub-adults (Blumstein and Henderson 1996); and adult, male *M. vancouverensis* marked more than adult females who marked more than 2-year olds (Heard 1977). In all species, scent-marking decreased seasonally (Koenig 1957, Heard 1977, Hébert and Prescott 1983, Rausch and Bridgens 1989, Shubin and

Table 13.2 Examples of the context in which scent communication occurs. Examples are from unedited field notes.

A. Chin-rubbing

Marmot Meadow Colony. Two-year-old female 2007 moves to a new perch rock. She sniffs the ridge, then runs her chin along the ridge. Then sits up.

Picnic Colony. Yearling male 2207 sitting on Split Rock. Leans down and sniffs and chin-rubs on the top of the rock and returns to sitting position. Then leans down and chin-rubs and cheek-rubs. Then sits.

River Colony. Adult female 431 sniffs at a rock, then sits. Gives two vigorous chin-rubs then walks away.

B. Cheek-rubbing during conflict between two females ($r = 0.0625$) from different matrilines. Space overlap was 10%.

Female 920 attacking female 301 who "shrieks" and lies partly hidden in the rocks. 920 mouth jabs at 301. 920 excited and cheek-rubs on underside of rock, pushes at 301 with her forelegs as if trying to dislodge her, pounces on her. Pushes rocks and dirt on 301. Cheek-rubs on another rock. At times 301 extended her head toward 920, but always pulled back when 920 returned. After 11 min, 920 moves downslope and cheek-marks on rocks. 301 cautiously gets up, sniffs at rock where 920 cheek-rubbed, then moves off upslope. Adult male arrived and sniffs rock where 920 cheek-rubbed.

C. Cheek-rub during male:female conflict.

Picnic Colony. Adult male 125 had contact with female 174 who "cried" and entered a burrow. The male cheek-rubs on the rocks. Female 301 comes to the burrow area. The male does more cheek-rubs.

Picnic Colony. Female 349 runs into Upper Slope Burrow as adult male 1 approaches across the talus. Male cheek-rubs vigorously on the rocks at the burrow.

Picnic Colony. New male 125 and female 301 look at each other. The male cheek-rubs on a large log. Female 301 moves toward him and he continues to cheek-rub. Gets down from log with tail-flagging, grapples with the female. Both on log, the female grooms the male, and they sit side-by-side.

North Picnic. Female and transient adult male greet on large rock. Female crouches as male tail-flags, then cheek-rubs. Male acts excited, then more cheek-rubs.

D. Scent-marking to indicate dominance.

Transient yearling male cheek-rubs on a rock then walks upslope. About 60 min later, adult male sniffs rocks where yearling marked, then cheek-rubs on same area on same rock. Male moves to another rock where yearling did not mark; sniffs, but does not mark. About 30 mins later, yearling returns and sniffs the rock, then cheek-rubs over the same area. Five min later, yearling cheek-rubs again.

E. Deterrence of an intruder.

Transient, unmarked (probably a yearling) carefully looking and exploring. Goes to Little Pinnacle and sniffs extensively at the peak where resident male cheek-rubs. Then moves on downslope. Not seen again.

F. Scent-marking as territorial marker.

Female 1434 moves four young from Little Pinnacle to Split Rock. She cheek-rubs on top of Split Rock. Repeats cheek-rub three times. Ten min later, she cheek-rubs on top of Split Rock. Cheek-rubs again one minute later. Then cheek-rubs again.

Intruding females were littermate sisters and cousins ($r = 0.0625$) of occupant.

Females 279 and 281 briefly mark rock near 1399's burrow. About 15 min later, 1399 sniffs at rock, then cheek-rubs on root protruding from the bank and cheek-rubs especially vigorously on the rock. Returns to roots and cheek-rubs. Also wiped forelegs over the body. Female 1399 continued to cheek-rub near her burrow for the next 6 weeks.

G. Head-wiping and chest-rubbing

Between females

Female 556 moves toward rock where female 855 sits. 556 suddenly stops and crouches. 855 rubs forelegs over head and chest rapidly and frequently. 556 tooth chatters. 855 resumes paw–head–chest-rubbing, moves with tail-flagging and circles to higher upslope position. 556 turns to face 855 who runs downhill and 556 follows. 855 stops, 556 approaches, they face off with tail-flagging. 855 moves off with tail up and 556 follows with tail up. They circle each other three times, rough and tumble fight occurs; 855 flees, 556 follows, then turns back.

Between males

Peripheral male 867 moves across the slope near territorial male 1084 who watches. 867 tail-flags and grooms by wiping head rapidly with forepaws and rubbing forepaws on chest. Then disappears into the spruce. Avoids 1084.

Table 13.3 Frequency analysis of scent-marking in yellow-bellied marmots. Expected values adjusted for the frequency of the age/sex class in the population. Adult males often tail-flag when cheek-rubbing.

	Observed	Expected
Adult male	10	4.62
Adult female	15	18.50
Yearling female	7	9.23
Yearling male	5	4.62

$\chi^2 = 7.78$, $p = 0.05$

Table 13.4 Frequency analysis of sniff and cheek-rubs for yellow-bellied marmots for all ages and sexes combined.

	Observed	Expected
Sniff-only	13	21
Sniff plus rub	13	21
Cheek-rub-only	37	21

$\chi^2 = 18.3$, $p < 0.001$

Spivakova 1993, Bel *et al.* 1995, Lenti Boero 1995, Blumstein and Henderson 1996, Brady and Armitage 1999).

Scent-marking typically occurs on rocks, but may occur on a dirt mound at the burrow and rarely on a dirt or shale bank. In yellow-bellied marmots the cheek-rub is usually brief, quick, and single. Sometimes cheek-rub follows an investigation of the surface that the marmot sniffed. The sniff-only and the sniff plus rub occur less frequently than expected and cheek-rub-only occurs more frequently than expected (Table 13.4). The two most common scent-marking sites are rocks (or tree roots) at the burrow site and rocks or fence posts used as perch or observation sites. Both adult males and females usually cheek-mark rocks where they sit or lie regardless of where they are located in the home range. Traps near the burrow or perch site may be marked as well as rocks along a pathway from the burrow area to the foraging site.

Adult, territorial male *M. marmota* patrol the borders of their territories. They walked conspicuously, tail-flagged, and scent-marked many times, but did not scent-mark at boundaries abutting empty areas. Scent-marking near the boundaries was far more common (80%) than at the burrow area (17%). Different individuals within the social group used the same marking places. New resident adults scent-marked at a higher rate than old residents (Lenti Boero 1995). In the French Alps, the resident adults also made "marking tours"; scent-marking occurred throughout most of the home range. Marking was more frequent at burrows and near the boundaries than in the central area (Bel *et al.* 1995). Adult *M. caudata aurea* generally marked (72.5% of cheek-rubs) within 10 m of the main burrow (Blumstein and Henderson 1996). Similarly, *M. vancouverensis* scent-marked more in the area of maximum use (Heard 1977), which always includes the home or nest burrow. Scent-marking rates of male and female adult *M. monax* did not differ; 96% of scent-marking occurred within 6 m of burrows (Ouellet and Ferron 1988). Scent-marking by *M. caligata* outlined territorial boundaries and may also express dominance over another marmot (Taulman 1990b).

The behavioral context in which scent-marking occurs provides the data for interpreting its function (Table 13.2). Scent-marking occurs during agonistic behavior in yellow-bellied marmots; a subordinate may escape the dominant by entering a burrow. The dominant female kicks dirt, sticks, and stones into the burrow and may even pick up a stone in her incisors and carry it to the burrow and drop it in (Armitage and Downhower 1970). This behavior was sometimes followed by the dominant moving excitedly to and fro in front of the burrow and vigorously cheek-rubbing on the rocks at the entrance. Cheek-rubbing often involves females from different matrilines (Table 13.2B).

Cheek-rub commonly occurred during conflict between adult males and females and was often associated with a new resident or transient male (Table 13.2C). These conflicts did not involve the typical chase of agonistic behavior, but is better characterized as sexual aggression (Armitage 1965, 1974). Cheek-rubs (scent-marking) may function to indicate dominance even when the protagonists have no direct contact (Table 13.2D).

Scent-marking may function as a territorial marker to prevent intruders from remaining in the area (Table 13.2E). Scent-marking may occur at a burrow when first occupied or after an intrusion by females from another matriline (Table 13.2F).

Head-wiping with chest-rubbing appears to be another form of scent-marking during conflict between females or between males (Table 13.2G). Adult males also performed this behavior during MIS. In all instances in the field, the subordinate individual did the head–chest-rub. This behavior was also reported for *M. menzbieri* (Bibikow 1996).

To summarize, observations of scent-marking are associated with territorial marking at burrow sites, perches or along pathways. During conflict, the dominant marmot scent-marks rocks or other objects. In effect, dominance is expressed by scent-marking.

Several studies explored the function of scent-marking experimentally by using wooden stakes or cotton balls with or without oral gland secretions. Woodchucks, yellow-bellied marmots, and golden marmots investigated stakes with unfamiliar smells longer than blank stakes or stakes with familiar scent (Meier 1991, Blumstein and Henderson 1996, Brady and Armitage 1999). Yellow-bellied marmots cheek-rubbed familiar and unfamiliar scents equally (Brady and Armitage 1999), whereas woodchucks marked unfamiliar scents more than their own scents (Hébert and Barrette 1989). In an experiment where dominance during paired encounters was determined, future subordinates marked stakes with unfamiliar scent more than future dominants (Hébert and Barrette 1989). These results suggest a role of scent-marking in dominance relationships and are consistent with studies of cheek-rubbing in captive woodchucks that reported a correlation between monthly rates of scent-marking and agonistic encounters. Furthermore, dominant individuals sometimes scent-marked after agonistic interactions (Hébert and Prescott 1989). In conclusion, both field observations and experiments indicate that scent-marking functions in dominance and territorial behavior.

Figure 13.1   Yellow-bellied marmot in the upright vigilant position.

Visual cues

Body posture serves as a visual cue to other marmots. In some cases, the meaning is evident. When marmots are vigilant (Fig. 13.1), they clearly face toward the stimulus, most often a potential predator. To the human observer, the behavior is obvious. The vigilant marmot faces the intruder and may shift position to maintain focus on the intruder as the intruder moves across the landscape. Several animals may become vigilant, but there is no evidence that when one marmot becomes vigilant, it stimulates other marmots to do so. Often marmots run to their burrow from a foraging area and become vigilant. But there is no evidence that vigilance induces other marmots to run to the burrow area. It is not possible to determine if all the marmots respond to the same stimulus or whether some marmots react when they see another marmot vigilant. However, often when a marmot is running toward the burrow area, other marmots see it and also run. A running marmot apparently is a cue that some "danger" is present.

Body posture clearly functions in agonistic encounters. Behaviors such as tail-flagging (Table 13.5A) or head-bob (Table 13.5B) characteristically occur. In many encounters, the two marmots lie and face each other in the "ready-alert" position (Armitage 1962) in which the body is typically curved with the head facing the other marmot.

Tail-flag or tail-flip is the most conspicuous visual signal among marmots. In Olympic marmots, tail-flip is significantly correlated with sex and age as follows: adult males > adult females > yearlings > young (Barash 1989: 106). The tail may also be raised and arched, especially during encounters (Armitage 1962, Heard 1977). Tail-flip, tail-up,

Table 13.5 Body posture during agonistic encounters.

A. Tail-flagging
   Marmot Meadow. Female 918 starts across the meadow and her niece 179 moving on a trajectory that will intersect pathway of 918 who turns toward 179. 179 stops then moves slowly with tail-flagging as 918 follows with tail-flagging. As 918 gets closer, 179 drops tail and runs faster, then slows and tail-flags. Finally 179 stops, rears up, and faces 918 who turns away and 179 dashes into a burrow.
B. Head-bob
   Picnic. Females 174 and her niece 571 alert to each other. 571 lies flat and still and stares at 174 who sits up and stares at 571. After 3 minutes, 571 turns head away at 45° angle. 571 turns back with slight head-bob. Head-bob repeated and then turns away as 174 remains motionless. 174 lowers body and faces 571 more directly, then resumes position. 571 raises fore part of body. Both females become alert at a disturbance. Then both depart area and return to their separate matrilines.

Table 13.6 The context in which a "shriek" or "scream" occurs in yellow-bellied marmots.

A. Between adult females
   Female 1194 attacks female 174 who "cries" as 1194 moves around and pounces at 174 who is down in the rocks. 174 flees uphill.
B. Between adult male and adult female
   Female 859 attacked three times by male 867. She "screams" but does not flee. She bares her teeth and is knocked over twice.

or tail-flag is used in agonistic encounters or to display dominance in yellow-bellied marmots (Armitage 1974), woodchucks (Bronson 1964), Vancouver Island marmots (Heard 1977), hoary marmots (Gray 1967), alpine marmots (Koenig 1957), and Olympic marmots (Barash 1973a).

Vocalizations

Marmots emit several kinds of vocalizations. The most prominent sound is the alarm call, variously reported as a whistle (Waring 1966, Gray 1967, Heard 1977, Lenti Boero 1992), a call (Taulman 1977), or cry (Bopp 1956). Whistle is a poor term as it implies that the sound is made with the lips. However, the alarm call clearly is vocal (Müller-Using 1955, Münch 1958). Marmots may growl, a brief, low frequency sound that may be expressed during play-fights or chases in Vancouver Island marmots (Heard 1977). I heard growls only from yellow-bellied marmots in their burrows (Armitage 1962).

A shriek, yelp, scream, cry or whine may be heard from Vancouver Island marmots (Heard 1977) or yellow-bellied marmots (Armitage 1962) during play and from hoary marmots during wrestling (also play?) (Gray 1967, Taulman 1977). In yellow-bellied marmots, a shriek may be emitted during conflict between adult females (Table 13.6A) or during male:female conflict (Table 13.6B). In all instances when shrieks are heard, the subordinate animal emits the scream.

Young (pups) yellow-bellied marmots may scream during handling. Screams sometimes brought the mother rushing from her burrow. The adult usually came to a sudden halt when she saw handlers and retreated to her burrow. The pup scream often contains nonlinear acoustic characteristics and is more than an order of magnitude longer than pup alarm calls (Blumstein *et al.* 2008b). In playback experiments elongated and average-length screams elicited higher level responses than shortened screams and average-length screams elicited higher level responses than adult alarm calls. Marmots significantly decreased foraging after hearing all playback stimuli and occasionally uttered alarm calls. Overall, marmots had a significantly greater response to the pup screams compared to the adult alarm call. What can we deduce from these responses? Obviously, mothers have fitness investments in their young and protecting that investment should be a high priority. That mothers respond to pup screams indicates that mothers attempt to defend young from predator or conspecific attacks. Observations of infanticide support this interpretation. A seized young screamed and her mother chased the adult female with the seized young, but failed to retrieve it (Brody and Melcher 1985). In another instance when a young was seized by an adult female, the mother attacked the perpetrator several times, failed to retrieve her young, and eventually retreated (Armitage *et al.* 1979). The mother would be at considerable risk if she attempted to protect her young from a large predator, such as a coyote. Her fitness would be greater if she retreated and lived to reproduce in another year. Such discretionary behavior probably explains why mothers retreated when responding to the pup's scream and encountering large, potential human predators.

Alarm-calling is by far the most common vocalization. Alarm calls have been recorded for 14 of the 15 species (Fig. 13.2). The number of calls varies from one to five. There is no obvious relationship between call structure and phylogeny. The variation in call structure possibly could be related to the habitat. If this hypothesis were correct, then a species call should transmit better in its habitat than in the habitat of other species. Although the acoustic environments of marmots vary (Blumstein and Daniel 1997) and the transmission properties of the call types significantly differ for *M. flaviventris*, *M. olympus*, *M. caligata*, and *M. monax*, there was no significant interaction between call type and habitat (Daniel and Blumstein 1998, also see Blumstein 2007). Thus, the acoustic environment may be rejected as an explanation for differences in call structure.

The spectrum of the alarm call may vary geographically. Eleven local populations of the steppe marmot were merged into four geographical populations that were separated by river valleys. The geographical populations coincided only partly with conventional subspecies (Nikol'skii 2002b). Topographic relief was suggested as a factor associated with the rhythmical structure of alarm calls. Populations inhabiting the plains produced evenly sequencing sounds whereas populations inhabiting areas with deep, irregular relief produced sounds in sets (Nikol'skii 2002c). Body size also affects the spectral frequency of steppe marmot alarm calls; larger size was associated with a lower signal frequency (Nikol'skii *et al.* 1994).

When accounting for transmission fidelity and home range size, the variation in alarm call repertoire size was explained by social complexity (Blumstein 2003, 2007, see Blumstein and Armitage 1997a, 1998 for the metric of social complexity). Thus, the more socially complex species, *M. caligata* (four alarm calls), *M. olympus* (four alarm calls), and *M. vancouverensis* (five alarm calls) were the only species with more than

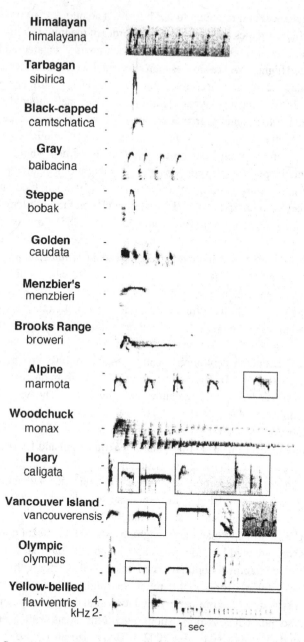

Figure 13.2   Spectrograms of the alarm-call repertoire of 14 species of marmots. Figure provided by D. T. Blumstein.

two alarm call types (Fig. 13.2). However, only 28.6% of the variation in alarm call repertoire size was explained by social complexity (Blumstein 2003). Interestingly, the large repertoires evolved only in the closely related Nearctic *Petromarmota* (Fig. 2.3). This pattern suggests a possible phylogenetic relationship in the evolution of large repertoires.

14 Alarm responses of the yellow-bellied marmot

Detecting and avoiding predators are critical behaviors that affect survival and fitness in free-ranging, terrestrial mammals. These behaviors are more effective if the mammal can distinguish between predatory and non-predatory intruders into its home range and take appropriate action. Presumably a mammal avoids undue risk when a predator is in the neighborhood and also avoids wasting time with inappropriate responses. With this general framework in mind, this chapter explores the origin and kinds of alarm stimuli, the nature of the responses to alarms, who emits alarm calls, to whom alarm calls are directed, and how and where alert behavior is expressed in the apparent absence of a predator.

Alarm responses

I recorded all alarm responses at three colonies during behavioral observations. Marmots respond to alarm stimuli by running to a burrow or rock, by becoming alert, or by emitting an alarm call. The run-in may be followed by an alarm call and the alert may be followed by a run-in or call (Table 14.1). These patterns suggest that marmots generally call from a safe place. A few times a marmot called, then ran or disappeared. Most alarm responses were alarm calls (Table 14.1). Sometimes marmots sat or stood and watched the alarm stimulus without calling or running. This response was most frequent at Marmot Meadow where humans often walked along an old dirt road that runs through the colony. About half of the time, the marmots responded by running to the burrow area, by being alert, or by watching. Marmots probably habituated to humans walking along the road and were cautious, but not highly alarmed. A similar situation occurred at River Colony where a road, frequently used by hikers (often with dogs), runs along the far eastern edge of the site (Fig. 8.2a). The road that marks the lower edge of the mountain slope occupied by Picnic Colony is more distant from the burrows and foraging areas than at the other colonies and the non-calling alarm responses occur at a lower frequency. In addition, a hiking trail passes close to major foraging and burrow areas; the nearness of people and dogs on the trail probably account for the higher frequency of calls in the alarm responses. Thus differences in the frequencies of different kinds of alarm responses among the three colonies are associated with the location (and nearness) of roads and their human traffic to main centers of marmot activity.

Table 14.1 The number of alarm stimuli and responses of yellow-bellied marmots at three colonies. The percent values in parentheses are the proportion of the total sample. The number of years of observation is given in parentheses following the colony name and the number of hours of observation is recorded below the name.

Colony:	Picnic (24) 1844 h	Marmot Meadow (21) 765 h	River (20) 1132 h
Alarm stimulus			
Unknown	337 (62.4%)	111 (51.2%)	276 (53.6%)
Mule deer	73 (13.5%)	9 (4.1%)	21 (4.1%)
Human	52 (9.6%)	35 (16.1%)	76 (14.8%)
Dogs	48 (8.9%)	48 (22.1%)	76 (14.8%)
Hawks	11	1	9
Coyote	7	3	3
Long-tailed weasel	3	1	5
Other birds	2		
Raven	1		6
Eagle	1		
Horse	1	3	25
Cattle	1	3	
Marten	1		
Airplane	1		2
Badger	1	1	2
Vehicles		2	11
Porcupine			2
Total	540	217	515
Alarm response			
Run-in only	48 (8.9%)	55 (25.3%)	49 (9.5%)
Run then alarm	8	8	4
Alert only	15 (2.8%)	26 (12.0%)	36 (7.0%)
Alert followed by run or call	6	2	1
Watches	17 (3.1%)	29 (13.4%)	33 (6.4%)
Alarm-calling	446 (82.6%)	97 (44.9%)	392 (76.1%)
Other marmots respond	187 (34.6%)	115 (53.0%)	141 (27.4%)
Responder unknown	341 (63.1%)	58 (26.7%)	224 (43.5%)

Origin of alarm responses

No one was present when alarm responses evolved, but their nature suggests possible causes. Obviously, a marmot aware of the nearby presence of a predator and runs to shelter is less likely to be predated than one who ignores or responds too slowly. It is easy to imagine that those other marmots that responded to the runner would also be more likely to survive. However, running silently to safety has a serious communication deficiency; it requires other marmots to see the responder. I have noticed that there is often an immediate response if other marmots are near the responder in a relatively open area. But frequently running by conspecifics lags by several seconds the initial response; the delay presumably increases their vulnerability. This scenario assumes that the first

Table 14.2 Conspecific combinations when the first named marmot fled the second after alarm-calling.

Conspecific combination	Number
Yearling male from adult female	4
Yearling female from adult female	5
Yearling male from adult male	1
Adult female from adult female	6
Adult female from adult male	4

runner is not running in order to communicate to conspecifics, but that conspecifics cue on each other's behavior. If there is some fitness advantage of communicating danger to conspecifics, a more effective means of communication should be strongly selected. The alarm call provides that advantage.

How did alarm-calling originate? Many mammals cry or shriek involuntarily when startled. A startle cry distracts, even if briefly, the predator or conspecific causing the reaction and provides an opportunity for escape. I observed this distraction effect in conspecific interactions in which a subordinate marmot alarm-called just before fleeing. The call seemed to give the caller a head start in fleeing from the aggressor, who was always an adult (Table 14.2). I assume that marmots that call have a better chance of surviving than those that did not. Thus natural selection would favor callers so that alarm-calling would become a basic component of marmot behavior.

Another factor that probably contributed to the evolution of marmot alarm-calling is the degree of stress or apprehension felt by a marmot. An alert marmot has an increased heart rate of 9.5% above that of sitting and heart rate increases in the absence of movement when a potential threat, such as a human or a dog, approaches (Armitage 2003b). Adult female yellow-bellied marmots that alarm-called in traps had significantly higher gluco-corticoid concentrations than non-callers (Blumstein *et al.* 2006c) and reproductive females encountering high predator pressure had higher fecal gluco-corticoids than mothers experiencing low predator pressure (Monclus *et al.* 2011).

Prolonged alarm-calling suggests the importance of stress in the evolution of the alarm response. Direct evidence for stress is lacking, but marmot behavior indicates stress. During prolonged calling, the marmot appears agitated, may look wildly in various directions while calling, and may jump down from a rock only to climb back up. Reproductive females are significantly more likely than other age/sex groups to engage in prolonged calling (Table 14.3). The most dramatic example of alarm-calling during stress occurred when an adult female called for 44 min after a coyote killed and carried off one of her four yearling daughters (Armitage 1982b). The remaining daughters assumed normal activity after several minutes and appeared to ignore the continued alarm-calling. Twice adult females groomed at the completion of the bouts. This self-grooming was similar to that infrequently observed during MIS, when the marmot is likely under some degree of stress, and may represent a displacement activity (Alcock 1989: 210).

Table 14.3 Instances of prolonged alarm-calling of 10 min or more. Value contributing significantly to χ^2 marked with an asterisk.

Caller	Number
Reproductive females	16*
Non-reproductive females	3
Adult males	2
Yearling female	2
Yearling male	1
Young	2

$\chi^2 = 38.3$, p < 0.001

The previous discussion speculated on the internal state that would generate vocalizations that natural selection could mold into an alarm call. What environmental or social conditions are more likely to be associated with the evolution of alarm-calling? An analysis of the evolution of alarm calls in rodents found significant associations between diurnality and sociality (Shelley and Blumstein 2004). Diurnality accounted for nearly three times as much of the variation as sociality in whether a species alarm-called. Phylogenetic tests revealed that the evolution of diurnality preceded the evolution of alarm-calling and that the evolution of diurnality and sociality were unrelated. These results are consistent with the hypothesis that alarm calls evolved to communicate to predators (Shelley and Blumstein 2004). Communicating to predators remains a major component of alarm-calling by yellow-bellied marmots (Armitage 2003g).

Alarm stimuli and responses

It is generally accepted that the function of alarm-calling is to warn kin of the presence of a predator (Sherman 1977). Although kin respond to alarm calls, so do any other marmots within hearing range. I recorded 30 instances of marmots at Picnic Colony responding to distant alarm calls, most of which apparently were emitted by marmots living at Boulder Colony (distance of about 250 m between the colony centers). Often the marmots at Picnic became vigilant and looked in the direction from which the sounds came. Before we can formulate any conclusions about the function of alarm-calling, we describe the situations that evoke alarm-calling, who calls, and who responds.

I identified 16 stimuli of alarm calls, which include 10 species of mammals, three of which are not predators (Table 14.1). The stimulus for over half of the alarm responses was unknown. The responder (i.e., caller, runner) was often unidentified, especially at Picnic Colony. Marmots apparently called while hidden in high vegetation or were behind boulders. Presumably the marmots would be difficult for a predator to find. An acoustic localization system verified that most alarm-calling occurs in positions of safety; i.e., near burrows (Collier et al. 2010). A predator attack on a calling marmot almost never occurs. Marmots sometimes sat or stood at a burrow and continued chirping at an advancing predator, a badger, dog, or human, until the predator approached closely whereupon the marmot quickly entered its burrow, often with a trill. I interpret these cases as special instances of "communicating with the predator," which will be discussed in greater detail later. Chipmunks and golden-mantled ground squirrels never elicited an alarm response, but yellow-bellied marmots and golden-mantled ground squirrels responded similarly to conspecific and heterospecific anti-predator calls (Shriner 1998). Woodchucks responded to conspecific and eastern chipmunk (*Tamias striatus*) alarm calls, but spent more time vigilant in response to conspecific calls (Aschemeier and Maher 2011).

The frequency of identified alarm stimuli varied among the three colonies. Deer, humans, and dogs accounted for 85.2%, 86.8%, and 72.4% of the identified stimuli at Picnic, Marmot Meadow, and River Colonies, respectively. The lower value at River is because horses and vehicles were important at that site (Table 14.1). Deer stimuli occurred more frequently at Picnic Colony because foraging deer often passed through the meadows surrounding the talus slope. Deer entered the meadows at Marmot Meadow infrequently and at River Colony deer typically foraged in meadows across the East River and rarely entered the meadows at the colony site (see Figs. 8.2, 8.7, and 8.8). Humans and dogs were frequent stimuli because they moved along trails or roads that passed through or near marmot colonies. The differences in the relative importance of these stimuli among colonies are attributable to the frequency and time of day of human and dog visitation. At River Colony, nearly every morning during peak marmot activity, hikers or cattlemen, often with dogs, passed by the colony and nearly always elicited a response, which was usually alert only, run-in, or watching; an alarm call was more likely when a dog was present. At Picnic Colony, hikers often appeared later in the morning when marmot activity was low. Horses were important as an alarm stimulus only at River Colony where local cattlemen pastured them in the meadow east of the gulley. Vehicles regularly passed along the road by Picnic Colony and never elicited a detectable response. Presumably, marmots were habituated to vehicular traffic. By contrast, vehicles were infrequent at River Colony and the marmots usually responded, mainly by becoming alert or running to the burrow area. Weasels and marten were nearly always chased when detected (Armitage 2003g). Responses to ravens, some hawks, and other birds were stimulated by birds swooping over the colony or by making sudden movements. Red-tailed hawks (*Buteo jamaicensis*) nested in the area and often soared above marmot colonies, but rarely attacked a marmot. Marmots became alert when a red-tailed hawk soared overhead and alarm-called about one-third of the time. In most encounters, the marmots watched the hawk and did not flee to a burrow (Armitage 2003g). Similarly, alpine marmots alarm-called in two of four interactions with Griffon vultures (*Gyps fulvus*) but did not hide in any of the four events (Heredia and Herrero 1992).

Unambiguous responses by other marmots or an overall general response occurred, on average, about one-third of the time (34.8%), with some variation among colonies (Table 14.1). I recorded the first responder; when other animals called or ran, they were included in the "other marmots respond" category. Multiple calls were recorded 27 times; including those in the "other marmots respond" category raised the frequency of responses to 36.9%. These numbers are underestimated because I did not include cases when two or

Table 14.4 Frequency analysis of alarm responses for yellow-bellied marmots when the responder was identified. Values contributing significantly to χ^2 are marked with an asterisk. O = observed. E = expected.

	Pre-emergence						Post-emergence					
	River		Marmot Meadow		Picnic		River		Marmot Meadow		Picnic	
	O	E	O	E	O	E	O	E	O	E	O	E
Adult male	12	9	5	5.1	9	11	16	16.9	1	11*	7	10.9
Adult female	63	42.8*	23	11.2*	43	35.8*	97	81.3*	31	25	54	35.2*
Yearling male	2	11.5*	3	8.5*	12	17.5*	2	5.1	7	4*	0	6.4*
Yearling female	6	19.6*	3	8.8*	16	15.6	12	23.7*	11	10	1	9.6*
	$\chi^2 = 27.8$		$\chi^2 = 20.0$		$\chi^2 = 3.55$		$\chi^2 = 10.7$		$\chi^2 = 12.9$		$\chi^2 = 28.5$	
	$p < 0.001$		$p < 0.001$		$0.05 > p > 0.02$		$0.025 > p > 0.01$		$p < 0.005$		$p < 0.001$	

three marmots were on a rock and one called, or one called and another marmot climbed up on the rock. Only when I could see a clear and general response did I include the response in the "other marmots respond" category. Many alarm calls were heard when I could not find any other marmot and often not the caller. Thus marmots alarm-call when no other marmots are above ground to be warned. I doubt that marmots call to warn conspecifics in their burrows; only once did I detect a marmot to emerge from its burrow and look around in striking contrast to marmots above ground who run to burrows, climb on rocks, or if lying or sitting, look up. These latter responses form the basis for the "other marmots respond" category. As I will argue later, marmots call when alone to communicate with the predator.

All age/sex categories give alarm responses (Table 14.4). Although young may alarm-call or run-in, I did not include them in the analyses. When young were included, chi-square values became greatly inflated, and I was more interested in the significance of the other age groups. I did determine that young call significantly less often than expected based on their frequency in the population. For example, young make up more than half of the population after their emergence, but are responsible for only 15.8% of the alarm responses. There is some evidence that young may respond differently to alarm stimuli than older individuals. Young steppe marmots gave an alarm call to almost every approaching object; they were intensely alerted, but resumed activity more quickly than older marmots. Marmots more frequently ignored juvenile calls and remained alert longer to adult calls (Nesterova 1996).

Because our time budget analysis revealed that alarm-calling occurred significantly more often in time periods following lactation (Armitage *et al.* 1996), I divided alarm responses into pre-emergence (before young appear) and post-emergence periods (young active above ground). Expected number of alarm responses was calculated from the frequency of each age/sex class in the population. The frequencies were adjusted for the disappearance of any members of each age/sex group. Finally, this analysis is based only on alarm responses when the individual giving the alarm was identified.

In all colonies for both pre- and post-emergence periods, the distribution of observed alarm responses differed significantly from the expected distribution (Table 14.4). Overall, adult males responded as expected; yearling males and females responded less often than expected, and adult females responded more often than expected. Yearling males and females did not differ in their frequency of alarm response (for the three colonies, χ^2 varied from 1.1 to 2.4, all p > 0.1). When yellow-bellied marmots are exposed to heightened risk in the form of alarm-call playback, sex and age differences persisted; males and yearlings responded more by reducing foraging than females and adults (Lea and Blumstein 2011).

Post-emergence alarm responses (mainly alarm-calling) were far more numerous than pre-emergence (χ^2 varied from 19.9 to 68.3; all p < 0.001). This difference could be a function of intruder pressure not a greater propensity to respond to intruders; predator pressure is expected to increase when vulnerable young are above ground as young are more easily captured (e.g., Thompson 1979, Armitage 2004a).

If one function of the alarm response is to warn vulnerable kin, then reproductive females should alarm-call more frequently than non-reproductive females. During the pre-emergence period, there was no overall trend for more alarm-calling from reproductive females; after young were weaned, alarm-calling by mothers was significantly more frequent than that of other females, which included adult sisters or daughters as well as unrelated adults (Table 14.5).

The importance of alarm-calling for warning kin was tested by experimentally inducing yellow-bellied marmots to call by exposing them to simulated predator attacks: walking toward a marmot, driving a radio-controlled chassis with a stuffed badger, walking dogs through marmot home ranges, and flying a radio-controlled glider over a social group. In all simulated attacks, the marmots that called were noted. The experiment was conducted in colonies where the relatedness among all individuals was known. The set of pair-wise r values for each individual was summed to calculate a total r, which reflected potential inclusive fitness benefits from alarm-calling (Blumstein et al. 1997). After pups emerged, females with pups called more than any other age/sex group. Over the entire study, 42% of the variation in the rate of calling was a function of whether the caller was a female that weaned pups. Total r explained no significant variation in the rate of alarm-calling. Thus, both experimental and long-term observational data support the interpretation that the rate of alarm-calling is a form of direct parental care and is not directed toward non-descendent kin. Callers are not at risk; alarm-calling is not altruistic (Barash 1975b). These results suggest that direct fitness,

Table 14.5 Statistical comparison of alarm-calling by reproductive and non-reproductive adult females.

Colony	Pre-emergence	Post-emergence
Picnic	$\chi^2 = 0.6$, p > 0.3	$\chi^2 = 29.6$, p < 0.001
Marmot Meadow	$\chi^2 = 2.5$, p > 0.1	$\chi^2 = 6.3$, $0.02 > p > 0.001$
River	$\chi^2 = 7.1$, $0.01 > p > 0.005$	$\chi^2 = 54.1$, p < 0.001

not indirect fitness, benefits maintain the higher frequency of calling by reproductive females (Blumstein and Armitage 1998b).

Function of alarm calls

The maintenance of rate differences in calling does not explain the basic function of alarm calls. The preponderance of evidence indicates that alarm-calling in yellow-bellied marmots is directed toward the predator. First, in nearly all alarm responses the marmots sit or stand and face the predator. Second, marmots shift position to keep the predator in view. Alarm-calling continues as long as the predator can be seen even when the predator is walking or running away from the marmots. Third, prolonged calling usually occurs when the predator remains in the area. For example, one adult female called for 34 min while standing and watching a coyote. Another female called for 30 min while facing dogs at the edge of her colony. Fourth, marmots may track a predator, such as a badger, as it moves through a colony. The marmot changes locations, often moving closer to the predator as the predator moves, and alarm-calls until the predator disappears (Armitage 2003g, 2004a). Alarm-calling sometimes appears to disturb the predator (Table 14.7).

I tested the "communicating with the predator" hypothesis by walking a dog toward a cabin under which a solitary marmot lived. The marmot immediately alarm-called and focused its attention on the dog. When the dog passed by the cabin so that the marmot could no longer see it, the marmot moved around the corner of the cabin, stood facing the dog, and continued to alarm-call. Alarm-calling continued until the dog crossed over a small hill and was no longer visible to the marmot. The dog–marmot encounter was repeated several times on different days with the same results. The marmot always alarm-called as long as the dog was in view and changed locations in order to keep the dog in view (Armitage 2003g).

"Communicating with the predator" is widespread among mammals (Dugatkin 1997: 94) and advantageous to both species. The predator is informed that its presence is known. Success is highly unlikely at this time and the predator is better off to seek prey elsewhere and return at another time. The potential prey benefits because it knows that the predator is no longer present. If a yellow-bellied marmot simply retreats to its burrow, it does not know if the predator is lurking nearby. The risk of not knowing if a predator is nearby is illustrated by observations of coyote predation. Predation was successful when marmots emerged from their burrow for the afternoon foraging period and failed to detect the coyote concealed in dense vegetation; the coyote leaped forth and grabbed a yearling female (Armitage 1982b).

Behavior during emergence from the burrow indicates that marmots are aware of possible ambush. When a marmot emerges, its head is visible in the burrow entrance as the marmot looks around. Then the body slowly emerges as the marmot continues to be wary (Fig. 14.1). When fully emerged, the marmot typically sits or stands by the burrow while it scans the environment (Fig. 14.2). Then the marmot generally moves to a boulder, fence post or a high mound, from which it continues to scan the environment. Only after this period of vigilance does a marmot move to a foraging area. Depending on the proximity of the burrow areas to the meadow, the marmot may stop at additional rocks,

Figure 14.1   Yellow-bellied marmot hiding behind plants as it emerges warily from its burrow.

boulders, or on a slight rise and scan the environment (Fig. 14.3). After the marmot moves into a meadow, it frequently scans the environment (see Fig. 8.3). Because these vigilance bouts follow the prolonged vigilance activity after emergence, they serve to check on the possibility that an intruder may be approaching or may have been missed during earlier vigilance. Such wariness characterizes marmot behavior above ground (Armitage and Chiesura Corona 1994). But as the above described predation event indicates, marmots do not always detect the predator. Because emergence from the burrow is a dangerous time, the marmot benefits by "telling" the predator to move elsewhere. Otherwise, the predator may simply conceal itself near the burrow and wait in ambush for an emerging marmot.

The pattern of marmot response to a predator strongly supports that alarm-calling evolved to communicate with the predator. The analysis of who calls suggests that natural selection has added additional functions to the alarm call. Clearly, the alarm call alerts other marmots to danger and the likelihood that a caller is a reproductive female indicates that selection favored the warning of direct kin; i.e., young offspring. These examples indicate that natural selection has acted on the caller. However, marmots produce individually distinct calls (Blumstein and Daniel 2004); thus, natural selection may act

Figure 14.2   Adult female yellow-bellied marmot vigilant at her burrow after emergence.

on the receiver; there is significant potential information about age, sex, and especially identity encoded in calls (Blumstein and Munos 2005, Matrosova *et al.* 2011).

Key variables in yellow-bellied marmot alarm calls that enabled discrimination of sex, age, and identity were repeatable, whereas variables that did not enable discrimination were less repeatable (Blumstein and Munos 2005). Marmot assessment of the reliability of callers assists them in deciding how much time to allocate to their own vigilance (Blumstein *et al.* 2004b).

The most important information contained in an alarm call is information about the predator. An alarm call indicates that a predator was detected, but the call could also indicate the kind of predator (terrestrial, aerial, or species) and the nearness (degree of risk) of the predator. Although referential calls, i.e., calls that indicate the kind of predator, were suggested for the alpine marmot (Lenti Boero 1992), the preponderance of evidence indicates that marmot alarm calls are not functionally referential, but convey the degree of risk (Blumstein 1995, 1999, Blumstein and Armitage 1997b, Blumstein and Arnold 1995). Risk is communicated by diverse mechanisms; e.g., in yellow-bellied marmots, high risk is conveyed by more rapid calling (Blumstein and Armitage 1997b). The variation in diversity of communicating risk is summarized by Blumstein (2007).

Figure 14.3 A marmot vigilant on a slight elevation before foraging in the area around the slight elevation.

Other alarm stimuli

In general, any intrusion into the marmot environment, including noise from vehicles or airplanes and non-predators such as goats and chamois (Lenti Boero 1992), elicits an alarm response. Yellow-bellied marmots also respond to the social sounds of coyotes, wolves, and golden eagles (*Aquila chrysaetos*) (Blumstein *et al.* 2008a). I observed yellow-bellied marmots become alert to red-tailed hawk calls, but I could not be certain that the marmots were not responding to the hawks' soaring above the colony. However, soaring hawks usually do not elicit a response; the marmots respond when the hawks swoop down over the marmots. Golden eagles were infrequently seen and yellow-bellied marmot responses were not observed. The eagles formerly nested on Gothic Mountain, which overlooks most of our study area, and we collected yellow-bellied marmot skulls from the nest site. A radio-controlled model of a golden eagle elicited alarm responses in hoary marmots (Noyes and Holmes 1979). When exposed to the model the focal animal called and ran simultaneously when more than 5 m from a refuge or called and visually tracked the model when less than 1 m from a refuge. The marmot never tracked the model from a non-refuge site. No animal entered a burrow and remained out of sight throughout a flight, but would climb on a rock at the refuge site and keep the model in view. These observations indicate that alarm-calling is not costly and that marmots attempt to track the activity of a potential predator, albeit from a safe location. The marmots responded less to coyote and wolf sounds than to marmot alarm calls and responded less to bird calls than to coyote and wolf calls, which suggest some level of sound discrimination (Blumstein *et al.* 2008a). Because coyotes do not vocalize while hunting (and presumably wolves would not either) and golden eagles attack suddenly at a low angle (Pigozzi 1989), the importance of responses to social sounds of predators is unknown.

Non-predators elicit marmot alarm responses. If there is a cost to the time allocated to alarm responses, marmots should not respond to non-predators. Alarm responses are probably not costly. On average, over all animal cohorts, yellow-bellied marmots throughout the summer spend about 4 min/day in alarm-calling. If the time spent alert and vigilant is added to the time spent calling, marmots spend as little as 15 min/day to as much as 80 min/day, depending on the animal cohort. This time is much less than the 200 min/day allocated to sitting and lying (Armitage *et al.* 1996); marmots have adequate time to meet essential daily needs. However, if a predator causes marmots to remain underground for significant time during the major time periods of above-ground activity, negative effects could occur. Potential negative effects on marmots have been investigated by monitoring human influences on marmot activity.

Human influences

The major source of human intrusions on marmots is hiking on trails or roads that pass through or near their habitats and can be a major cause of alarm responses (e.g., Table 14.1). In addition, most species of marmots are or have been hunted (Bibikow 1996, Armitage 2003a), perhaps for thousands of years (Tomé and Chaix 2003), which could lead to marmots considering humans to be a threat.

The response of alpine marmots to various levels of hiking activities in France and Switzerland verified that marmots respond to humans by modifying their activity (Table 14.6). Marmots usually immerged if hikers had dogs (Mainini *et al.* 1993). In highly frequented areas, alpine marmots habituate to the presence of hikers by reducing flight distance and the time spent in their burrows (Neuhaus *et al.* 1992, Mainini *et al.* 1993, Gibault *et al.* 1996, Neuhaus and Mainini 1998, Louis and Le Berre 2000). Resource availability affected marmot responses to high human frequency; more marmots foraged where forb abundance was higher (Table 14.6) (Gibault *et al.* 1996).

Alpine marmots adjust their responses to human intrusion based on experience. Hikers affected behavior more in hunting areas (Louis and Le Berre 2002) and marmots located their burrows near hiker's paths (Table 14.6). As marmots foraged closer to the

Table 14.6 Response of alpine marmots to various levels of hiking and hunting activities.

Human activity	Marmot response
Hikers on trails	Rarely immerged
Hikers crossed through marmot territories	Immerged about one-half of the time
Hikers pass through main burrow area	Always immerged
Highly frequented area with high forb abundance	Higher vigilance, remained closer to burrows, more foraging marmots
Low frequency area	Fewer foraging marmots
Hunting permitted	Significantly more time idle when hikers present
Non-hunting area	Hikers did not affect amount of time spent idle
Hiker paths	More secondary burrows near paths
Disturbance (hikers present)	Marmots move away from the paths and closer to their secondary burrows

paths they were closer to their secondary burrows, thus reducing the risk of foraging (Semenov *et al.* 2002).

In general, marmots adjusted to human disturbance by shifting their daily activity cycle. They foraged much less during times of disturbance, which usually were during midday, and foraged more early in the morning and later in the afternoon (Neuhaus *et al.* 1992, Franceshina-Zimmerli and Ingold 1996, Louis and Le Berre 2002). Spending more time below ground in areas of high hiking pressure may not be costly as marmots allocate considerable time to sitting or lying and remaining in the burrow under low or no human disturbance, especially during midday (Armitage *et al.* 1996).

The response patterns of alpine marmots suggest that human disturbance may have demographic effects. This possibility was investigated in Olympic marmots (Griffin *et al.* 2007b). Although some behavioral differences, such as looking-up and flight behavior, differed between marmots at high-visited and low-visited sites, the overall pattern of daily activity was not affected. Furthermore, there were no differences in litter size, reproductive and survival rates, and body condition between marmots at the two sites. Similarly to alpine marmots, Olympic marmots evidenced some habituation to human intrusions (Griffin *et al.* 2007b).

Vigilant behavior

The basic response of marmots to an alarming stimulus is an alert or vigilant response. Vigilance behavior is characterized by phenotypic plasticity. This plasticity is expressed in time spent vigilant and flight distances; e.g., yellow-bellied marmots increased the time spent vigilant when the presence of motorized vehicles was high and reduced flight distance as disturbances from vehicles or pedestrians increased (Li *et al.* 2011). Alpine marmots more than doubled the number of bouts of vigilance/minute in a closed site (where the main burrow averaged 81.2 m from the forest edge) than in an open site (where the main burrow averaged 239.6 m from the forest edge) but mean duration of vigilance bouts did not differ between the sites. Red fox were observed much more frequently at the closed site (Ferrari *et al.* 2009). Both male and female alpine marmots spent more time vigilant than foraging when young were present than in a year when young were absent (Lenti Boero 2003b). Vigilance behavior in the yellow-bellied marmot had a low heritability ($h^2 = 0.08$) further emphasizing the high environmental variability (Blumstein *et al.* 2010). Plasticity in this trait permits marmots to adjust their behavior on both short- and long-term time scales.

Marmot pups (young of the year) apparently learn what is dangerous. Steppe marmot pups alarm-called with a higher probability to any approaching object than did nearby adults. Only the pups called when hares or cows were present. Pups also allowed human intruders to approach much more closely then did adults (Nesterova 1996). By contrast, the reactions of recently emerged alpine marmot pups were slight and similar in remote areas and in areas highly frequented by hikers. By late summer, their reaction increased significantly but to a much larger extent in remote areas, which suggests that young also habituate to intruders (Neuhaus and Mainini 1998). The probable greater risk of intruders to pups may be compensated in part by the behavior of adults; the presence of pups outside the burrows increased adult alert and flight distances and delayed re-emergence (Louis and Le Berre 2000). These results suggest that young marmots may be more vulnerable to predation (see Chapter 17). Recently weaned yellow-bellied marmots may emerge from their burrow when a human sits only a few meters distant. They are "nervous," watch the human, quickly retreat at any movement, but soon reappear. By late summer they are more responsive and flight distance increases markedly. Some marmot researchers catch recently weaned young by lying behind the burrow entrance and grabbing a young as it emerges. The response patterns of young suggest that alarm stimuli may operate on two levels: (1) a response to any intruder; and (2) subsequent discrimination of the degree of danger.

Predator discrimination

The wide variety of stimuli that elicit alarm responses suggests that the fundamental reaction is to respond to any intrusion. If one eliminates irregular stimuli, such as vehicles or humans, the basic terrestrial stimulus is the elongate, four-legged shape (Armitage 2003g). A basic response to this shape can account for the frequent alarm responses to non-predators (Table 14.1, Lenti Boero 1992, Armitage 2003g). To this basic response, marmots probably added a second level of discrimination to identify predators.

Although cattle and deer are non-predators, they often elicit prolonged responses. I observed yearling males and females and adult females alarm call from 7 min to 20 min while facing a foraging deer. In contrast to a coyote, a marmot may face and follow a deer at close range (Armitage 2003g). A major difference between alarm responses to deer and coyote is that yellow-bellied marmots are far more likely to alarm-call than not to coyote than to deer ($G = 4.6$, $p < 0.05$). By contrast, marmots do not call more frequently to dogs than to deer ($G = 0.4$, $p > 0.5$). One confounding factor is that deer are usually foraging in meadows and dogs are walking along a road bordering the colony, whereas the coyotes, when detected, were within the marmot colony. The few times dogs approached marmots closely, usually while running, the marmots trilled. Thus, the more frequent alarm-calling to coyotes may represent degree of risk rather than predator identification.

Cattle appear in late summer and many of the alarm responses are given by young. At Marmot Meadow, cattle may graze throughout the meadow and close to the burrow areas; the marmots frequently remain in their burrows. However, if cattle approach the burrow areas when marmots are active, the marmots alarm-call. The calling seems to divert the cattle away from the burrow area (Table 14.7) and alarm-calling also causes deer to retreat. Horses elicit the full range of responses, run-in, alert and watch, and alarm-calling. Alarm-calling was more likely when the horses ran through the meadow or when they approached closely to the burrow areas. I never observed prolonged alarm-calling when horses were present. Usually the horses were distant from the main area of marmot activity and did not elicit responses; possibly the marmots habituated to the horses under these situations.

Table 14.7 Examples of disturbed behavior by intruders resulting from alarm-calling.

Predator response

Marmot Meadow. Dog into the meadow. Marmots run to Aspen Burrow. Then a yearling male begins calling and
 yearling female also calls. When the dog gets close, the yearling male and two yearling females call rapidly
 while facing the dog. Three female adults at Main Talus watch but do not call. Dog seems disturbed, barks;
 alarm-calling continues as dog turns back and leaves the meadow. Calling ceases.

Large, non-predatory mammal response

Alarm call. Young immerge but adult female sits and watches. Chirping and female moves toward the cattle and
 they detour around the burrow area.

Figure 14.4   The long-tailed weasel (*Mustela frenata*), left, and the American marten (*Martes americana*),
right, are small predators, especially of young, that are vigorously chased by yellow-bellied
marmots.

There is some evidence for predator discrimination which may be based, in part, on
size (Cooper and Stankowich 2010); marmots respond differently to small predators
and to the large deer and cattle. Yellow-bellied marmots vigorously chase long-tailed
weasels (Fig. 14.4, left) and marten (Fig. 14.4, right) and rarely alarm-call (Armitage
2003g). They also chase red foxes (Blumstein *et al.* 2009a), and I observed a steppe
marmot chase a red fox. The red fox preys on the young steppe marmots but apparently
is not a threat to the adults. There is some evidence for discrimination between
dogs and coyotes. Although marmots alarm-called 61% of the time when dogs
and 85% of the time when coyotes were detected, the difference was not significant
($G = 2.4$, $p > 0.1$). Lack of statistical significance was likely affected by the proximity
factor as previously discussed for deer, dogs, and coyotes. Also, the sample size for
coyotes is small and a larger number of similar observations could reach statistical
significance.

Although marmots alarm-call and watch dogs, coyotes, and badgers, they never chase
these predators nor move as close to them as they do to deer and cattle. Clearly, marmots
recognize that the three large predators represent considerable risk. Recognition of
risk is also evident in behavioral changes following predation by coyotes or badgers.

Yellow-bellied marmots shift their foraging patterns for several days to avoid the areas where predation occurred (Armitage 1982b, 2004a).

Do yellow-bellied marmots remember predators that they have not encountered for many generations? The multipredator hypothesis states that anti-predator behaviors evolve together and thus prey may respond to long-absent predators as long as they experience other predators. This hypothesis was tested by creating life-size photographic models of the wolf and mountain lion, absent for >70 years, of current predators, the red fox and coyote, and of the gray duiker, an ungulate that marmots have never seen (Blumstein *et al.* 2009a). Overall, there was no significant stimulus effect on low vigilance (lie-look and stand-look), foraging, locomotion, or time out of sight. There was an overall stimulus effect on high vigilance (stands with head fixated) and the combined proportion of time allocated to high vigilance and out of sight (in the burrow). There was no clear pattern of quantitative significant differences among the five species. The results support the multipredator hypothesis. However, the multiple-predator response may simply be a function of the basic alarm stimulus: an elongate, four-legged animal rather than through experience with other predators. Also, the similarity between dogs and wolves may explain some of the qualitative differences between wolves and the other predators.

Summary

Yellow-bellied marmots express alarm responses to intruders as large as or larger than a weasel. Marmots alarm-call and watch intruders as long as they are present and enter their burrows only when the risk of capture becomes great. This pattern of behavior enables marmots to keep track of intruders and encourages the intruders to leave. As a consequence, marmots maintain above-ground activity and decrease the risk of being ambushed (e.g., by coyotes) when they emerge from their burrows. Adult females are more likely to alarm-call and reproductive females call more frequently than non-reproductive females after the young are weaned. Alarm-calling, which functions to "communicate to the predator" has evolved additional functions such as warning closely related kin, especially offspring.

Part IV

Reproductive success

Male reproductive success depends on the ability of a mature male to locate and mate with reproductive females. Adult male marmots have three possible strategies for associating with mature females: (1) remain in the natal environment; (2) disperse and obtain residency with females elsewhere; or (3) wander as transients seeking mating opportunities. A potential problem with the first strategy is that it likely leads to inbreeding and avoidance of inbreeding is often considered the ultimate factor responsible for male bias in the dispersal patterns of mammals (Greenwood 1980, Bowler and Benton 2005). However, inbreeding avoidance may be a consequence rather than the ultimate cause of male-biased dispersal. The major problem with the dispersal strategy is the increased mortality, especially of long distance dispersers (Frey-Roos 2005), in comparison with philopatric individuals (Van Vuren and Armitage 1994a). The transient strategy also has the likelihood of increased mortality plus the difficulty of moving through a snow-covered landscape. This strategy is not available for those species that mate before emerging from hibernation.

In the relatively asocial *M. monax*, where all individuals typically disperse in their first year, males must find a place to settle, establish a territory, and mate with females whose home ranges overlap his. Dispersal of both sexes appears to be a consequence of agonistic behavior between the adult female and her offspring (Barash 1974c) and the subsequent dispersion of dispersers reduces the likelihood of inbreeding. However, in a dense population where acceptable areas of settlement are few, many juveniles remain philopatric and settle within their natal home range (Maher 2009a). These philopatric individuals were more closely related than dispersers and their close proximity made inbreeding possible. Multiple paternity occurred in 63% of the litters and there was considerable variation in male success; about 57% of the males were not assigned paternity and the number of offspring to which males were assigned paternity varied from 0 to 8. Although genetic analysis revealed mating with close relatives, there was no evidence of inbreeding depression (Maher and Duron 2010).

Restricted family

In marmot species with restricted family social organization (Table 5.2), the general pattern appears to be male dispersal and immigration into sites with resident females. Some males may settle in unoccupied habitat and form new families, e.g., *M. olympus*

(Barash 1973a), *M. vancouverensis* (Bryant 1998). Dispersal may not occur from some small populations, but it is unclear whether males remain in their natal area and become the dominant, reproductive male. No replacements were reported in *M. caligata* in Alaska where pairs remained together for up to 4 years (Holmes 1984a) and one succession occurred in a Canadian population when a peripheral male replaced a male that disappeared (Kyle *et al.* 2007). Much remains to be learned about male reproductive strategies in this group of species.

Because annual male reproductive success is directly related to the number of females with whom a male mates (Armitage 1986c), males should seek matings. When family groups are in close proximity, a male may make excursions to seek matings in other families, a process named gallivanting (Barash 1981). Gallivanting occurs during the mating season and all gallivanters were repulsed by the resident male. However, if the resident male were removed, the gallivanter successfully copulated. Monogamous males were more likely to depart from a female and polygynous males were more likely to mate-guard. Males gallivanted more when females were non-fertile. In general, males attempt to maximize their copulations and to prevent females from multi-mating (Barash 1981).

Extended family

When several or many families occur on a habitat patch and several males occur within a family, there are more opportunities for extra-pair copulations by neighbors or transients and for more replacements either within the natal family or in neighboring families.

In *M. baibacina*, *M. bobak*, *M. caudata*, and *M. menzbieri*, a male may form a new family when he settles on a new site and is joined by a female or alternatively, he may join a neighboring family especially when there is no territorial male or subordinate adult male (Mashkin 2003). This exchange of males among families probably reduces the likelihood of inbreeding, but a quantitative analysis of male reproductive success is lacking. Because *M. bobak* families typically have two or three adult males and one adult female, polyandry and competition among males for reproductive success are expected (Nikol'skii and Savchenko 2002a).

M. marmota males frequently make excursions into neighboring territories during spring. The intruder may be fiercely attacked and even killed by the resident male (Ferrari *et al.* 2012), but many males take over a territory while others establish residency in vacant sites (Arnold 1990a). In some cases, males form pairs with half or full sibs; thus, inbreeding was possible in 22% of newly established pairs. In addition, polyandry may involve inbreeding and extra-pair mating, including close kin; e.g., between son and mother (Arnold 1990b). A male may succeed to territorial status when a monogamous pair evicts the former residents (Lenti Boero 2003a). When a new male or new pair occupied a territory, no reproduction occurred in that year (Lenti Boero 1994). No male became territorial in his natal family; territorial males were replaced either by a male from an adjoining family, by a satellite male, or by an immigrant (Lenti Boero 1999). This pattern of male residency suggests that the ultimate factor in the process of obtaining territorial status is competition among males, not the avoidance of inbreeding. Males remain in their natal family, which increases their survivorship until they are large enough to compete for the dominant position (Armitage 1996b). Dispersal to seek opportunities to reproduce should occur when the probability of success is higher than it would be if the male dispersed at an earlier age but before philopatry reduces lifetime reproductive success (LRS) (Arnold 1990a, Armitage 1996b).

Because the mating system of *M. marmota* is polyandrous, territorial males face competition within the family. These males should behave in a way that maximizes inclusive fitness. One tactic is reproductive suppression of the adult subordinate males by the dominant territorial male (Arnold and Dittami 1997). Physiological depression of reproduction occurs in all 2-year-old subordinates and among older subordinate non-sons. Sons had androgen levels as high as those of the territorial male whereas non-sons had significantly lower levels. As a consequence, any successful mating by subordinate males will be those most closely related to the territorial male, which results in higher inclusive fitness than if non-sons successfully mated with resident females.

Male alpine marmots may gain fitness by impairing reproduction of the dominant female (Hackländer and Arnold 1999). If male takeover of a territory occurred during the mating period, there was no effect on female reproduction. However, if male takeover occurred after the mating period, females failed to reproduce despite signs of pregnancy; e.g., enlarged nipples and late molt. Females failing to wean young had higher reproductive success the following year. By impairing reproduction in females with whom the male did not mate, the females gain sufficient mass for successful reproduction the following year. On average, the body mass of successful females is 141 g larger than in years when the same females failed to reproduce and vernal body-mass averages 177 g lower in the year after reproduction than in the year in which young were weaned. Female reproductive investment also can be terminated by infanticide (Perrin *et al.* 1994). In at least one instance, the female whose young were killed by an invading male produced five young the next year (Coulon *et al.* 1995). In brief, a male by terminating the female's reproductive process provides the female the additional time needed to gain sufficient mass to both survive hibernation and reproduce the following year.

The competition among adult males in the alpine marmot family illustrates the trade-off between fitness-related behaviors. Males are the primary helper during social thermoregulation. When helpers are absent, the offspring sex ratio is biased toward the helping sex (males); mothers with helpers produce unbiased sex ratios (Allainé 2004). Sex ratio within litters was not affected by the mother's body condition, territory exposure to the sun, presence of a litter the previous year, litter size, or change in the dominant male. Juvenile survival during winter increased precisely with the number and proportion of subordinate males in the hibernating group, but not with the number of subordinate females (Allainé *et al.* 2000). The significant increase in survival increases the fitness of both parents. However, extra-pair copulations occur (Cohas *et al.* 2007a) and reduce the territorial male's fitness.

In studies of asocial and family-group marmots, individually identified males were not followed through their lifetimes to determine the factors contributing to lifetime reproductive success (LRS) and its variance among males. LRS probably is related to the number of females with whom a male mates; annual reproductive success of male yellow- bellied marmots was directly related to the number of females in a male's harem (Armitage 1986c).

Male residency in yellow-bellied marmots

The patchy habitat occupied by yellow-bellied marmots and the matrilineal organization provide an opportunity for polygyny. A male that successfully occupies a habitat patch with one or more matrilines has an opportunity for a high LRS (Armitage 1998). The male must establish cohesive relationships with the resident females; as a consequence, amicable greatly exceeds agonistic behavior in adult male:female relationships (Armitage 1974).

The primary mechanism whereby males gain access to females is by dispersing from their natal sites and seeking residency elsewhere; 92.2% of resident males were immigrants (Table 15.1) and only 35.2% were born at one of our study sites. The amount of movement by adult males is exemplified by the number of transients, animals that are present at a site for several days at most before disappearing. Transients formed 30.8% of the adult males. No transient was known to become resident with adult females.

The probability that a male successfully becomes resident with females can be estimated from life table statistics (Schwartz *et al.* 1998). Only 15.6% and 8.3% of males live to ages 2 and 3 years, respectively, the ages when residency is most likely. However, the proportion of 2-year-old residents is only 0.34 compared with 0.66 for males aged 3 years or older. Some unknown proportion of these males was older than 3 years, but, because adults first trapped beyond age 2 cannot be accurately aged, the proportion remains unknown. The mean age of residency for the 10 males that became resident at their birth sites was 3 years. Because 18.5% of the resident males were known to be aged 4 years or older, I estimated the age of residents by assigning the minimal age at which residency was achieved. The mean age of residency was 2.96 years; this age is an underestimate as some unknown number of males was older than the assigned minimum. Both approaches for estimating age of residency indicate that a mean age of 3 years is a workable estimate. This estimate raises the question of why most males disperse as yearlings (Armitage 1991a) when the death rate is high ($q_x = 0.702$) and remains high for 2-year olds ($q_x = 0.475$), both rates much higher than those of females, whose residency at the natal site is much higher (Schwartz *et al.* 1998).

Table 15.1 Residency characteristics of adult males aged 2 years or older.

Number of males	224
Number of residents (colonies or satellites)	128
Age of philopatric residents	3 aged 5 years, 1 aged 3 years, 6 aged 2 years
Number whose birth site known	45
Number of transients	69
Number living peripherally 1 or more years	34

Yearling male dispersal

The major reason yearling males disperse is competition with adult males. Social behavior of adult males is primarily agonistic; amicable behavior between adult males has not been observed and although some amicable behavior between adult and yearling males was observed, social behavior between these cohorts is overwhelmingly agonistic (Armitage 1974, Armitage and Johns 1982). More than a third of the agonistic behavior was chases. I observed a yearling male watch an adult male emerge from his burrow; the yearling immediately dispersed from the site even though no chase occurred. This observation indicates that for some yearling males the presence of an adult male is sufficient to induce dispersal. The timing of male yearling dispersal is related to rates of agonistic behavior; the higher the rate the earlier the dispersal (Downhower and Armitage 1981).

Despite the agonistic environment in which yearling males find themselves, not all yearling males disperse. A critical factor in delaying dispersal is the ability of the yearling to avoid the male; e.g., moving out of an area when the adult approaches (Armitage 1974). I compared space overlap between adult males and yearling males that either dispersed or were philopatric, most of whom returned as 2-year olds. Mean space overlap of residents did not differ from that of dispersers (Table 15.2); seven yearlings that had no overlap with the adult male (not included in the analysis) were peripheral (living at the edge of the site) and six returned as 2-year olds. Because yearlings and adults may use the same space at different times, space overlap does not necessarily indicate the amount of contact between the yearling cohorts and the adult male; agonistic behavior between the adult male and resident yearlings was significantly less than that between adult males and dispersers (Table 15.2).

Multiple-factors affect yearling male residency; residency occurs when yearlings can avoid the adult male, when adult male density is low, when a new or an old adult male is present, when multiple adult males are in conflict, and when agonistic behavior is avoided (Table 15.3). New residents often appeared late in the active season (usually after

Table 15.2 Space overlap and agonistic behavior characteristics of philopatric (resident) and disperser yearling males. Philopatric males are those who were present into August, most returned as 2-year olds. Adult females are non-mothers of the yearlings.

	Philopatric	Dispersers	Statistics
Space overlap with adult males	18.7%	21.6%	$t = 0.8$, $p > 0.4$
Agonistic behavior (#/animal/h) with adult males ($x \pm$ SE)	0.018 ± 0.004	0.062 ± 0.013	$t = 10.8$, $p < 0.001$
Space overlap with adult females	32.7%	23.9%	$t = 2.7$, $p = 0.01$
Agonistic behavior with adult females	0.014 ± 0.005	0.018 ± 0.007	$t = 0.45$, $p > 0.5$
Space overlap of yearlings with adult females vs. adult males	Significantly greater with females	No difference	$t = 4.5$, $p < 0.001$ $t = 0.62$, $p > 0.5$
Agonistic behavior between yearlings and adult females vs. adult males	No difference	Significantly less with females	$t = 0.64$, $p > 0.5$ $t = 3.1$, $p < 0.01$

Table 15.3 Characteristics associated with the failure of yearling males to disperse when an adult male was present. N = 53 yearling males.

Number of yearlings	Characteristic
9	No adult male present or disappeared in early summer. At RMBL, three 2-year olds formed separate territories
17	An area available as a refuge where adult male did not visit
8	Only one male present when the colony typically had two or three resident males
4	New male present
5	Old male (6-years old or older) present
1	Multiple males in conflict with each other, no interactions with yearlings
1	Transient 2-year-old disappeared, yearling dispersed the next year when an older male immigrated
5	Rates of agonistic behavior with adult male (0.016–0.019/male/h) significantly ($t = 2.31$, $p < 0.05$) less than that of yearling dispersers
3	No obvious pattern

the breeding season) and typically have a higher rate of social interactions with the resident females than do returning resident males (Armitage 1974). Possibly the time a new male allocates to the resident females subtracts from the time spent in conflict with yearlings which allows them sufficient time for foraging without harassment.

The data support the interpretation that male dispersal is conditional and depends primarily on the level of conflict with the resident, territorial male. The importance of a male in initiating dispersal is illustrated by the following. At Marmot Meadow in 1995 no adults were present and three male and three female yearlings engaged in play or amicable behavior. When an adult male immigrated in late July, agonistic behavior between the adult male and the yearling males was five times greater than amicable (the reverse prevailed between the male and yearling females); all male yearlings disappeared within 1 week and all females returned as 2-year olds.

Dispersal was also linked to the voluntary separation from an opposite sex parent (Wolff 1993). I tested this relationship by determining whether the mother was present when 51 yearling males did not disperse and were recaptured at their natal site as 2-year olds, thus, verifying their resident status. The mother was far more likely to be present than absent (37 to 14, $\chi^2 = 10.4$, $p < 0.005$). Thus, male marmots do not disperse to avoid their mothers.

However, non-mother adult females could affect male dispersal. Space overlap between these females and resident male yearlings was greater than that of dispersers (Table 15.2), and residents had greater space overlap with adult females than with adult males; but space overlap of dispersers did not differ between adult males and females. These differences in space overlap indicate that adult females do not have a major role in male dispersal and this interpretation is supported by rates of agonistic behavior. There is no difference in rates of agonistic behavior between adult females and resident or disperser yearlings (Table 15.2). Furthermore, agonistic behavior of resident male yearlings did not differ between adult males and adult females, whereas agonistic behavior was significantly greater between dispersers and adult males than between dispersers and adult females (Table 15.2). I conclude that male yearling dispersal is a consequence of mate competition and is induced primarily by agonistic behavior with the adult, territorial male. There is no evidence that adult males discriminate between sons and unrelated yearlings.

I conclude that dispersal of yellow-bellied marmot males is not deterministic, but is conditional and is related to the level of agonistic behavior the potential disperser receives from the resident, territorial male, who may be his father. Yearlings that avoid agonistic behavior may remain until 2-years old and a small number achieve residency in their natal sites. Inbreeding avoidance is a consequence of, not the reason for, dispersal (see Moore and Ali 1984), which is caused by competition for mates.

Age of male dispersal and immigration in yellow-bellied marmots

Most male yellow-bellied marmots disperse as yearlings, the year before maturity. This dispersal pattern characterizes marmot species except those in which dispersal is delayed 1 or more years (Table 5.1). Marmots should disperse when age and size increase the probability of successful immigration; a larger 2-year-old male would have a greater chance for success than a yearling. The age of dispersal must also be considered from the viewpoint of the resident adult. Undoubtedly, a 2-year old represents a greater threat than a yearling to the continued residency of the adult territorial male. Thus, the resident adult should engage in conflict with potential rivals when he has the best chance of winning. As expected, yearling males always lose in contests with adult males.

We can estimate the age of successful immigration (the resident territorial male) by examining the fate of dispersing males trapped at our research sites. The number of potential immigrants is a minimal estimate because some dispersers disappeared before they could be trapped and others must have moved through a site when we neither trapped nor observed. Of 36 yearlings trapped, 22.2% became residents (Table 15.4). A territorial adult male was known to be present for 11 of the 13 yearlings whose colony of origin was known,

Table 15.4 Age and residency of male yellow-bellied marmots.

Age (years)	1	2	3 or older	Statistical comparisons
Resident	8	45	83	Residency increases with age G = 23.8, p < 0.001
Transient/peripheral	28	57	46	
Proportion resident				
Satellite sites		0.34	0.34	
Colonial sites		0.66	0.66	
Proportion transient				
Satellite sites		0.73	0.27	No difference in proportion by age at the two sites G = 1.94, p > 0.1
Colonial sites		0.54	0.46	

which supports the interpretation that dispersal occurs to escape conflict. Yearlings became resident where no adult male was present or at large sites where unoccupied areas were available (Table 15.3). In only one of the successful immigrations was a philopatric yearling male present, he disappeared (possible badger prey). We introduced nine yearling males into localities where a colonial male resided; all yearlings emigrated (Armitage 1974). Thus, successful immigration of yearling males is most likely when there is no competition from resident yearlings, no territorial adult male present, or unoccupied space is available.

We never observed a colonial male to be displaced by an intruder. Three 2-year-old and three 3-year-old males were introduced into localities with a resident male; all emigrated. Typically, the introduced male fled at the approach of the resident (Armitage 1974). One 4-year-old male, who was introduced in one colony and fled from the resident male, appeared at a nearby colony 2 weeks later. After a series of encounters with the resident, he disappeared and lived in a peripheral area and became resident in part of the locality as a 5-year old (Armitage 1974). When territorial males were removed, immigrants soon occupied the site (Armitage 1974, Brody and Armitage 1985). Because nearly all males became territorial in the year that the former resident did not return, we will examine the residency status of adult males.

In addition to residents and transients, some males lived as peripherals. The proportion of trapped males becoming resident increased with age (Table 15.4). The proportion of 2-year olds did not differ from that of 3-year olds at satellite vs. colonial sites; or that were transient at satellite or colonial sites (Table 15.4). Thus, males are more likely to become resident as they get older, but their residency and movements prior to residency (the transients) do not differ between the two major habitat types.

A male may spend one or more years (mean = 1.44 years) peripheral to a site before becoming the resident territorial. Once territorial, the male may be resident for one to eight years; the overall mean is 2.63 years (Table 15.5). Males spend significantly more time as residents than as peripherals but length of residency does not differ statistically between satellite and colonial males. The lack of a difference between settlement patterns and length of residency indicates that males achieve territorial status whenever and wherever the opportunity occurs.

Of the 34 males who were peripheral for 1 or more years, half became resident at that site. Spending 1 or more years peripherally confers a reproductive advantage during the first year of residency. About one-third of new resident males do not reproduce in their first year, although first-year males are more likely to reproduce than not ($\chi^2 = 10.6$, $p < 0.005$). A

Table 15.5 Residency times of adult, male yellow-bellied marmots.

Mean length of residency	Ranged from 2.09 to 3.75 years among colonies 2.63 years overall
Mean time resident vs. peripheral	1.19 years more as residents, $t = 6.1$, $p < 0.001$
Mean time resident: satellite males vs. colonial males	Satellite (2.31 years), and colonial (2.78 years) males do not differ statistically, $t = 1.66$, $0.1 > p > 0.05$

greater percentage (84%) of males known to have been peripheral the year before residency reproduced than males (50%) who were not present the previous year (Armitage 2004c).

Multiple-male associations

There was no instance of multiple-male territories that persisted throughout the active season and males contributed little to the number of adult residents at the study sites (Armitage 1991a). There were several instances when two or more males were present in the same territory or site. Three modes of activity characterized the multiple-male associations.

One mode was characterized by the dispersal of all 2-year-old males when an older male was present. In each case, the 2-year olds were full brothers or paternal half-brothers/ maternal cousins (average $r = 0.375$) and sons of the older, resident male. Space overlap was minimal, social interactions were agonistic and all 2-year olds dispersed before young were weaned (Table 15.6).

Table 15.6 Activity patterns associated with multiple-male groups.

Colony	Year	Activity pattern
2-year old dispersal		
North Picnic	1988	Six 2-year olds dispersed early; no social behaviors observed; no space overlap with 6-year-old resident male or among the 2-year olds.
River	1989	Three 2-year olds dispersed, space overlap between adult father and 2-year olds was 5%; among 2-year olds space overlap was 4%. All social interactions between adult male and 2-year olds agonistic (0.05/animal/h). Only amicable behavior between adult females and 2-year olds (0.042/animal/h).
Picnic	1992	Four two-year-olds dispersed by mid-June. Their space overlap with adult male averaged about 3%, all social interactions with adult male agonistic (0.063/animal/h).
Formation of independent territories		
North Picnic	1971, 1972	No space overlap and no observed behaviors between the two adult males. Only social behaviors were sex-grapple (Armitage 1974) of adult females by the males.
Picnic	1999	No social interactions and no space overlap among the males.
Two or more males present		
Picnic	1989	No social interaction between two immigrant males at Lower or between immigrants and two resident males (one at Middle, one at Upper). Immigrants had few amicable and agonistic behaviors with adult females. Space overlap was 5% between immigrants and zero between immigrants and resident males.
Marmot Meadow	1980	Male 374 active at both major burrow areas, chases 372. Male 372 establishes residency at Aspen Burrow, the two males have distinct foraging areas. Male 519 chases 372, occupies Aspen Burrow, then moves to Main Talus. Disappears after 3 weeks. Agonistic behavior seven times more frequent than amicable among males; agonistic behavior twice as frequent as amicable between males and females.

The second mode was characterized by the formation of independent territories in a site with a large area (Table 15.6). Two male immigrants into North Picnic formed independent territories; one territory was located at the burrow area usually occupied by females and the other coincided with the presence of adult females at locations that were usually unoccupied. In neither year was there any indication that one male attempted to expand his territory by excluding the other male. Similarly, at Picnic two immigrant males and a 3-year-old male born at the site partitioned the space to form three territories that coincided with the three major burrow areas (Upper, Middle, and Lower). Each male maintained an independent, individualistic existence.

The third mode was characterized by the presence of two or more males in the same area within a site. One response pattern was characterized by mutual avoidance. In 1989, two adult male immigrants at Lower Picnic had no contact with each other or with two male residents (Table 15.6). Both immigrants were killed by a badger in late July and the male at Middle occupied the Lower area the next year.

The second response pattern was characterized by conflict and shifting patterns of space use. In 1980, three immigrant males lived at Marmot Meadow for varying lengths of time. Dominance relations among the males were complex. Male 374 was the first to appear; when male 372 appeared, he established residence at Aspen Burrow (Table 15.6). Subsequently, the males wrestled, chased each other, allo-groomed and at times lay together at Aspen Burrow (Frase and Armitage 1984). Male 519 appeared 4 days later and became resident at Main Talus where the three adult females lived. While 519 was present, 374 shared a foraging area with 372 around Aspen Burrow; after 519 disappeared, 374 returned to Main Talus and 372 remained at Aspen Burrow. Agonistic behaviors characterized male social interactions; females tended to avoid the males. The next year, only 372 was present and his territory included both major burrow areas. To summarize, the presence of multiple-male residents is characterized by mutual avoidance or a combination of intense conflict followed by mutual avoidance. There is no evidence of cooperation among males; e.g., 374 and 372 did not collaborate to exclude 519.

In 2001, the yellow-bellied marmot population began an abrupt increase and tripled by 2008. This increase in population is associated with an increase in the number of resident males in 1 or more years at nine sites. As many as 11 males were present in a behavioral social group for the duration of the active season or in a reproductive social group during spring. The males in these groups were considered mutually tolerant in the sense that a male was not forced to leave a group (Olson and Blumstein 2010). Despite the large number of males present in a group, male behavior was characteristic; i.e., over 8 years at all the sites, only 80 male:male social interactions were observed; 68.8% were agonistic. Although males were not forced to emigrate, male movement was restricted; the proportion of burrows used by an individual male was smaller for males in groups with a greater number of males. There was no evidence for cooperative alarm-calling and male reproductive success decreased with more males per female in a group (Olson and Blumstein 2010). In conclusion, apparently the presence of multiple males in a social group is a result of the increased number of males in the population coupled with the inability of a male to obtain exclusive use of a site and does not represent cooperative behavior between males.

Table 15.7 Male composition at Picnic Colony.

Male	Year(s) present	Status
1575	2001, 2002	Long-term resident
17	2001	2-year old born at Lower Picnic
2727	2001, 2002	Adult immigrant at Upper Picnic
178	2001	Adult immigrant at Upper Picnic, never observed after being trapped
2630	2001	2-year-old recruit, never observed, dispersed and trapped at Boulder
2772	2002	Adult immigrant at Upper Picnic
2774	2002	Adult immigrant at Middle Picnic
2786	2002	Adult immigrant at Lower Picnic
2738	2002	2-year-old immigrant at Lower Picnic
2773	2002	Adult transient, later present at Cliff Colony

The distribution of males at a site includes both spatial overlap and areas of exclusive use. At Picnic Colony in 2001, initially five males were present (Table 15.7). Most space overlap occurred between widely wandering male 17 and the two resident adults (Fig. 15.1 upper). Only rarely were two adult males recorded in the same grid and never simultaneously. The only social interactions were agonistic (0.018/animal/h) when 17 fled from 1575. In 2002, seven males were identified, but only four were observed within the site (Fig. 15.1 lower) as the activities of two males at Upper Picnic were obscured by high vegetation and another male was a transient. Again, only rarely were two different males recorded in the same grid and never during the same census. Most space overlap occurred between 1575 and 2786 (Fig. 15.1 lower). Both males were present in 2003; 2738, who ranged widely in 2002, settled at Upper Picnic where neither male from 2002 was present in 2003. In conclusion, multiple-male residency is characterized by some degree of space overlap, a high turnover of individual males, and a notable lack of social interactions.

Lifetime reproductive success (LRS)

A high LRS requires that a male be capable of maintaining a territory in which one or more adult females live. I observed 19 encounters between a resident and an adult intruder male. Intruders typically are chased by the resident (Table 15.8A).

The frequency of intrusion is low in May. Typically the landscape is snow covered and there is little movement away from the burrows until patches of vegetation appear. Movement is dangerous because predators such as coyotes and golden eagles are active and most escape or flight burrows are snow covered; the only refuge is the hibernaculum. The frequency of intrusion is high in June and early July, then declines rapidly (Salsbury and Armitage 1994b); intrusions rarely occur after mid-August. Presumably, marmots focus on preparing for hibernation. Transients typically move through a site; space overlap with the adult male averages 9.4%. Peripheral males often make several incursions into the resident male's territory; space overlap averages 14.9%.

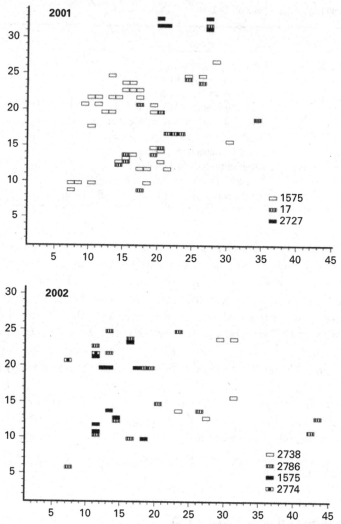

Figure 15.1 Patterns of space use by adult male marmots at Picnic Colony in 2001 (upper) and 2002 (lower). The grid sizes are 9 × 9 m. Two symbols connected vertically indicate that two males were recorded at least once in the same grid. The degree of space overlap may be visualized by observing the intermingling of locations between two males; e.g., 17 and 1575 and none between 1575 and 2727 in 2001. The lack of space-sharing is exemplified by the frequency with which a male was recorded in a particular grid. For example, 48 (42.5%) of the 113 location records of 1575 were at grid 23–16; 17 was never recorded at that grid. Fourteen (37.8%) of the 37 location records of 17 were at grid 24–24; 1575 was also observed there, but for only 7.0% of the locations.

Sighting intruders is rare; in 4909 hours of observation at four colonies, only 55 intruders were seen. This number is a slight underestimate as 68 transients or peripherals were trapped at these sites. Not all sighted intruders were trapped. If one assumes that the two samples are independent and combine the two estimates, the number of intruders remains small, on the order of 0.025/h. On average, during summer, resident males spend

Table 15.8 Responses of resident males to intruders.

A. Typical response

Male 125 chases intruder male from Split Rock to Upper Picnic. Intruder moves on upslope as male 125 dashes through upper area. Intruder moves toward Upper-Upper while male 125 at Upper with tail-flagging. Interaction took 11 min. Intruder not seen again.

B. Incursion by peripheral male

Male 321 standing near burrow where 401 entered. 401 emerges and runs up hill, 321 pursues and catches him. Vigorous fighting. Chase and wrestling back downhill. Much tail-flagging and mouth-sparring, standing on hind legs and pushing with fore legs; apparent grasping of head or jaw and shoulders. Wrestle and tumble for at least 3 min. 401 breaks away, flees, 321 pursues. Both disappear.

C. Corticosteroid concentration

	ng/ml	N	Statistical comparison
Other males present	4.4 ± 1.01	10	$t = 1.85$, $p = 0.1$
Other males absent	2.3 ± 0.54		
Conflict males	5.04 ± 0.38	10	$t = 2.66$, $p < 0.02$
Non-conflict males	3.54 ± 0.41		

about 9 h above ground, significantly more than any other age/sex cohort (Armitage et al. 1996). Thus, a male could expect to encounter an intruder about once every 4 days, on average. However, the frequency of intruders is not spread evenly over the summer; about 76% of the intrusions occur during the peak periods of June and July, with about 7% occurring in May and 17% after mid-July. This pattern of intrusion is reflected in energy expenditures of territorial males.

Metabolic rate of males is maximal after mating when the rate of intrusions is highest and then declines linearly (Salsbury and Armitage 1994a, Armitage 2004b). FMR can be predicted from body size and compared with the measured FMR to look for rates that depart from the predicted. There was considerable variation in measured FMR and the highest rates occurred when each male was in conflict with other adult or yearling males. Mean measured FMR of conflict males was about 66% larger than their predicted FMR (Armitage 2004b). When FMR of non-conflict males was measured, it was significantly lower than predicted FMR. Neither body mass nor season accounted for the difference in measured FMR of the two groups. Conflict males maintained a higher FMR at any time during the season when intruder males were present. Otherwise, FMR declined significantly with time; i.e., season (Armitage 2004b). The seasonal decline in metabolic rates and FMR is associated with a seasonal decline in fecal corticosterone metabolites (FCMs) (Smith *et al.* 2012), which suggests that adrenocortical activity may function in the seasonal regulation of metabolism. Specific FMR (FMR/body mass) did not differ between adult males and females when all values were included or when reproductive females and conflict males were removed from the analysis nor did FMR differ between conflict males and lactating females. These results suggest there is no difference in the metabolic reproductive costs of males and females. The nature of the cost differs between the sexes; for males, it is territorial conflict and costs of defending females especially when females are widely dispersed (Salsbury and Armitage 1995); for females, it is lactation. However, total annual costs may be greater for males. Females have high costs for about 30 days whereas males may have a high FMR for up to 60 days (Armitage 2004b).

There is abundant evidence that male yellow-bellied marmots undergo considerable stress. For example, several adult males went into shock and died while being handled. All were in conflict with other males at their sites. Adult male FCMs are higher than those of yearling males and yearling and adult females (Smith *et al.* 2012) and males have higher rates of mortality and a shorter life span than females; no known-aged male lived beyond 9 years (Schwartz *et al.* 1998). Since that demographic analysis, one male lived to age 12 years. Social behavior influences corticosteroid concentrations (Armitage 1991b). When male 401 lived peripherally at Picnic Colony, he made incursions into the territory of resident male 321, who patrolled the entire colony; male 321 was dominant (Table 15.8B, note some behaviors similar to those of play). During this time, 401 had the highest bound corticosteroid concentration and the second highest total corticosteroid concentration of males whose status was known. When 401 became the resident, territorial male the following year, and no other adult males were present, his corticosteroid concentrations were among the lowest. Two other lines of evidence indicate that male:male competition is associated with increased corticosteroid concentrations. Total corticosteroid concentration was higher in males when other males were present or when males were in conflict (Table 15.8C).

Analyses are confounded by seasonal changes in corticosterone concentrations (Fig. 15.2). There is a trend for concentrations to increase through the lactation period and the first 2 weeks post-lactation, then to decrease. A similar pattern occurs in FCMs (Smith *et al.* 2012). This trend is influenced by the social status and/or conflict of individuals. For example, the highest concentration occurred in a 2-year-old male during the second post-lactation time period. This male was in conflict with the resident male and subsequently dispersed. Similarly, two males with high values during gestation soon disappeared; a third was a new male, who also disappeared. The possible seasonal change in corticosterone concentrations coincides with the FMR values, which suggests that males must cope with metabolic and social stress. Although evidence supports both a seasonal trend and social effects on corticosteroid concentrations, a systematic study of the role of corticosteroids in male marmots is needed.

LRS, measured as the number of young fathered, was determined for males living at six colonial and two satellite sites. The number of years a male was territorial, the number

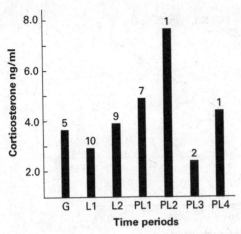

Figure 15.2   Seasonal changes in corticosterone concentrations in adult male yellow-bellied marmots. Season time periods: G = gestation; L1 = the first half of lactation; L2 = second half of lactation; PL1 = first two weeks of post-lactation; PL2, PL3, PL4 = subsequent two week periods following lactation. N is indicated by the number above each bar.

of adult females aged 2 years or older resident on the territory, mean age of adult females, mean matriline size, and the number of young weaned were tabulated for each male. Because an adult female could be present for 2 or more years, the number of adult females refers to the number of female-years (a female-year is one female resident for 1 year) and not the number of individual females. Of the five variables, only mean female age differed significantly among sites (Armitage 2004c).

When the number of young was regressed against the other variables, significant relationships occurred with number of females, mean matriline size, and number of years resident. Subsequent analyses (linear model ANOVA) revealed that only the mean number of females was significant:

$$\text{No. young} = 0.563 + 2.23 \text{ No. females, } p = 0.0001, \text{ adj-}R^2 = 0.76.$$

Thus, adult males should attempt to include as many adult females in their territories as possible. However, the size of male territories is not related to the number of females, but to their dispersion (Salsbury and Armitage 1994a). Some males defend widely spaced, isolated females and have large territories that include these females, whereas clumped females are readily defended within small territories. The variation in territory size and number of territories in large colony sites is consistent with the premise that males include as many females as possible within their territories.

Because a yearling has a greater probability of living to adulthood than a young individual, LRS was measured as the number of yearlings produced over a male's lifetime. The general linear model ANOVA with the variables colony, number of females, number of years, mean matriline size, and the number of young identified two significant variables, the number of young and mean matriline size (MMS):

$$\text{No. yearlings} = -0.985 + 0.475 \text{ No. young} + 0.78 \text{ MMS.}$$

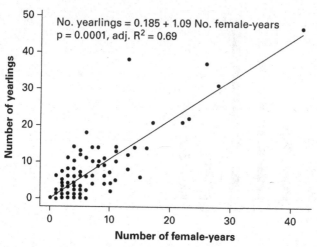

No. yearlings = 0.185 + 1.09 No. female-years
p = 0.0001, adj. R^2 = 0.69

Figure 15.3   The relationship between the number of yearlings and the number of adult females (female-years) in an adult male's territory.

Thus, the number of yearlings increased as the number of young and mean matriline size increased. Although MMS was significantly related to the number of yearlings, the relationship explained little of the variation between these variables (adj-R^2 = 0.067). Because the number of yearlings is related to the number of young weaned, the number of yearlings produced over a male's lifetime is significantly related to the number of adult females (Fig. 15.3).

The significance of MMS suggests that social factors within matrilines affect the number of young surviving to become yearlings. Both the number of young produced and the number surviving as yearlings increase as matriline size increases (Armitage 1986c). Therefore, male LRS is affected by the social organization of adult females; male LRS would be greater if females lived in matrilines of two or three, but mean matriline size is 1.39. There is no indication that an adult male affects the social organization of adult females and its effects on female reproduction. However, I observed a male break up a female chase; such activity could help maintain female residency and reproduction.

Summary

For yellow-bellied marmots, MRS is determined by his ability to disperse, survive, and obtain a territory, to form cohesive associations with resident females, and to mate with as many females as possible, but is affected by levels of cooperation and competition in female groups. The ability to obtain a territory by inheriting the natal territory, by takeover of an occupied territory, or by settling in a territory without a dominant male occupant probably is the key factor for all species of marmots. The transient strategy has little or no success in yellow-bellied marmots, but provides some success in alpine marmots and probably in woodchucks where some males travel widely. Because transient male alpine marmots were not identified, their LRS is unknown. Is the transient strategy temporary until the male can obtain a territory or does it represent the best that some individuals can do?

There are clear costs and benefits to male territorial residency in the extended family. The presence of multiple adult males in the family provides a benefit to the territorial male as survivorship of young is highly dependent on social thermoregulation provided by the subordinate males. The territorial male incurs a cost through EPP. Overall, the system increases the fitness of the territorial male as he fathers most of the offspring (see Chapter 5). Numerous factors; e.g., age/sex composition of the family, length of residency, affect the fitness of male alpine marmots, but a quantitative analysis of the contribution of these factors to the LRS of individual males has not been calculated.

16 Female reproductive success

Reproductive success of female marmots is determined by the number of their reproductive descendents. Several environmental factors impose constraints on potential success of female yellow-bellied marmots and some of these likely apply to other species of marmots. First, large body size and the short active season permit only one litter per year. If for any reason a female cannot wean young in a given year, she incurs a reproductive loss that reduces her potential fitness. Second, if she weans a litter late in the active season, there is a high probability that the young will not survive hibernation (Table 16.1; Armitage et al. 1976). Third, in one or more years there may be no resident adult male present during the mating season. This situation cannot be rectified by seeking a male elsewhere; habitat patches typically are too far apart for such travel to be practical in a snowy landscape where predators may be present. Fourth, prolonged snow cover may cause her to deplete fat reserves and be unable to muster the energy required for reproduction (Andersen et al. 1976).

The importance of fat reserves for reproduction is critical for all species of marmots. Woodchucks use fat during the spring breeding season (Snyder et al. 1961) and marmot species that regularly skip one or more years of reproduction do so because there is insufficient time after weaning to accumulate sufficient fat to both survive hibernation and reproduce the following year (Armitage 2000, Armitage and Blumstein 2002). Delaying reproduction until forage becomes available or moving to a location where snowmelt has exposed forage are not viable strategies for achieving reproductive success (Andersen et al. 1976). Thus, the first step in reproductive success is to occupy a good habitat site. Female reproductive success of golden marmots was related to food sources the previous year, which indicates that females that hibernate in good condition have a higher probability of reproducing (Blumstein and Foggin 1997). Similarly, alpine marmot females in good condition produced more juveniles (King and Allainé 2002).

A major consequence of the constraints on reproduction imposed by the environment and the delay in the age of first reproduction (Table 5.1) is that for most species of marmots, on average, less than 50% of adult females in any year breed (Table 16.2). The number breeding may vary from year to year; e.g., from 55.6 to 81.5% for *M. monax* (Snyder and Christian 1960). Females that live in social groups apparently pay a reproductive cost by reproducing less frequently than the solitary *M. monax*.

Table 16.1 Survival of young yellow-bellied marmots weaned early (before July 15) or late (after July 15). Early- and late-weaning females were chosen from the same colony in the same year or the same female weaning early or late in different years to limit variation due to different habitats or to female variability. G = 18.8, p < 0.001.

	Died	Survived
Early	24	56 (70%)
Late	50	18 (26.5%)

Table 16.2 The percentage of adult females that reproduce. References are in Armitage (1996b, 2007) unless otherwise noted.

	Age of first reproduction	% reproducing
M. monax	1	72
M. flaviventris	2	52 (Colorado) 46 (Oregon)[a]
M. baibacina	2	28
M. marmota	2	45
M. vancouverensis	3	53
M. olympus	3	45
M. caligata	3	45

[a] Mean for 4 years (Thompson 1979).

Sociality as a reproductive strategy

The basic social unit of yellow-bellied marmots is a mother with her offspring. This association may lead to the formation of adult kin groups that persist through time as matrilines (Armitage 1998). Matrilines form when 2-year-old daughters remain in their mother's home range; this process produces matrilines consisting of mothers:daughters: sisters (Fig. 16.1). Of 83 matrilines, 52 (62.7%) were mother:daughter groups, 23 (27.7%) were sister:sister groups (derived from mother:daughter groups), and the remaining 8 (9.6%) were groups that persisted after older adults died. Sisters could be either littermates or non-littermates who could be either full sisters (same father) or half-sisters (different father). Conflict is more likely among non-littermate sisters regardless of the degree of relatedness. Matriline formation characterizes colonies but is poorly developed at North Picnic Colony and at satellite sites (Fig. 16.2). Approximately 71% of satellite and 66.5% of North Picnic adult females were immigrants compared to 15.4% at Picnic Colony. Although matrilines formed at seven satellite sites, only five matrilines of two adults occurred in 109 satellite years (a satellite-year is one site studied in 1 year). Thus mean matriline size is much smaller than that of colonies (Table 8.1). Most matrilines become extinct, but one matriline at Picnic Colony persisted for at least 42 years (Fig. 17.3).

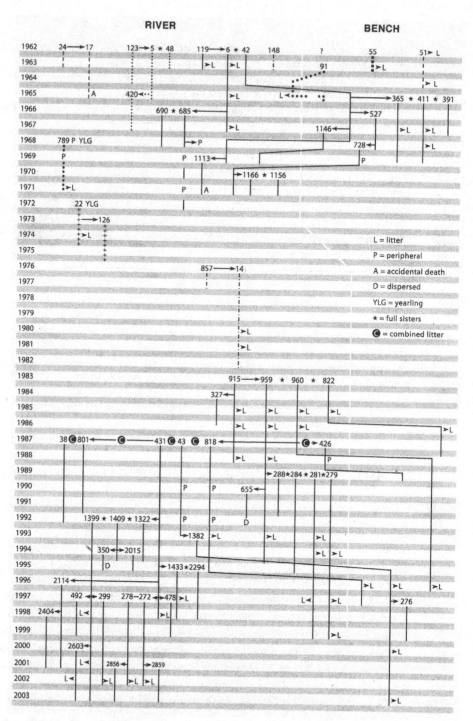

Figure 16.1 Matrilineal organization at River/Bench Colony. Each individual is identified by its left ear-tag number in the year of immigration or birth. Vertical lines indicate the years of residency. A horizontal line connects the vertical line of the mother with her recruits in their year of birth. Litters from which there were no recruits are indicated by L. A horizontal line that crosses two or more vertical lines indicates the recruits are from litters that intermingled before young could be assigned to a particular adult; e.g., 1987.

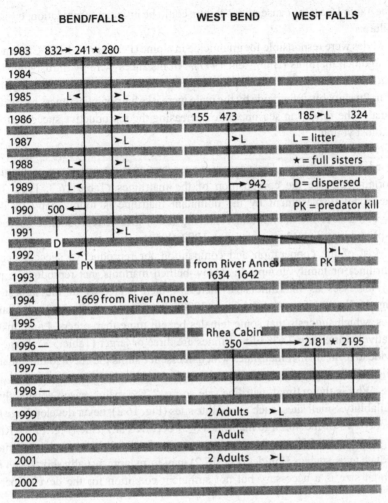

Figure 16.2   Matrilines at the Bend Falls complex of satellite sites. Lines and symbols as in Figure 16.1.

Large matrilines provide a competitive advantage in conflict between matrilines. When adjacent matrilines each consist of one female, an average of 1.17 to 1.36 yearling daughters are produced yearly. However, if an adjacent matriline consists of two or more females, annual production of yearlings is reduced by one-half because the single female weans a litter about half the time; litter size is not reduced (Armitage 1998). The failure to wean a litter may occur because of infanticide. The extent of infanticide in yellow-bellied marmots is unknown, but I observed two instances of an intrusive, adult female enter the burrow of the resident in a different matriline. In each case the resident female was reproductive as indicated by enlarged nipples and the frequency of carrying grass to the nest burrow. Following the intrusion, grass-carrying ceased, nipples regressed, and no young were weaned. Infanticide may be cryptic; the likelihood of observing the intrusion is very small. There were numerous instances of lactating females failing to wean (see Table 16.5); the cause of this failure could be infanticide, predation, or physiological failure.

Males were responsible for infanticide in alpine (Perrin *et al.* 1994, Coulon *et al.* 1995) and golden marmots (Blumstein 1997). All instances were by intrusive males or males that had taken over the territory. In some but not all instances the females bred the next year. Presumably infanticide by the male increases his potential fitness by mating with the females the following season while decreasing the reproductive success of the females that lost young.

A numerically dominant matriline may acquire resources by displacing the females of an adjacent matriline. The process may occur over several years as recruitment and mortality change the composition of the matrilines (Table 16.3, Figs. 16.1, 17.3). Successful displacement by the dominant matriline is followed by its descendents spreading over the habitat and achieving much greater reproductive success than the displaced matriline (Armitage 1989, 1992, 2002).

Most of the cooperation and competition for reproductive success occurs within matrilines or family groups. In yellow-bellied marmots, net reproductive rate (NRR, R_0) increases as matriline size increases then decreases in the largest matrilines (Fig. 16.3) (Armitage and Schwartz 2000). R_0 varies considerably in the larger matrilines, which suggests that some females benefit more than others. Large matrilines are relatively rare; only 7.6% of matrilines are three or larger (Table 5.3). Large matrilines divide to form smaller matrilines which in essence shifts the curve to the left and restores a high NRR. The optimal matriline size is three, but only 4.3% of matrilines were of this size. Why is the optimal matriline size so rare? The major reason apparently is habitat availability. Small sites, such as satellite sites (Fig. 16.2), never develop large matrilines. Twenty-two of the 40 large matrilines (≥ 3) occurred at Picnic (Fig. 17.3), River/Bench (Fig. 16.1), and Marmot Meadow, among the largest habitat patches. A site of intermediate area, such as Boulder (Fig. 17.2) may form large, usually short-lived matrilines. Habitat size is a necessary but not sufficient condition for the development of large

Table 16.3 Female displacement by a numerically dominant, adjacent matriline.

Colony	Time period	
River	1994–97	1399 and her mother 431 occupy prime burrow area. Sisters 279 and 281 in a different matriline in the adjacent area (Fig. 16.1).
	2000	431, 279, 281 no longer present. Three 4-year-old sisters form a new matriline and displace 1399 and her daughter 2603 to the adjacent area.
Picnic	1978	Littermate sisters 301 and 349 occupy central area. Mother 1194 and daughter 920 at a less favorable site.
	1981	1194 and daughters 503 and 489 occupy prime burrow area. Female 301 displaced to secondary area.
	1982	Female 301 recruits daughter 573; 489 alone, the two groups occupy same areas as the previous year.
	1983	High mortality reduces population to 573 and new immigrant 855.
	1986	573 displaced from talus slope after 855 recruits daughters.

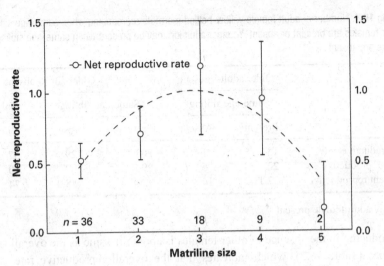

Figure 16.3 The relationship between matriline size and net reproductive rate. n = number of years that matriline size occurred. Values are mean plus 95% confidence intervals. $Y = 0.158 + 0.751 X - 0.122 X^2$, $p = 0.003$, $R^2 = 0.12$, where $Y = R_o$ and X = matriline size. Figure modified from Armitage and Schwartz (2000).

matrilines. The three sites with most of the large matrilines also have numerous burrow sites (Svendsen 1974) and large meadows with abundant forage (Kilgore and Armitage 1978, Frase and Armitage 1989). By contrast, North Picnic, the largest site, is characterized by widely spaced, small resource patches and has low density and small mean matriline size (Table 8.1). The importance of habitat quality is also apparent in the variation of R_o values; matrilines of similar size had markedly different values in different colonies (Armitage and Schwartz 2000).

Another reason why large matrilines are uncommon is that competition, as well as cooperation, characterizes marmot societies. R_o of some large matrilines may be lower than that of smaller matrilines (Fig. 16.3). Matrilines of one consistently have low R_o, in large part because most immigrants form matrilines of one and do not reproduce in their first year of settlement. Also, matrilines of one are subject to reproductive suppression by adjoining numerically dominant matrilines.

One factor that reduces the reproductive success of a matriline is the effect of the presence of an older female on reproduction of a younger female. When an older female is present, a younger female is significantly less likely to wean a litter ($G = 53.4$, $p < 0.001$) (Armitage 2003f). This relationship changes when additional adults are present. When same-age siblings are present, an adult female is more likely to reproduce than not when an older female is present (Table 16.4, $G = 5.8$, $0.025 > p > 0.01$). When older adults are absent, the effect of the same-age sibling increases ($G = 7.3$, $0.01 > p > 0.005$). The presence of younger kin affects reproduction; when both same-age kin and older adults are absent, the presence of younger kin (Table 16.4) increased the percentage of adult females that reproduced by 15.6% ($G = 4.4$, $0.05 > p > 0.025$). The effect of younger kin was essentially the same as the presence of same-age kin. The percentage

Table 16.4 Number of adult female yellow-bellied marmots reproducing when same-age siblings and older females are present or absent. Younger adult kin may be present when same-age siblings and older adults are absent.

	Older adults present		Older adults absent		
	Same-age siblings		Same-age siblings		Only younger
	Present	Absent	Present	Absent*	Kin present
Reproductive	13	1	80	129	57
Non-reproductive	22	6	29	92	20
Percent reproductive	37.1	14.3	73.4	58.4	74

* Only adult female present

reproducing in the absence of other females is about the same as the overall rate for the species (Table 16.2), which indicates that the overall reproductive rate is strongly affected by the high frequency (72.7%) of matrilines of one. The increase in the likelihood of reproduction by the presence of same-age or younger siblings in the matriline supports the occurrence of cooperative breeding in yellow-bellied marmots.

Reproductive skew

Reproductive skew (Hager and Jones 2009) is probably universal in the social marmots. Skew is clearly indicated by the frequency of reproduction by marmots of different ages. It is important to distinguish reproductive frequency related to environmental harshness (Armitage 2000) from reproductive frequency related to social factors. The mean age of first reproduction is typically older than the age of reproductive maturity (Table 5.1). In the woodchuck, only about 20% of the yearlings reproduce. This low rate may be a consequence of population density as the percentage increased to 42% when the proportion of adults in the population decreased (Snyder 1962). There is no evidence of age-related reproduction in woodchucks; a higher proportion of adult females reproduce than in any other marmot species (Table 16.2).

For all of the social marmots, from 28% to 53% of adult females breed (Table 16.2). In the extended family group of the alpine marmot, female dominance and experience strongly affected reproductive success (King and Allainé 2002). Experience may contribute to the increasing percentage of yellow-bellied marmot females that reproduce with increasing age (p = 0.009, Fig. 16.4). A similar pattern occurs in the gray marmot; however, for some species such as the Siberian marmot, the relationship may be curvilinear (Bibikow 1996). For most marmot species, age of females was unknown, but reproductive skew is widespread (Blumstein and Armitage 1999).

Reproductive skew in yellow-bellied marmots occurs because many females fail to wean a litter. Some females initiate reproduction, as evidenced by nipple development, but fail to wean young (FR), others are non-reproductive (Table 16.5). The number of FR is underestimated as not all females were trapped at the correct time for evaluating

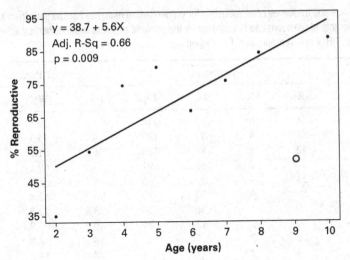

Figure 16.4 The relationship between age of adult females and the percentage of females that weaned litters. Data are for females living within matrilines at River, Marmot Meadow, Picnic, and North Picnic Colonies. Total N = 385. Age 9 was not included in the regression because of small sample size, but the mean is shown as a O. When age 9 is included, the regression is not statistically significant (p = 0.12) but adj-R^2 = 0.21 which indicates that the regression explains a substantial proportion of the variation.

reproductive status, but the relatively small numbers of FR indicate that the error is small. FR on average occurs in less than 10% of the adult females, but varies among age groups (Table 16.5). FR is especially high in older females and coupled with low rates of weaning may exemplify reproductive senescence. If one excludes 2- and 3-year olds where reproductive suppression is frequent (Armitage 2003f), and combine FR with weaning, 70.2% of adult females (age ≥4) initiate reproduction annually. Thus, about 30% of these females do not attempt to reproduce in 1 or more years.

Frequency of reproduction does not differ significantly between philopatric females and immigrants, in part because of the lack of 2-year-old immigrants (Table 16.5), but does vary with social structure (Table 16.6). Failed reproduction is most frequent when other non-matriline females are present, is much lower when females live in the same matriline, and is lowest when females are solitary. This pattern indicates that competition is higher among unrelated (or distantly related) females from different matrilines and that competition within matrilines reduces the reproductive success of females. The low rate of FR in solitary females likely reflects the absence of competition and indicates that litters are lost for other reasons, such as predation or infanticide. Assuming the same rate of loss for non-social reasons in the two social groups, the evidence points to social conflict as a major factor producing reproductive skew. Blumstein and Armitage considered marmots to be cooperative breeders as they express the following attributes proposed by Solomon and French (1997a): (1) delayed dispersal; (2) some form of alloparental care; and (3) reproductive suppression. Alloparental care occurs during hibernation in some species (see Chapter 7). The best evidence for cooperative breeding and reproductive skew comes from the occurrence of reproductive suppression.

Table 16.5 Age and frequency of reproduction for yellow-bellied marmots from all sites from 1962 to 2002. Age of immigrants of unknown age is indicated by the minimal possible age followed by a "+" sign. WL = weaned litter, NR = non-reproductive, FR = failed reproduction.

Philopatric	Number				Percent		
Age	NR	FR	WL	Total	FR	WL	FR + WL
2	141	10	63	214	4.6	29.4	34.1
3	85	8	77	170	4.7	45.3	50.0
4	42	8	80	130	6.2	61.5	67.7
5	21	7	63	91	7.7	69.2	76.9
6	23	4	38	65	6.2	58.5	64.6
7	10	6	27	43	14.0	62.8	76.7
8	9	1	26	36	3.8	72.2	75.0
9	6	2	13	21	9.5	61.9	71.4
10	2	2	7	11	18.2	63.6	81.8
11	4	2	1	7	28.6	14.3	42.9
12–15	5	0	1	6	0	16.7	16.7
Totals	348	50	396	794	6.3	49.9	56.2
Immigrants							
3+	26	6	33	65	9.2	52.8	60.0
4+	12	5	18	35	14.3	51.4	65.7
5+	5	0	11	16	0	68.8	68.8
6+	2	1	4	7	14.3	57.1	71.4
7+	1	0	1	2	0	50.0	50.0
Totals	46	12	67	125	9.6	53.6	63.2

Table 16.6 Statistical analysis of the frequency of failed reproduction (FR) for solitary females (OFP = only female present) and for two social groupings: the colony (ONMF = other non-matriline females present) and the matriline (MF = other matriline female present). $G = 13.0$, $df = 2$, $p < 0.005$.

	OFP	ONMF	MF
Reproductive	106	114	211
FR	5	27	30
FR (%)	4.5	19.1	12.4

Reproductive suppression

Reproductive suppression is widespread in mammals (Wasser and Barash 1983) and probably is universal among marmot species. Reproductive suppression is suggested in the asocial woodchuck: more yearling females breed at low population density (Snyder 1962) and adult females act agonistically at higher rates with adult daughters than with younger offspring (Maher 2009b). Much of the evidence for reproductive suppression is based on the low or no reproduction by the younger adults in a group and their higher rates of breeding in reduced populations and reduced reproduction in high-density populations (Armitage 1996b). For example, 2- or 3-year-old gray marmots frequently reabsorb embryos when an older adult female is present, but suppression is not universal (Mikhailuta 1991). However, density effects may be a consequence of the direct action of food resources rather than by social mechanisms. Also, marmot reproduction is affected by habitat differences and weather patterns (Bibikow 1996). Therefore, I will concentrate on those studies where density, habitat, and weather patterns can be excluded because the research was conducted on the same populations over several years and individual marmots were identified.

Most evidence implicating reproductive suppression is demographic. Subordinate, usually younger, females fail to wean a litter in the presence of a dominant, usually older female (Table 16.7). In yellow-bellied (Armitage 2003f) and alpine marmots (Hackländer *et al.* 2003) high reproductive skew is not due to differences in body mass or capacity for reproduction. The dominant and subordinate adult alpine marmots did not differ in plasma estradiol concentrations; all females had access to the territorial males and some subordinate females were observed to copulate with adult males (Hackländer *et al.* 2003).

In both alpine and yellow-bellied marmots, nipple development was observed in some subordinates, none weaned a litter, failure to wean was related to low and declining plasma levels of progesterone. Progesterone levels in yellow-bellied marmots (Armitage and Wynne-Edwards 2002) and woodchucks (Sinha Hikim *et al.* 1991, 1992) decline in

Table 16.7 Evidence of reproductive suppression in marmots.

Species	Evidence	References
Alaskan marmot	Only the dominant pair bred in a 6-year period in a captive population when other mature individuals present.	Rausch and Bridgens (1989)
Hoary marmot	In bigamous families, females may skip an additional year when both expected to breed; if both breed, subordinates produce one-half as many young and one-fifth as many yearlings.	Wasser and Barash (1983)
	Polygynous groups include sexually mature non-reproductive adults.	Kyle *et al.* (2007)
Olympic marmot	Second non-parous female may be present in the family.	Barash (1973a)
Alpine marmot	Territorial, but not subordinate females wean litters in stable families. Subordinate females fail to wean litters in years when dominant female fails to breed.	Arnold (1990a), Perrin *et al.* (1993b), Lenti Boero (1999)
Yellow-bellied marmot	2- and 3-year-old females significantly less likely to breed when older females are present in the matriline. Suppression acts mainly to delay age of first reproduction.	Armitage (2003f, 2007)

late gestation then increase during lactation. Low progesterone levels indicate that reproductive suppression occurs before parturition. There was no evidence of infanticide in alpine marmot stable families. Agonistic behavior directed from dominants toward subordinates occurred primarily during gestation; nearly all the amicable behaviors were initiated by subordinates. The movement of a pregnant female, who was the recipient of agonistic behavior from an older female, to another family where she achieved residency (the adult female of that family disappeared) (Goossens *et al.* 1996) likely enabled her to escape reproductive suppression and successfully wean a litter.

Relatedness influenced social interactions in alpine marmots; amicable and neutral interactions predominated between a dominant and her daughter and agonistic interactions characterized behavior between dominants and unrelated subordinates. In all cases, agonistic behavior increased during gestation (Hackländer *et al.* 2003). An increased number of subordinates decreased the probability of a dominant successfully reproducing. Only once did a subordinate female, the adult daughter of the dominant female, wean young. Conflict continued between the females and the daughter eventually evicted her mother (Hackländer *et al.* 2003). The patterns of conflict indicate that individual marmots strive to maximize direct fitness and that suppression of a young by an older female is not always successful.

To summarize, reproductive skew in marmots is a consequence of reproductive suppression of younger, subordinate females by older, dominant females. Why do younger females remain in the family or matriline in the face of reproductive suppression? I hypothesized elsewhere that marmots remain in their natal area because of likely habitat saturation (Armitage 1996b). Although survival is enhanced by remaining in the social group, a female must eventually seek opportunities to reproduce either at home or by dispersing and successfully immigrating elsewhere. The extensive forest-edge habitats of woodchucks and the numerous small habitat patches of yellow-bellied marmots scattered throughout their montane area (Svendsen 1974, Armitage 1991a) provide opportunities for dispersers to survive and reproduce. But where habitat is restricted to an alpine meadow (e.g., alpine, Olympic, and Vancouver Island marmots) or when an extensive steppe is fully inhabited (e.g., the steppe marmot), a marmot has little choice but to remain at home until a nearby opportunity for reproduction occurs or to eventually leave the natal area to seek reproductive success elsewhere. The choice should be one that potentially has the greatest fitness benefits. The trade-offs between these two choices are analyzed for yellow-bellied marmots in Chapter 18.

Sex-ratio variation

There is considerable evidence for significant deviations from 1:1 sex ratios in some species of mammals (Clutton-Brock 1985). Sex-ratio bias should be directed by a female to the sex that contributes most to her LRS (Clutton-Brock and Albon 1982, Armitage 1987b) or to the sex whose future reproductive success can be affected by the female.

In alpine marmots, subordinate adults participate in social thermoregulation, which increases survival (Arnold 1993b). Sex ratio at weaning is male-biased and juvenile survival increased precisely with the number and proportion of subordinate males, but not with the number of subordinate females, in the group (Allainé *et al.* 2000). A given mother produced a greater proportion of males (sex ratio = 0.65) when helpers were absent in the family group than when they were present (sex ratio = 0.46) (Allainé 2004). Because helpers increase survivorship of the young, sex-ratio bias in this species is directed toward future fitness benefits.

Sex-ratio bias in yellow-bellied marmots is related to the recruitment of daughters. Small litters were not sex-biased, and stressed females did not produce sex-biased litters. Young adult females (age below population mean) living in a matrilineal group when other matrilines were absent, weaned significantly more daughters than sons and recruited significantly more daughters per litter and a greater proportion of the young daughters than did young females living alone or when other matrilines were present (Armitage 1987b). Among old (age above the population mean) females, there was no sex-ratio bias and females living alone recruited significantly fewer daughters per litter and a significantly smaller proportion of their daughters than expected regardless of whether another matriline was present. A female is more likely to produce male young as the number of offspring produced by older females increases (Nuckolls 2010). In general, females appear to produce the sex conveying the highest fitness gains (Armitage 1991a).

For both alpine and yellow-bellied marmots sex-ratio bias is not genetic but is facultative and responsive to the social environment. Individual alpine marmot females do not bias sex ratios when helpers are present, but do so when they are absent. Young adult yellow-bellied marmot females produce female-biased litters when the probability of recruitment is high but wean non-biased litters when they become older. For both species, adjustment of sex ratios is best understood as a measure to increase fitness.

This description of sex-ratio variation suggests that it is adaptive, but the outcomes may be a fortuitous consequence of the sex composition of litters. In particular, fetal position may affect life-history traits, a female adjacent to one or more males *in utero* may be masculinized (Vom Saal 1989). It is not possible to determine fetal position in wild mammals, but the possible affects of sex ratio on life-history traits can be examined.

In alpine marmots, females born in a male-biased litter had a significantly higher probability of becoming dominant and more likely to become dominant by aggression (Hackländer and Arnold 2012). These females did not suffer any loss in total reproductive output. In yellow-bellied marmots, masculinized females, those with a large AG distance, were less likely to survive their first hibernation, more likely to disperse, and less likely to become pregnant and wean young (Monclus and Blumstein 2012). These two examples indicate that male-biased litters can have either positive or negative effects on reproductive success and that the effects are likely related to the life history of the particular species. These examples raise the question of whether the effects extend beyond the first generation. This question was explored in yellow-bellied marmots.

I determined the sex ratio for 139 litters which had at least one female young. Sex ratio (proportion of males) varied from 0 to 0.8. The number of female yearling recruits decreased significantly as sex ratio increased (Fig. 16.5); the proportion of female recruits also decreased as sex ratio increased, but the relationship was not significant (p = 0.24, adj-R^2 = 0.002). However, the number of yearlings recruited was significantly and directly related to the proportion of yearlings recruited (p = 0.015, adj-R^2 = 0.035). The low amount of variation explained by these relationships suggests that other factors, such as age and social status of the adults, markedly affect the outcomes of sex ratio bias (also see Monclus and Blumstein 2012).

I analyzed the effects of their natal sex ratios on the LRS of 96 females who lived to be at least 2-years old (Table 16.8). Only one of six yearlings and no adults were recruited

Table 16.8 Lifetime reproductive success of adult females recruited from litters with different proportions of males at weaning. All values are mean ± standard error. N = number of females, PNR = proportion of females that never reproduced. Because of small sample sizes, litters with sex ratios of 0.17 and 0.25 were combined with 0.2; 0.43 was combined with 0.4, and 0.67 was combined with 0.6. Values rounded to nearest hundredth.

Sex ratio as a proportion of males in natal litter	Number of young per year			Number of years resident	N	PNR
	Male	Female	Total			
0	0.78 ± 0.17	0.96 ± 0.21	1.73 ± 0.36	4.05 ± 0.70	20	0.35
0.2	0.41 ± 0.13	0.39 ± 0.14	0.8 ± 0.26	2.93 ± 0.63	15	0.53
0.33	0.66 ± 0.27	0.66 ± 0.26	1.32 ± 0.48	3.78 ± 1.1	9	0.44
0.4	0.95 ± 0.27	1.20 ± 0.32	2.14 ± 0.47	3.0 ± 0.52	13	0.23
0.5	1.34 ± 0.20	1.02 ± 0.20	2.36 ± 0.37	4.28 ± 0.68	25	0.28
0.6	0.97 ± 0.19	0.64 ± 0.15	1.61 ± 0.32	4.07 ± 0.69	14	0.29

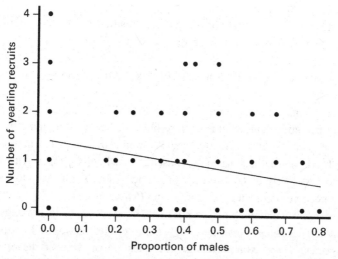

Figure 16.5   The relationship between the number of female yearling recruits and the proportion of males in the mothers' natal litters. Number of yearlings = 138. NYR = 1.37 − 1.08 Proportion of males; p = 0.004, R^2 = 0.054.

from litters with sex ratios of 0.75 and 0.8 which indicates a possible negative effect of high sex ratio on females, but may be a chance effect of small sample size (only six litters). Regression analyses found no relationship between the proportion of males in their natal litters and various measures of LRS (Table 16.8): number of male young ($p = 0.21$), number of female young ($p = 0.90$), total number of young ($p = 0.47$), length of residency ($p = 0.68$) or PNR ($p = 0.32$). Sex ratio increased as the proportion of males in the female's natal litter increased ($p = 0.14$, adj-$R^2 = 0.33$). This relationship was strongly affected by the sex ratio of young when the sex ratio of the mother's litter was 0.6; it was the only sex ratio that deviated significantly from 0.5 (Table 16.8). In conclusion, there is no transgenerational effect of sex ratio of a female's natal litter on her subsequent reproductive performance with a possible exception that females from highly male-biased litters produce male-biased litters.

Inbreeding

Inbreeding by yellow-bellied marmots is not avoided (Armitage 2004c, Olson *et al.* 2012). There appears to be no selection against inbreeding. The probability of inbreeding can be calculated from the likelihood of a male becoming resident in his natal colony, the probability of living to age 3 years, and the probability that an adult female relative is reproductively mature. The probability of a male mating with his mother or sister is less than 25 matings in 10 000; brother:sister mating, about 5 matings in 10 000; father:daughter matings are more likely, about 2 per 100 (Armitage 1974). These estimates are too low; of 780 potential matings, inbreeding was possible in 79 (10.1%). In years of high population growth and survival when the number of males in a colony increased, relatedness among reproductive pairs was higher than expected from random mating; father:daughter pairs followed by paternal half-sib pairs were the most frequently inbred (Olson *et al.* 2012). Inbred pairs weaned litters at a significantly lower rate than outbred pairs (Armitage 2004c) and survival of young through their first winter was significantly lower in inbred litters (Armitage 2004c, Olson *et al.* 2012).

Nevertheless, both the adult male and female gained fitness (measured as the number of surviving young). If this annual breeder fails to breed in a given year, her reproductive output is reduced by about 27% on average. Because a male on average is resident for fewer years than the average female, his reproductive output is reduced by about 42%. When the loss due to inbreeding is subtracted, the female and male on average gain 1.7 young over not breeding (Armitage 2004c); thus when choice is limited, inbreeding is more fit than non-breeding.

There could be a fitness cost if the detrimental effects of inbreeding are passed on to the offspring of inbred litters. There is no evidence that fitness loss passes from mother to daughter when inbred and outbred litters were fathered by the same male (Table 16.9). I conclude that inbreeding has short-term but no long-term fitness loss and that inbreeding, when no alternative exists, increases fitness over not breeding.

Table 16.9 Reproductive success of females born into inbred or outbred litters that were fathered by the same male. Mean values are followed by standard error and (N).

	Outbred	Inbred	Significance test
Number of yearling recruits	10	14	G = 0.03, p > 0.5
Proportion of females reproducing	0.7	0.54	G = 0.42, p > 0.1
Mean litter size (number litters)	3.8 ± 0.25 (21)	4.1 ± 0.41 (16)	t = 0.62, p > 0.4
Lifetime total number of weaned young:			
Mean per adult (number females)	7.9 ± 2.7 (10)	5.3 ± 2.0 (13)	t = 0.76, p > 0.4
Mean per reproductive adult	11.3 ± 3.1 (7)	9.9 ± 2.8 (7)	t = 0.34, p > 0.5

Sociality and recruitment

Recruitment of females into their natal site is the major mechanism by which resident yellow-bellied marmot populations increase (Armitage 1996b). Recruitment benefits both the adult (resident) and yearling (recruit) female and is significantly more likely if the mother is present, but is not affected by the mother's reproductive status (Armitage 1984). The presence and reproductive status of other females in the matriline enhanced rather than inhibited recruitment. Recruitment was not affected by whether a new adult male was present, but was significantly unlikely if a newly immigrant female was present.

Many adult females fail to recruit daughters, and, of those that do, the number of recruits varies widely among females (Figs. 16.1 and 16.2, also see Figs. 17.2 and 17.3). Because recruitment is a major factor that determines a female's fitness, we examine the factors that distinguish recruiters from non-recruiters. Recruiters were resident for a longer time, weaned more litters, and produced more female young and female yearlings than non-recruiters. However, there was no difference in mean litter size between the two groups (Armitage 1984). It would be expected that long residency would be associated with weaning more litters and thus producing more young and yearling recruits. Therefore, the demographic data were submitted to stepwise discriminant analysis. Three variables entered the model: number of litters, of female young, and of female yearlings. The number of female yearlings by far was the most important variable. This result is consistent with life table statistics; the life expectancy of female yearlings indicates a high probability that a yearling daughter will live to become a reproductive adult (Schwartz *et al.* 1998).

Individual behavioral phenotypes

The demographic characteristics of adult females characterized as social or asocial were similar except social females recruited about twice as many yearling daughters as the asocial females. This variable was the only demographic factor that discriminant analysis identified to distinguish between the two groups (Armitage 1984). The importance of

Table 16.10 The relationship of behavioral phenotype to lifetime reproductive success of adult females classified into one of three phenotypic groups.

No significant relationship	Significant relationship
Number of weaned young	Number of female yearlings
Number of yearlings	Number of female yearling recruits
Number of young or yearlings per year of residency	Number of 2-year-old resident daughters
Number of female young	

individual variability to LRS was determined for 19 adult females who were classified into one of three phenotypic groups. Phenotype was not related to measures of production of young or yearlings, but was significantly related to the production and recruitment of daughters (Table 16.10). Females in the sociability group had higher values for each of the three variables (Armitage 1986a).

Amicable but not agonistic behavior was correlated with measures of LRS; e.g., females with high lifetime rates of amicable behavior produced higher numbers of yearlings. Adult females recruited more female yearlings if they were amicable with female yearlings and with adult female kin, if they initiated amicable interactions, if they had a high rate of amicable compared to agonistic behaviors, and if they had high rates of total social interactions. Amicable behavior was strongly biased to close kin ($r = 0.5$). Thus, LRS was greater for those females who maintained an "amicable or cohesive environment" with adult and yearling kin regardless of which behavioral phenotype they expressed (Armitage 1986a).

Age of first reproduction

LRS, and thus fitness, varies among individuals. One factor that affects fitness is age of first reproduction, which ranges from 2 to 6 years. A later age of first reproduction could be compensated for by higher survival, an extension of maturity to older ages, or larger litter sizes. However, the number of reproductive events, mean litter size, and post-maturity survival were unrelated to age of first reproduction (Oli and Armitage 2003). Thus, delay in the age of first reproduction (a) resulted in a loss of individual fitness (λ^m):

$$\lambda^m = 1.83 - 0.12a, R^2 = 0.24.$$

This relationship indicates directional selection for early maturity (Oli and Armitage 2003). Individual fitness did not differ among colonies, but was influenced by the total number of adult females and the number of reproductive females. These variables explained 14.1% of the variation in λ^m. This loss of individual fitness is a consequence of the imposition of reproductive suppression on young adult females and does not represent a "choice" to forego reproduction.

Possibly the loss of individual fitness; i.e., direct fitness, does not result in a loss of inclusive fitness because of fitness benefits gained from assisting kin; i.e., indirect fitness. Indirect fitness benefits would be especially important for those young adult females that die before they reproduce (see Figs. 16.1 and 16.2 for examples of young adult females that failed to reproduce). Direct fitness declined significantly as the age of first reproduction increased ($p < 0.0001$); there was no relationship between indirect fitness and age of first reproduction (Oli and Armitage 2008). Females that failed to wean a litter had greater indirect fitness than those that did, but the inclusive fitness of the reproductive females was 2.3 times greater than that of the non-reproductive females. Indirect fitness benefits do not compensate for the loss of direct fitness when females fail to reproduce or when the age of first reproduction is delayed (Oli and Armitage 2008).

Given these fitness differences, it is most unlikely that delayed age of first reproduction could evolve by means of natural selection. Why do females join a matriline and face possible reproductive suppression? The benefits to the older adult are expressed in the increased NRR as the matriline increases in size (Fig. 16.3). The imposition of reproductive suppression by the older female contributes to her fitness; it is her young that will use the resources available to the matriline rather than the young of another female even though that younger female is likely to be her daughter. The decision of the younger female cannot be divorced from the alternative of dispersal. Because of the central role played by dispersal, the reasons and conditions for remaining at home will be fully discussed in Chapter 18.

Briefly, the major reproductive strategy of female yellow-bellied marmots is to form a matriline with daughters or sisters as both survival and reproduction, and thus fitness, are enhanced.

Factors affecting reproductive success

The previous discussion emphasized the role of sociality as the major strategy for achieving reproductive success. However, within that sociality framework, other factors, such as the habitat site affect reproductive success (Table 16.11). The inability to form

Table 16.11 Variation in reproductive success among yellow-bellied marmot sites. Reproductive success was measured as number of yearlings divided by the number of weaned young. Sites are arranged from the smallest to the largest. Modified from Armitage (1988).

Site	Mean reproductive success
Boulder	0.170
Cliff	0.467
Marmot Meadow	0.458
River	0.477
Picnic	0.405
North Picnic	0.329
Kruskal-Wallis $H = 9.38$, $0.1 > p > 0.05$	

persistent matrilines at Boulder and North Picnic is associated with reduced reproductive success.

Within a matriline, females may nest communally; i.e., females wean litters in the same burrow. In 178 colony-years, 233 litters were weaned; 56 (24%) were communal. About 38% of the communal associations were littermate sisters, about 33% were mother:daughter pairs; about 24% consisted of a mother:daughter:sister trio, and about 5% were formed by non-littermate sisters. The benefits of these communal associations, as distinct from matriline formation, are unknown, but there is some evidence to indicate that communal nursing occurs (Armitage and Gurri-Glass 1994). Also, members of a communal group and their young share space and have low space overlap with other resident females. Thus, all members of the social group share resources and cooperate in defense against predators and conspecific intrusions. Several times I observed a communal group of females together rebuff an intrusive female from another matriline. By contrast, I observed single females retreat or eventually shift to another burrow as a response to intruders.

There is no evidence that adult females discriminate among the young; an advantage may be gained through the dilution effect in which the probability that any specific young will be predated is reduced with the increase in group size. Possibly females sequester their young from infanticide. This benefit would be most important after the young emerge as at that time they are most vulnerable to infanticide (Brody and Melcher 1985). Because females apparently accept any young into the social group up until the time of emergence (Michener 1974), transfer to or raising young in the communal nest eliminates the likelihood of infanticide by the adult females in the group. Finally, communal nesting may occur because of a shortage of quality nest burrows. The highest percentage of communal nesting occurred at two sites with limited nesting burrows (Armitage and Gurri-Glass 1994).

Is communal nesting costly; e.g., by reducing recruitment? I examined whether communal nesting enhances recruitment for 119 litters produced by females living alone and for 68 litters weaned by 31 communal groups. The communal groups recruited significantly more yearling females (1.52 per group) than single females (1.13 per female, $G = 15.2$, $p < 0.001$). However, the communal females recruited only 0.691 yearling females per litter, a rate significantly lower ($G = 4.6$, $0.05 > p > 0.025$) than that of single females. This difference in recruitment success is related to the social structure of the matriline. Recruitment is most likely when the mother or no adult female is present (see Chapter 18). If recruitment occurs from the litter of a female living alone, there will either be no adult female present (if the female died) or the female will be the mother. In a communal group, the mother may be absent or females may not recruit when parentage is uncertain and the yearling female may disperse. Single females compensate in part for their much lower gross and NRRs (Armitage and Schwartz 2000) with a recruitment rate about 61% higher than that of communal females. When recruitment occurs, a single female forms a mother:daughter matriline and gains the survivorship and reproductive advantages of matrilineal organization.

Although we argued that the major reproductive strategy of yellow-bellied marmots is sociality, i.e., matriline formation, cooperation and competition occur within the matriline and multiple factors affect the reproductive success of an adult female. Nuckolls (2010) used Hierarchal Linear Models, also known as Linear Mixed Models, to examine the effects of 38 environmental, demographic, and "intrinsic to the female" variables on annual and LRS. The following discussion is based on the analyses performed by Nuckolls.

The multivariate analysis of annual reproductive success revealed that three variables increased and two variables decreased the probability of reproducing (Table 16.12). For each covariate, the shape of the response is expressed when all other significant covariates are set to their means. Because partial effects can be difficult to interpret, it is instructive to compare univariate analyses with the multivariate analysis. In general, the univariate analyses agreed with the multivariate analysis for all of the covariates (Table 16.12); however, slopes differed (Nuckolls 2010). The univariate analysis for the effect of age revealed that the likelihood of reproduction increased to age 5, then remained constant. However, when partial effects of the other significant covariates were set to their means, the probability of reproducing increased to a maximum at about age 7, then decreased (Fig. 16.6). Note that the probability of reproducing was higher if the female reproduced the previous year. The univariate analysis indicated that reproduction by same-aged females always resulted in a probability of reproducing of 1.0, but the multivariate analysis revealed a gradual increase in the probability of reproduction as the number of offspring produced by same-aged females increased (Fig. 16.6). This relationship probably represents a situation in which conditions favored reproduction and all or nearly all females reproduced. The different shapes of the univariate and partial effects (multivariate) curves emphasize that the effect of one variable depends on the status of the other variables.

Table 16.12 Factors that significantly affected reproductive success of female yellow-bellied marmots. MA = multivariate analyses, UA = univariate analyses, + indicates the variable increases and − indicates the variable decrease the probability (from Nuckolls 2010).

Variable	Annual probability of reproducing		Lifetime probability of reproducing at least once		Total number of lifetime offspring	
	MA	UA	MA	UA	MA	UA
Age	+	+				
Presence of new male	−	−				
Reproduced previous year	+	+				
Number in matriline of:						
same-aged females	−	−				−
younger adult females		+	+			+
offspring of same-aged females	+	+				
older adults		−		−		−
Life span			+	+	+	+
Resident of natal colony			+	+		
Percent of years with resident male			+	+		

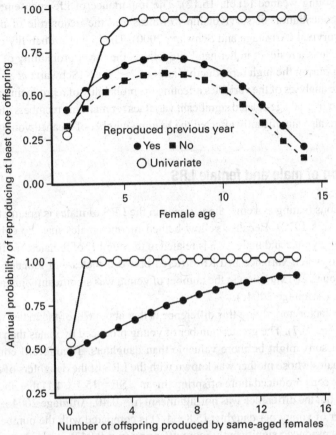

Figure 16.6 A comparison of univariate and multivariate (partial) effects of age and the number of offspring produced by same-aged females on the annual probability of reproducing. The curve for partial effects (solid symbols) represents the variable when the other four significant covariates are set to their means. Redrawn from figures in Nuckolls (2010).

The univariate analyses identified two additional significant factors: the probability of reproducing decreased as the number of older matriline adults increased and increased as the number of younger matriline adults increased (Table 16.12). Different analyses confirmed the effect of younger adults in the matriline (Table 16.4), but found no significant effect of the number of older females (Armitage 2003f). The effect of increasing the number of older females is slight. Many of these variables are correlated; e.g., number of older adults and number of same-aged adults (Nuckolls 2010). When entered into the multivariate model, number of same-aged females had a significant partial correlation and may act as a proxy for other variables. None of the environmental variables significantly affected the probability of annual reproduction.

The probability of reproducing at least once during a female's life span was significantly increased if she was a resident in her natal colony, associated with an established male, and had a long lifespan; lifespan was the major factor that determined the total number of young weaned (Table 16.12). The importance of life span emphasizes the significant contribution of survivorship to fitness and the major role of matrilines in increasing survival (Armitage and Schwartz 2000). The higher probability of reproducing if a female is a resident in her natal site reflects the lower probability of immigrants reproducing during the high reproductive ages of 2–7 years (Schwartz *et al.* 1998).

Univariate analyses of the variables affecting the probability of reproducing at least once added two additional age-related significant variables for matriline members (Table 16.12). Age structure also significantly affected the lifetime number of weaned young.

Comparison of male and female LRS

For polygynous mating systems, the variance in the LRS of males is greater than that of females (Trivers 1972). Because yellow-bellied marmot males may be either monogamous or polygynous and male LRS is related to the number of females, we investigated the degree to which variance in LRS differed between males and females. The mean number of young and variation in the number of young was significantly greater in males than females (Armitage 2004c).

The major factor underlying this difference is that about 6% of the males were highly successful (Fig. 16.7). The greater number of young produced by males than by females suggests that sons might be more valuable than daughters. Therefore, I compared the LRS of 24 males whose mother was known with the LRS of the daughters of those same mothers. The sons produced more offspring (mean $\pm$ SE = 15.1 $\pm$ 4.1) than the daughters (10.5 $\pm$ 3.3), but the difference was not significant (p = 0.40; Armitage 2004c). However, if the number of young per daughter (5.8 $\pm$ 1.7) is compared with the number of young per son, sons produced significantly more offspring (p = 0.047).

Both the higher LRS of males and the greater number of offspring per son than per daughter suggests that sons are more valuable than daughters. Theory suggests that the sex ratio of offspring should be biased toward the more valuable sex, i.e., the sex that would have the greatest effect on the mother's fitness (Clutton-Brock and Iason 1986). Because yellow-bellied marmot males produce more offspring than females, sex ratio should be biased toward sons; however, the sex ratio at weaning is 1:1 (Schwartz *et al.* 1998). Survivorship of males and females probably affects sex allocation. The best estimate of the mean age of first reproduction for both males and females is 3 years. It could be older for males as the sample size of known-age males was much smaller than that of females. However, by age 3, adult females outnumber adult males by 2.16 times; a number similar to 1.79 times more young produced by males (see Fig. 16.7). A female must produce two sons for each daughter to have the same probability that each sex will reproduce, and produce two surviving daughters for each surviving son to produce the same number of grandoffspring. When survivorship and reproductive output are combined, two daughters are equal to two sons and both sexes contribute equally to their mother's fitness and the population sex ratio is stable at 1:1.

Reproductive success of each sex is affected by the social dynamics and demography of the other sex. Female success decreases when no adult male is present or arrives late in

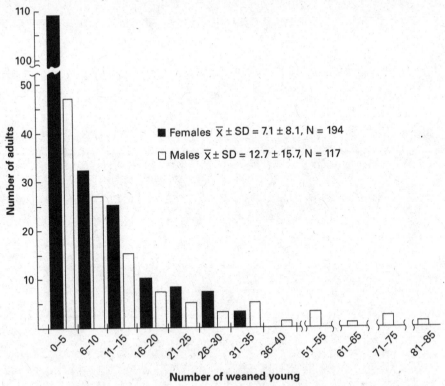

Figure 16.7   The relationship between the number of weaned young and the number of adult females or males producing that number of young. Young are grouped in intervals of five.

the active season and increases when the same male is present in subsequent years. Male success increases when female recruitment increases the number of reproductive females and decreases when reproductive suppression reduces the number of breeding females. Neither sex can affect these influences of the other sex.

Summary

Female yellow-bellied marmots are highly individualistic in that they strive to maximize direct fitness. Their fundamental strategy is to recruit daughters to form a matriline with its attendant benefits of increased survivorship and reproductive success. The benefits are enhanced if a female has a long life span and occupies a high quality habitat. The cooperative effects of matriline formation are countered by competitive effects which mainly are expressed in the age structure of the population and reproductive suppression of younger by older females.

Part V

Population dynamics

17 Basic demography

The numerous factors that affect demography have been variously studied among marmot species. Therefore, this chapter considers marmots broadly and the yellow-bellied marmot is emphasized for those topics for which extensive data sets exist. One major factor is social structure which determines population structure.

Age structure

Age structure in marmots is seldom stable because of variation in habitat quality among sites and years and demographic changes associated with changes in social structure. The basic demographic unit is the family or matriline; within this unit, age structure is an important variable. The proportion of young and adults varies among species (Armitage 1996b) and for any species varies among years or among geographic populations (Table 17.1). Over a 6-year study of the steppe marmot, the proportion of young and adults varied by 100%; the proportion of yearlings was inversely related to the proportion of young (Shubin 1991); age structure was similar in a black-capped marmot population (Yakovlev and Shadrina 1996).

Family composition varies in a population and differs between species (Table 17.1). The black-capped and steppe marmot distributions differ primarily in the presence of 2- and 3-year olds. The variation in age and family structure is likely a consequence of demographic processes that are based in the family and that are not correlated among families; e.g., black-capped marmots rarely have young in successive years. Thus, a stable age structure is unlikely (Shubin 1991, Schwartz and Armitage 1998).

Habitat quality affects age structure. In the gray marmots, mean age (juveniles excluded) is lower and more young-aged, mature adults occur in the favorable habitat (Table 17.1; Mikhailuta 1991). This difference in age structure probably relates to breeding opportunities for young adults; a higher proportion of families on the favorable habitat have litters, litter sizes are larger, and family size is about 50% larger. Probably younger adult females are more likely to disperse from families on unfavorable habitat to seek reproductive opportunities elsewhere.

Table 17.1 Variation in the population structure of marmots.

Age structure	Proportion (%)	
Age group	Among marmot species	Steppe marmot
Young	14–48	25.1–50.6 (x = 34.7)
Adults (3 years or older)	28–52	39.4–67.3 (x = 51.5)
Family composition	**Steppe marmot**	**Black-capped marmot**
Adults and young	44.7	53.8
Adults and yearlings	28.5	15.4
Adults, yearlings, and young	19.5	7.7
Adults with 2- and 3-year-old offspring		15.4
Adults with non-offspring 2- and 3-year old		7.7
Habitat quality	Mean age (years)	Proportion (%) over 5-years-old
Favorable	2.9	15.9
Unfavorable	3.2–3.9	26–31

General characteristics

The *M. flaviventris* and *M. marmota* (Farand *et al.* 2002) have Type II survivorship curves (Krebs 1985), which implies a constant survival rate after age 1 year. Partial survivorship curves are available for *M. olympus*, *M. vancouverensis*, and *M. caudata*. These curves are more similar to a Type I curve; all curves show higher survivorship than *M. flaviventris* at all ages greater than 1 year (Armitage 1998). This difference in survivorship curves is a consequence of delayed dispersal to age 2 or older in the three species with a family social structure. There is a trade-off between reproduction and survival. Whereas *M. flaviventris* reaches reproductive maturity at age 2 but has higher mortality associated with earlier dispersal (Van Vuren and Armitage 1994a), the other species with higher survivorship do not reach reproductive maturity until age 3 (Table 5.1). There is no difference in survivorship between the sexes in the three family species in contrast to the higher survivorship of female than of male yellow-bellied marmots (Blumstein *et al.* 2002).

For yellow-bellied marmots, life expectancy at birth was 1.4 years for males and 1.9 years for females. Generation time of females at 4.49 years was 2.4 times greater than life expectancy at birth. As can be deduced from the survivorship curve, sex ratio at birth is 1:1, but is female-biased at all ages of 2 years or older (Schwartz *et al.* 1998, Schwartz and Armitage 2003). A stable age distribution, a common assumption in population modeling, was not present in the East River Valley population (Schwartz and Armitage 1998). A chi-square analysis revealed that proportions in age classes were not equal. For example, the proportion in the 0 (young) age class varied from 10.6% to 49.4%. When the proportion of young was high, such as in 1980, a high proportion of yearlings occurred in 1981; of 2-year-olds in 1982, of 3-year-olds in 1983, and of 4-year-olds in

1984. In other words, the effects of reproduction in a given year could be traced through the age classes in subsequent years. A very poor reproductive year could result in zero values in subsequent years for ages 2 years and older.

Change over time

Yellow-bellied marmots maintain relatively stable average numbers through time (Armitage 1991a), but numbers fluctuate among years and differ among sites (Fig. 17.1). The number of adult females is strongly affected by the number of recruited 2-year-old females which is significantly correlated with the number of female yearlings (Armitage and Van Vuren 2003); recruitment is significantly related to the number of young produced (Armitage 1984). In effect, the number of female young born at a site in a year strongly influences the number of yearlings the next year which, in turn, affects the number of 2-year olds in the third year and ultimately, the number of adults.

The mean population numbers for each of the four age groups were always larger at Picnic and always smaller at Boulder (Table 17.2). All differences were significant: Picnic > River > Boulder, except the mean number of yearling and 2-year-old females did not differ significantly between Picnic and River. The sequence of mean numbers is identical to the sequence in areas. This result is not surprising; larger areas are expected to have more residents as long as there are no major differences in habitat quality. Large area is a necessary but insufficient requirement for larger population size; e.g., North Picnic, the largest locality, has fewer adult females and smaller matrilines than much smaller areas (Table 8.1).

Variation among localities

Population age composition varies considerably among sites and years (Fig. 17.1).The number of individuals in any of the age groups is not correlated among sites (r varied from −0.130 to 0.383, all p > 0.05); population numbers at a locality change independently of changes at other sites (also see Armitage 1977). Mechanisms that affect population numbers also vary among sites; e.g., the lack of a resident male affects the production of young and a high number of yearlings in one year results in an increase in the number of 2-year olds the following year (Table 17.3).

The proportion of individuals in one age class present in the next age class varies among the three localities. The smallest proportion of young survives to become yearlings at Boulder and the largest proportion of young to survive as yearlings occurs at River. There is little difference among the three sites in the ratio of yearlings to 2-year-olds and in the ratio of 2-year-olds to older females, although there is a progressive increase in the ratio from Boulder to Picnic to River. These differences suggest that demographic processes vary among sites and these differences will be explored primarily in Chapter 19.

As a consequence of the low number of yearlings and 2-year-olds at Boulder, the average proportion of young at Boulder (65.8%) is much higher than the proportion of young at

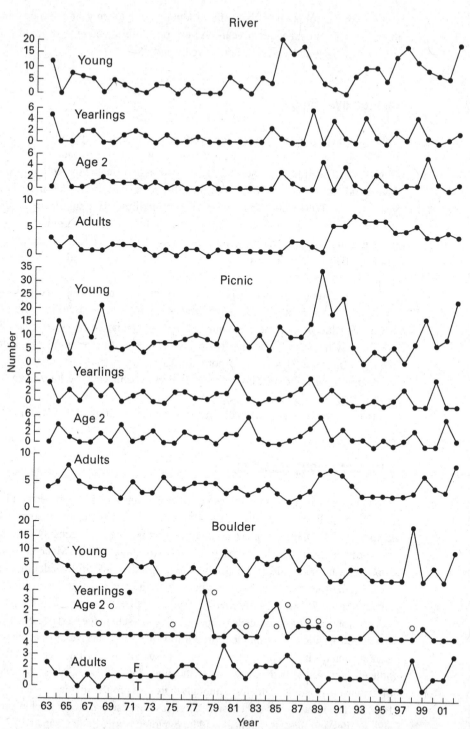

Figure 17.1   Population changes through time for residents at a down-valley colony (River), an up-valley colony (Picnic), and a satellite site (Boulder). Because population dynamics is primarily a function of the change in the number of females, except for young, only female numbers are reported.

Table 17.2 Mean population numbers of residents at three marmot localities. Values are mean ± standard error, N = 41 years for all sites.

Age group	River	Picnic	Boulder
Young	6.3 ± 0.93	10.4 ± 1.2	4.0 ± 0.72
Female yearlings	1.2 ± 0.26	1.4 ± 0.22	0.38 ± 0.14
2-year-old females	1.0 ± 0.24	1.4 ± 0.23	0.40 ± 0.13
3 years or older females	2.8 ± 0.36	4.5 ± 0.26	1.3 ± 0.15

Table 17.3 Variation in the timing among colonies of mechanisms that affect population numbers.

Colony	No resident male	Late male
River	1991	1964, 1977
Boulder	1967, 1968, 1970, 1991	1975
Picnic	Never occurred	Never occurred
	High number of yearlings	Increase in number of 2-year-old residents
River	1984	1985
	1988	1989
	1990	1991
Boulder	1978	1979
	1985	1986
Picnic	1981	1982
	1988	1989
	2000	2001

River (55.7%) and Picnic (58.7%). The proportion of young in these three localities is generally higher than that reported for other species (Armitage 1996b) and more similar to the values reported for the steppe marmot (see above). Most reports of age composition are a snapshot in time and likely do not represent long-term means. As indicated by their standard errors (Table 17.2), there is considerable year-to-year variation (also see Schwartz and Armitage 1998). Only long-term demographic studies of the same populations can reveal the variation and means of the age composition of marmot populations.

The frequency of female reproduction affected age distribution. At River in 1968, there were no young; two 2-year-old females were present, but neither ever reproduced even though they lived to ages 3 and 6. In 1978 and 1979, the resident female did not reproduce, but in 1980 when a new male was resident, she weaned a litter in that and the subsequent year. Why some females never reproduce or reproduce rarely (see females 843, 976, and 931 in Fig. 17.3) is unknown.

Immigration (see Chapter 18 for immigration rates) affected the age composition primarily by affecting the number of 2- and 3-year-old adults. At Boulder, 2-year-old immigrants were present in the year after a year in which there were no yearling residents; 1975, 1998 (Fig. 17.1). In 1968, a 2-year old was a late immigrant; there were no yearlings the previous year. Boulder is characterized by a high level of immigration (Fig. 17.2); 53.6% of resident adult females were immigrants. By contrast, only 15.4% of resident adult females at Picnic were immigrants (Fig. 17.3).

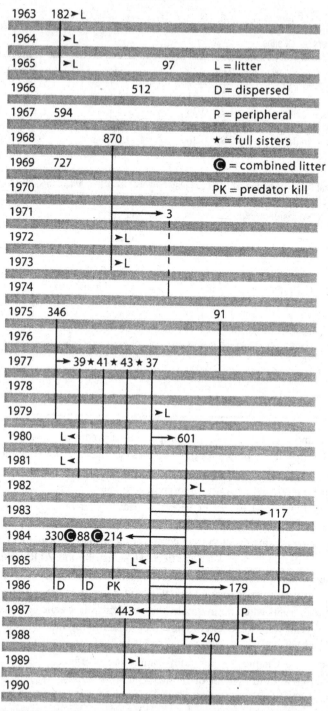

Figure 17.2 Residency chart for Boulder. The marmot identification number is recorded in the first year of residency. When a female was resident for more than one year, the years of residency are indicated with a vertical line. Female recruits are recorded in the year of birth; a horizontal line that crosses two vertical lines indicates that the litters were intermingled (combined) and the young could not be assigned to a specific female.

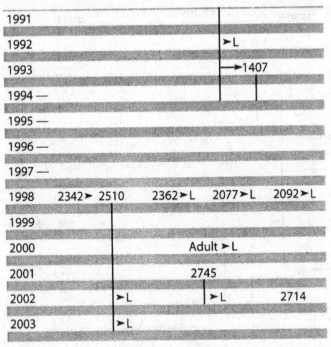

Figure 17.2   (Cont.)

Population change affects average relatedness among female colony members. For example, at Picnic in 1991 when several matrilines were present and population numbers were high, average *r* was 0.27; when the population declined in 1996 to one matriline, average *r* increased to 0.46.

Predation and the timing of mortality

Changes in all marmot populations are similar to those affecting mammalian species in general. In yellow-bellied marmots the resident female population is affected by recruitment and immigration (Figs. 17.2, 17.3) and by mortality. One potential source of mortality is social conflict. Between 1982 and 2001, we recorded 14 individuals with wounds, such as large cuts, slashes, or punctures. Seven individuals survived and seven disappeared and were presumed to have died. A marmot may also lose all or part of its tail during conflict; of nine individuals with partial tail loss, four disappeared (died or dispersed) and five were successful residents. Four individuals with limb injuries failed to survive through the subsequent hibernation. Infections were detected in four individuals; two survived. These sources of mortality have little demographic effect, but obviously impact individual fitness.

The primary causes of mortality are predation and death during hibernation (Armitage 1996b). Marmots appear to minimize predation risk by performing most behaviors close to refugia, especially the home burrow area. Overall, 50% of all activity is close to a

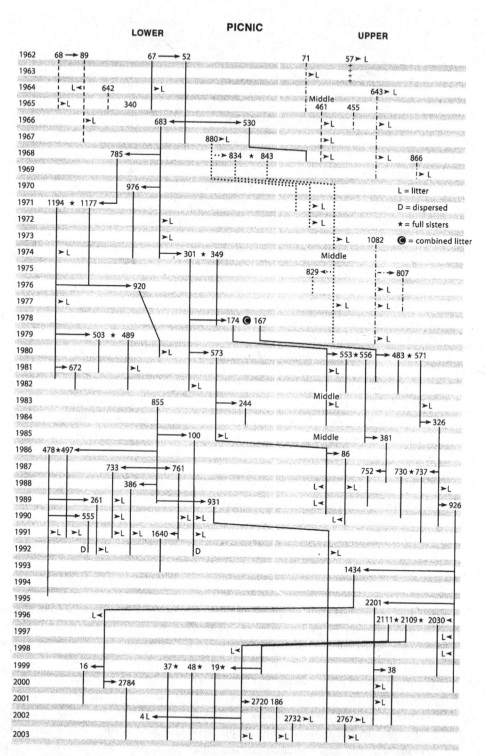

Figure 17.3   Residency chart for Picnic. See Fig. 17.2 for explanation of symbols and lines.

burrow (Blumstein 1998). Foraging is the riskiest behavior and marmots reduce risk through alert behavior and by foraging as close to a burrow as possible. For that purpose, marmots dig flight burrows close to or in foraging areas.

It is difficult to discriminate between death during hibernation and undetected predation when it occurs late in the season prior to immergence or early in the season shortly after emergence. For example, we reported high hibernation mortality in yellow-bellied marmots (Schwartz *et al.* 1998) in contrast to low hibernation mortality in the Olympic marmot in which no radio-implanted animal 1-year old or older died during hibernation (Griffin *et al.* 2008). In the Vancouver Island marmot predation probability was highest in August and September prior to immergence and probability of death during hibernation was very low (Bryant and Page 2005). Peak consumption (determined by fecal analysis) of Olympic marmots by coyotes, a major predator, were in September (Witczuk *et al.* 2013). Hence predation mortality may be underestimated and hibernation mortality, overestimated because late season predation was undetected.

The timing of disappearances and mortality for resident yellow-bellied marmots for all known cases was determined for the entire study period. Over the 41-year period, 140 mortalities and 13 disappearances were detected (Table 17.4). An additional 32 predation mortalities occurred among dispersers. The time of many predations was not explicitly recorded or was reported as midsummer, early spring, etc. However, predations occurred before late summer except for four badger predations. All dispersal mortalities and the 17 hibernation mortalities were detected by radiotelemetry (Van Vuren 2001).

The cause of death for those found dead (N = 29) is unknown except for an adult female, who had nerve damage that was probably a result of social conflict

Table 17.4 Time of mortality of yellow-bellied marmots. D = found dead; Dis = disappeared; I = infanticide. All other active season mortalities were caused by predators: P = predator unknown, B = badger, Br = bear, C = coyote, E = golden eagle, M = marten, W = long-tailed weasel. Bold = dispersers. The number before a letter is the number of mortalities.

| | Adult | | Yearling | | Young | | |
	Male	Female	Male	Female	Male	Female	Unknown
May	P	C, P					
June 1–15	D						
Spring or early summer	**C, P** P, D	3P, C		P			
June 16–30	P	3B, C	2C, B, D	4B, 2C, W	2B	2B	W
July 1–15				B, 3C	D	2D	
Midsummer	2P, E	2Dis	C, 2D		Br	Dis	
July 16–31	2B, D	3B, D		B	3B, 4D, Dis	9B, C, 4D, Dis	I, 2B, 6D
August 1–15		D		3P, Br	2D, P		M
August 16–31		3C	2D, P				
Late summer	**2P, B**	2B			12B, C, 2P, 5Dis	7B, P, 3Dis	
September		B		**Br**	2B	B	
Summer, date	**4P, Br**	P, C	**P, 4P, B**	**P, 8P, B**			
Unknown	B, C, 2Br		**5C, E, M**	**4C, 2B, M**			
Hibernation	2	1, 4	2, 3	2, 2		1	

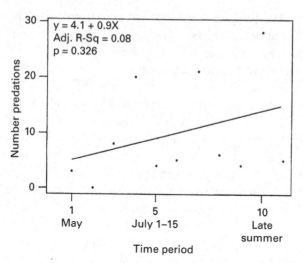

y = 4.1 + 0.9X
Adj. R-Sq = 0.08
p = 0.326

Number predations

1 May   5 July 1–15   10 Late summer

Time period

Figure 17.4   The relationship between the number of predation events and time.

which resulted in loss of use of her hind legs, and a young, who was injured by a dog seen chasing a group of young. An adult female emerged from her burrow, walked several meters, and dropped dead. Disappearances during the active season are likely caused by predation; several occurred when predators were known to be active in the area and predation was confirmed for other members of the group. However, two adult females disappeared when they were involved in social conflict; both were wounded and may have dispersed to avoid further conflict or died from their wounds.

There is no indication of an increase in predation late in the summer except for a major incident of badger predation (Armitage 2004a). The regression of predation mortality against time is positive, but not statistically significant (Fig. 17.4). Predation did not change with time for adults ($p = 0.8$, adj-$R^2 = 0$) or yearlings ($p = 0.77$, adj-$R^2 = 0$), but increased with time for young ($p = 0.15$, adj-$R^2 = 0.13$). Thus, most of the increase in mortality after mid-July can be attributed to predation on young who are active above ground primarily from mid-July through August. Furthermore, an analysis of coyote scats revealed that marmots were most susceptible during July (Van Vuren 1991). Total mortality was highly correlated with predation mortality for the entire population ($r = 0.94$, $p < 0.01$), and for each of the age groups (all $r > 0.95$, $p < 0.01$). These relationships strongly support the interpretation of Van Vuren (2001) that most mortality during the summer is caused by predation.

The predators

Both aerial and terrestrial predators utilize marmots as prey. Golden eagle (Fig. 17.5) predation was observed or documented for *M. flaviventris* (Van Vuren 2001), *M. vancouverensis* (Bryant 1996a), *M. marmota* (Pigozzi 1989, Lenti Boero 1999), *M. caligata* (Murie 1944) and probably occurs wherever eagle and marmot distributions overlap (Bibikow 1996). Yellow-bellied marmots were the most important food of nestling

Figure 17.5   The golden eagle (*Aquila chrysaetos*) preys on marmots wherever the two species co-exist.

eagles (Marr and Knight 1983) and may be more important in eagle diet when populations of other prey are depressed (Knight and Erickson 1978). Other raptors, such as the
red-tailed hawk, the steppe eagle (*Aquila nipalensis*), and the rough-legged buzzard
(*Buteo lagopus*) as well as some owls also may prey on marmots (Bibikow 1996). No
predator can specialize on marmots and some predators, such as wolves, may turn to
marmots when their primary prey is scarce (see Armitage 2003, for further discussion of
marmot predation in North America).

Wherever marmots occur there generally is at least one canid, such as the coyote, one
felid, such as the cougar, and mustelids, such as marten, weasels, and polecats. Yellow-
bellied marmots form the major item of coyote diet in the high ranges in Teton National
Forest (Murie 1940) and probably are an important part of coyote diet wherever the two
species co-occur (Van Vuren 1991). In North America, the badger is a major predator.

Figure 17.6   A red fox (*Vulpes vulpes*) carrying a freshly killed young yellow-bellied marmot (*Marmota flaviventris*). Photo courtesy of D. Inouye. (See plate section for color version.)

The red fox (*Vulpes vulpes*) seems to occur wherever marmots occur and can cause significant mortality on local populations. In recent years a red fox family established residency at the RMBL, which is also home to a marmot colony. The fox caused significant mortality among young marmots (Fig. 17.6). Wolves prey on marmots (Fryxell 1926), but their importance currently is minor because of their widespread extirpation in North America. However, in the Kazakhstan Mountains, some wolf families primarily feed their cubs on marmots (Bibikow 1996).

When environmental conditions change, the impact of predators on marmot populations can increase and cause extinction of local populations. Predation is the major cause of mortality of the endangered Vancouver Island marmot where wolves, cougars, and golden eagles caused a precipitous decline in the population (Bryant and Page 2005). The predation impact on the marmots was probably facilitated by landscape changes created by modern forestry practices. These predators are a major obstacle in the efforts to reestablish the marmot populations. Introduced marmots suffer higher predation mortality than wild-born marmots; as a consequence, survival of captive-born marmots was much lower than that of wild-born marmots (Aaltonen *et al.* 2009). Olympic marmot populations have also declined recently. Many of the colonies described by Barash (1973a) are extinct; recent surveys indicate that local extinctions are widespread and no recolonizations were detected (Griffin *et al.* 2008). The major cause of the population decline is increased predation on adult females by coyotes (Griffin 2007). Coyotes likely reached the higher elevations of the Olympic Peninsula in the 1940s; hence it is a relatively recent predator of Olympic marmots. The invasive coyotes are widespread and prey on the marmots throughout their active season (Witczuk *et al.* 2013). The coyote population is maintained by an abundance of alternative prey; thus coyote predation on Olympic marmots is substantial, is responsible for the drastic decline in numbers of Olympic marmots, and could drive this species to extinction (Griffin 2007, Witczuk *et al.* 2013).

Major weather events

Although weather variables affect survival and reproduction of various age classes of yellow-bellied marmots, the adult age class is less vulnerable to annual fluctuations in temperature and precipitation (Schwartz and Armitage 2005). However, severe weather events not only affect reproduction and survival of all age classes, they also affect social organization and dynamics that are expressed over several years (Armitage 2003c).

The grouping of study sites into DV and UV areas was chosen to coincide with patterns of snowmelt, which occur later UV (Van Vuren and Armitage 1991). For each of the two areas four demographic variables were calculated (Table 17.5). The total number of individuals includes immigrants and transients, because their numbers should be related to regional reproduction and survival.

The mean values of the demographic characteristics for 1985–90 are very similar in the two areas and serve as a baseline for evaluating subsequent demographic patterns. After 11 years of population change, only the mean number of adult females differed

Table 17.5 Demographic variables for Down Valley and Up Valley populations following major weather events. N = number, L = litter, f = female.

	Down Valley				Up Valley			
		Number of adult	Young			Number of adult	Young	
Year	Total N[a]	females	N/f	N/L	Total N	females	N/f	N/L
1985–90[b]	70	15	2.2	4.0	79	15	2.3	3.7
1991[c]	64	22	1.1	2.8	83	15	2.6	4.9
1992	88	20	2.4	5.3	46	11	1.0	3.7
1993	79	20	1.4	3.1	23	7	0.7	2.5
1994	51	22	0.8	3.6	25	6	1.8	5.5
1995[d]	57	18	1.2	4.4	21	4	0.8	3.0
1996	82	15	3.0	4.5	31	6	3.2	3.8
1997	92	18	2.9	3.8	40	8	2.0	4.0
1998	82	19	1.9	4.5	75	15	1.8	3.0
1999	53	16	0.9	4.7	77	12	3.5	4.7
2000	43	12	1.4	4.3	71	10	3.1	5.2
2001	33	9	1.7	2.5	62	14	1.6	4.6
1991–2001[b]	65.8	17.4	1.7	3.96	50.4	9.8	2.01	4.08

[a] Includes all age groups, residents, dispersers, and transients
[b] Mean values
[c] Year of late summer drought
[d] Year of late snow cover

Table 17.6 Differences in mean demographic variables and their correlations between the Up Valley (UV) and Down Valley (DV) areas. For r, all $p > 0.05$.

Variable	r	Differences between areas
Total number	−0.228	$t = 0.8$, $p > 0.4$
Number adult females	−0.204	$t = 4.4$, $p < 0.001$
Young/female	0.030	$t = 1.1$, $p > 0.2$
Litter size	−0.202	$t = 0.3$, $p > 0.5$

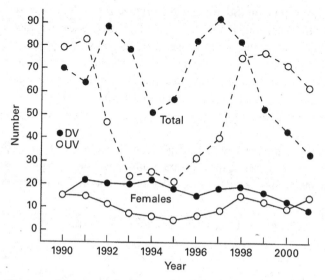

Figure 17.7 Demographic patterns for down-valley (DV) and up-valley (UV) sites following major weather events.

significantly between the two areas (Table 17.6). Although there was little difference in mean demographic values, there was considerable difference in the pattern of demographic change (Fig. 17.7). These patterns are a consequence of two unusual weather events interacting with other typical events that affect demography.

A late summer drought occurred in 1991. Population parameters were typical except reproduction was low DV because no male was present at one of the two major colonies. The drought caused a significant decline in overwinter survival UV of reproductive females and young (Armitage 1994), which resulted in a decrease in total population and adult female numbers (Table 17.5). Cliff Colony went extinct and was settled by an adult male in 1992 who was the only marmot there until he dispersed in early 1995. The drought affected the UV marmots because young were weaned on average 8 days later than DV and the young and reproductive females had less time to grow before the drought-induced quality of the vegetation declined (Fig. 17.8). Late summer dryness is common, but in most years a sufficient number of patches of green vegetation remain to meet marmot requirements (Fig. 17.9). Thus the young and reproductive females weighed much less than non-reproductive marmots at the time of hibernation

Figure 17.8   A meadow foraging area in late summer when drought greatly reduced the amount of palatable vegetation. (See plate section for color version.)

Figure 17.9   A patch of green vegetation, mainly cinquefoil, in late summer.

Figure 17.10   Picnic Colony on June 19, 1995, showing extensive snow cover.

(Armitage 1994, Lenihan and Van Vuren 1996a); non-reproductive animals, who began mass gain about 3 weeks earlier than the young and reproductive females, survived.

Population decline continued because of poor reproduction; e.g., the high number of adult females DV included numerous non-reproductive 2- and 3-year olds, who are especially vulnerable to reproductive suppression (Armitage 2003f, 2007). Some females UV never reproduced or lost litters prior to weaning, thus the number of young/female remained low (Table 17.5). Because the two areas experienced somewhat different causes of demographic change, none of the four demographic variables was significantly correlated over 11 years (Table 17.6).

Prolonged snow cover in the spring of 1995 led to higher mortality and reduced reproduction. In that year, June 10 was the latest date on record for 50% snow cover DV; a precise date was not determined for UV, but it was more than 50% snow covered on June 19 (Fig. 17.10). The DV sites were moderately affected; two satellite sites (Bend and River Annex) went extinct; only 28% of the females reproduced. Although another site (Falls) went extinct the following year, the population increased markedly (Fig. 17.7, 1996) because of a high rate of reproduction (Table 17.5).

Snow cover affected UV demography similarly; four sites (Beaver Talus, Boulder, Cliff, and North Picnic) went extinct. Recruitment of young females at Picnic and Marmot Meadow Colonies was responsible for recovery of the UV population. The low rate of reproduction in 1998 and 2001 occurred because a large number of non-reproductive 2-year-old recruits maintained a high number of adult females but made little contribution to the total number of marmots, which declined slightly despite the large contribution from Boulder (Fig. 17.1). This pattern further emphasizes the degree to which individual sites act independently in the larger metapopulation.

The drought of 1991 and late snow cover of 1995 affected matriline structure at Picnic (Armitage 2003f). In 1991, five matrilines occupied the entire habitat patch. One female emigrated and was trapped at Cliff Colony the next year, but soon disappeared. Emigration in response to drought may be common (Anthony 1923, Rudi *et al.* 1994). By midsummer 1995, two matrilines remained, a single female that never reproduced and a mother:daughter pair (926 and 1434, Fig. 17.3) that occupied a small area within the site. In the absence of immigrants, these two females and their daughters spread over the entire site over the next 6 years and eventually split to form two matrilines (Fig. 17.3). These females were all descendents of a female who was resident in 1962. This 41-year genealogy on one habitat patch is the longest known for any marmot species.

The UV and DV populations followed different trajectories. Although the UV population increased and maintained high numbers after the year of late snow cover, the DV population, which suffered only a minor decline and quickly recovered, declined after 1997. This pattern raises the question of why the population declined in the absence of any obvious external factors. Several stochastic events converged to promote the decline. First, reproduction was poor in 1998 and 1999 when potential reproduction by a large number of young females was likely suppressed by the older females. Second, no male was present at a major colony, Gothic Townsite (RMBL), in 1999. Third, badger predation reduced the number of adults, yearlings, and young in 1999 and 2001. Badger predation and the absence of an adult male are unpredictable, but these two factors depressed the DV population when it otherwise was expected to increase or remain stable.

A major characteristic of the recovery process was the failure of marmots to recolonize several sites. In part, this failure occurred because virtually no yearling female dispersed from the major colonies. Bend/Falls, River Annex, and Beaver Talus remained uncolonized in 2001, a reproductive female did not appear at Cliff until 2001, and Boulder and North Picnic were not colonized by reproductive adults until 1998 and 1999, respectively. Recolonization of the extinct sites was slow which suggests the population decline in 1995 was widespread. The slowness of recovery has important implications for possible effects on marmot populations stemming from climate change (see Chapter 21).

Demographic mechanisms of growth and decline

The total number of marmots in the UV population was significantly correlated with the number of adult females ($r = 0.92$, $p < 0.01$) and the correlation was positive in the DV population ($r = 0.51$, $r_{0.05} = 0.58$). These correlations suggest that adult females are excellent indicators of demographic change. However, numbers of adult females do not indicate which life-history variables are important determinants of population growth. Therefore, it is important to identify the demographic factors that contribute to population change.

We used prospective and retrospective perturbation analyses to determine the relevant demographic mechanisms of population growth rate. Prospective methods; e.g., sensitivity and elasticity analyses, explore the functional dependence of projected population growth rate on demographic variables and quantify likely changes in population growth rate in response to proportional (elasticity) or small absolute (sensitivity) changes in demographic variables. These procedures quantify potential influences of demographic variables on population growth rate but do not consider observed changes. On the other hand, retrospective analyses, especially the analysis of life table response experiments (LTRE), ask how variation in demographic variables expressed themselves in observed changes in population growth rate. (An extensive discussion of these procedures and methods of analysis are available in Oli and Armitage (2004).)

The four demographic variables, age of first reproduction, juvenile survival rate, adult survival rate, and fertility rate, exhibited substantial temporal fluctuations (Oli and Armitage 2004). As a consequence, projected population growth rate varied; it was positive (> 1) in 12 years and negative (< 1) in 25 years; the overall mean was 0.96 ± 0.20. Delayed age of first reproduction generally characterized years of negative growth rate.

In the elasticity analyses, age of first reproduction and annual juvenile survival rate were the most important factors affecting population growth rate and their relative magnitudes did not differ between years characterized by negative or positive population growth. In the LTRE analyses, demographic variables contributed to population growth in opposite directions; e.g., age of first reproduction could be positive when the other variables were negative. As a consequence, projected population growth rate remained essentially unchanged. Overall, age of first reproduction and annual fertility rate made the largest absolute contributions to growth rate. As suggested by the pattern of population changes (Fig. 17.1), pairs of colonies within a temporal period had substantially different population growth rates. This result emphasizes that the local populations follow their own trajectories, but of great interest is that these trajectories are influenced by the same demographic variables. Only this long-term study could capture the complexity of the multiple factors acting on population dynamics.

The contributions of the demographic variables to population growth differed in the two analyses: elasticity analysis identified juvenile survival as making the largest contribution whereas LTRE identified age of first reproduction and fertility as making the largest contributions. And importantly, LTRE distinguished between population growth and decline. Fertility followed by age of first reproduction made the largest contributions to decreases in population growth and fertility, juvenile survival, and age of first reproduction made almost equal contributions to increases in population growth. Of special significance is that the age of first reproduction, which is mainly determined by social behavior, is a critical factor in population growth or decline. High population growth occurred only in years characterized by early reproductive maturity; large increases or decreases in population growth were frequently associated with substantial changes in the age of first reproduction. This analysis provides a direct link between social behavior in the matrilineal structure of marmot populations and population dynamics. In addition, the inclusion of fertility indicates that reproductive variables rather than survival (adult survival had almost no affect on population growth) drive population growth (Oli and Armitage 2004).

Role of disease

A wide range of diseases and parasites were reported for marmots, but in most cases there was little evidence that disease affects the size of marmot populations. Their incidence and distribution were summarized by Young and Sims (1979), Bibikow (1992, 1996), and Bassano (1996). Viral and bacterial diseases are little known and often seem to be localized. Viral hepatitis is patchily distributed, woodchucks from north-central Maryland, central New Jersey, and southeastern Pennsylvania harbored the virus, but it was not detected in New York or New England. The virus was not found in woodchucks less than 3 months old, but infection rate was similar to that of adults by 7 months (Tyler *et al.* 1981). Significant genetic heterogeneity was detected between woodchucks from New York (virus absent) and Delaware (hyperepidemic) at two loci. Allele 2 for peptidase with glycyl leucine-4 was much higher in the Delaware population and two alleles for phosphogluconate dehydrogenase were present in the New York population, but only one allele was detected in the Delaware population. Because only these two of 22 loci significantly varied between the two populations, these differences may be a response to hepatitis infection (Wright *et al.* 1987). The virus was not detected in yellow-bellied marmots (Armitage 2003h) nor in alpine marmots (Cova *et al.* 2003), but a hepatitis virus was reported in the steppe marmot (Pole 2003). Leptospirosis antibodies were detected in 9% of 153 woodchucks in central New York (Fleming *et al.* 1979) and rabies was reported (67 cases from 1971–84) in woodchucks during an epizootic of rabies in raccoons in the mid-Atlantic states (Fishbein *et al.* 1986).

Two questions arise from the reported incidence of these diseases. First, is the local distribution and limited number of pathologies a consequence of marmot insensitivity to bacterial and viral infections or a lack of epidemiological investigations? A lack of investigations is considered the more credible hypothesis (Bassano 1996). Second, does disease incidence have a major impact on marmot demography? For example, white muscle disease was reported in a woodchuck population in central New York where 13 of 24 individuals trapped in September showed various degrees of the disease (Fleming *et al.* 1977). This paper, as in virtually all reports of disease, provides no data on mortality. The absence of viral hepatitis in some marmot populations and the low infection rate of plague in marmots (the average number in a focus does not exceed 2–3% of the examined animals, but the proportion may reach 22% of the animals procured from burrows harboring infected animals, Bibikow 1996) suggest that disease does not have a major demographic impact.

More likely disease impacts are localized. Four Vancouver Island marmots hibernating in a subalpine meadow died from a bacterial infection (Bryant *et al.* 2002). A survey of pasteurellosis in eastern Mongolia reported finding 300 corpses over 50 km stretch of territory (Bibikow 1996), but Bibikow does not report the population size of the tarbagans where the mortalities occurred. Although high mortality may be associated with infections, the infections may have been important because of other factors, such as lack of food (Bibikow 1996). Plague epizootics occurred in alpine gray marmots after periods of low summer temperature and surplus rainfall; in arid high mountains, epizootics occurred after several dry summer seasons. Both of these weather conditions decrease food availability, which suggests that marmots in poor condition are more susceptible. When 108 gray marmots were experimentally infected with plague in August, 21% died before hibernation, another 6% showed infection in the following spring, and 73% survived. The marmots were more resistant during hibernation. Resistance seems to be associated with mass gain; i.e., the marmots are in better condition (Bibikov 1992). There is just enough tantalizing evidence to suggest that bacterial and viral infections may be associated, in part, with environmental stress. Clearly, the role of disease in marmot biology requires considerable additional research.

Marmots and plague

Some insight into the importance of disease in marmot biology can be gained from the studies of marmots and plague. The plague organism *Yersinia pestis* occurs in discrete geographic foci. The natural focus is an area in which a set of interspecific relationships evolved that consists of the disease agent, its vectors, and the animal recipients (susceptible species) and reservoirs (highly resistant species) (Nelson 1980). The most extensive studies of plague foci and marmots were conducted in Asia (Bibikow 1996).

Plague in Eurasian marmots affects their demography and human health. Four species of marmots are the major hosts of plague: *M. caudata*, *M. bobak*, *M. sibirica*, and *M. baibacina* (Bibikow 1996, Pole 2003). In the late nineteenth and early twentieth centuries, severe plague epidemics were attributed to human contact with infected marmots. Local people had long associated plague epidemics with marmot hunting. Anti-plague stations were opened and long-term inspection of marmot populations in the high mountainous areas of Central Asia and Kazakhstan led to the description of plague foci: Tien Shan, Alai, Gissar, and Talasi (Pole 2003). Further foci were found, such as the Kokpak mesofocus, an area of 100 000 ha. The epizootic developed in house mice when marmots were hibernating and plague was found later in the gray marmots and their fleas as well as in 8 mammal species and 10 species of arthropod parasites (Ageev *et al.* 1997).

Several species of fleas were described from marmots (Bassano 1996); *Oropsylla silantiewi*, *Rhadinopsylla liventricosa*, and *Ceratophyllus lebedewi* transmit plague and maintain the plague organism during winter. Fleas infected with plague can survive for up to 420 days in abandoned marmot burrows at low temperature (Bibikov 1992). Although as many as 8–21 species may be found in a plague focus, generally, one species of flea is predominant. *O. silantiewi* was the dominant flea on the long-tailed and gray marmots from 2600 to 5000 m in the Pamir and Tien Shan Mountains, whereas *C. lebedewi* predominated on long-tailed marmots between 2250–3800 m in the Alai and Gissar regions. *Pulex irritans* was abundant along with *C. lebedewi* only in Talas between 2250 and 3200 m (Ageev and Pole 1996). The distribution of the fleas was more likely explained by their environmental requirements than by marmot populations.

Plague in marmots may not be a serious threat to humans. Apparently humans become infected by contact with infected marmots, but the disease can be transmitted to other humans. One epidemic in 1942 resulted in eight deaths from pneumonic plague. The epidemic started when a horse herder became ill (horses and other livestock frequently graze in meadows where marmots live). He decided to go home and a friend traveled with him. The horse herder died before reaching home and his infected companion continued the journey, became ill and infected two women who cared for him. All three died as well as four others with whom the sick persons had contact. Eight died in an outbreak in 1942, but there were only seven reported cases in central Tien Shan in the next 54 years (Aikimbayev *et al.* 1997).

Epidemics in humans were common; 190 were recorded in Mongolia from 1866 to 1957 and 84 for Transbaikalia from 1863 to 1930 (Pole 2003). Although most plague epidemics were associated with minor loss of human life, the Manchurian plague of 1910–11 caused 60 000–100 000 deaths. Because marmots were considered the main source of infection, the basic efforts of suppression were directed to the elimination or extensive reduction of marmot populations. The technique was tried for the first time in Transbaikalia where by 1956 the tarbagan was virtually exterminated (Pole 2003). An epizootic in marmots was not reported for 15–20 years. However, after 10 years, plague was found in a population of *Spermophilus dauricus*, which had become the dominant species in the biocenosis. In the Tien Shan and Alai plague foci, red and gray marmots were destroyed on an area of 8200 km^2. Marmot numbers were reduced to 2–5 animals/km^2, but the plague epizootic was suppressed for only 5–15 years.

The next phase of suppression used DDT to kill fleas in marmot burrows over 19 700 km^2, which was about 80% of the area of the focus. The number of fleas on marmots and in their burrows was greatly reduced. After 10–12 years, fleas reached 20% to 50% of their initial numbers. Fleas recovered much more slowly in the alpine zone. Plague epizootics occurred after 8–23 years in various foci. In the Kokpak focus, gray marmot burrows were disinfected over 812 km^2. Excavations of 35 nests and plugs of 550 burrows revealed no fleas; only one plague epizootic occurred in 15 years, by which time the major flea species, *O. silantiewi*, was completely restored (Pole *et al.* 2003). Since the dissolution of the USSR and the formation of the Commonwealth of Independent States (CIS), observations of plague foci were greatly reduced because of a lack of financing.

The reduction of *M. baibacina* populations to control plague markedly affected population structure. The population was reduced to 5–150 individuals/km^2; this reduction resulted in isolated, small settlements of 5–30 families (Pole 1992). During recovery many new families were formed and litter size increased. About 25% of the 2-year-old females reproduced, which was 5–8 times greater than normal. In effect, the reduction of mean age from 3.1 years to 2.5 years allowed the young females to escape reproductive suppression (Armitage 1996b). The new families, on the "periphery" of established families, had a younger age structure, which indicates that the younger, not the older, females moved to form these families.

There is some evidence that marmots have genetic resistance to plague. Sensitivity to plague in *M. baibacina* in the Tien Shan plague focus fluctuated with four blood groups determined by hemo-agglutination methods (Aikimbaev and Pole 1996). The fourth phenotype was associated with high sensitivity to plague; the second phenotype, with immunity. The first and third phenotypes were associated with relative resistance. The plague epizootic occurred when 51.5% of the marmot population had phenotype four; plague was absent when only 39% of the marmots had the phenotype. Populations of *M. sibirica* were categorized as one of three groups: (1) high plague intensity; (2) plague sometimes occurs, but not for 15 years; (3) plague free, not reported for 100 years (Batbold 2002a). Mean frequencies of observed alleles of four polymorphic loci varied among the three populations. Low plague and plague-free populations had comparatively higher frequencies of Tf^L and both populations were very similar in genetic structure. Plague was a strong selective force; population density decreased by 76.6% and male mortality was twice that of female. After infection, Hp^S, Tf^M, Tf^K, and Al^B allelic frequencies decreased and Tf^L, Al^A, and Hp^F allelic frequencies increased. Selection acted most strongly on the transferrin locus. After the epizootic, genetic structure changed slowly and after 8 years became similar to the plague-free population. This change suggests that plague resistance has a cost such that the plague-resistant genetic structure of serum proteins is less fit in the absence of plague.

The first case of plague in the western United States was recognized in 1900 and was probably brought by rats to ports such as San Francisco. Plague quickly spread to ground squirrels and was reported in marmots in northeastern Oregon in 1929 (Eskey and Haas 1940). In 1934, an epizootic in marmots infected a camp in Oregon; if not for this one case, the epizootic would not have been detected. The role of marmots as a plague reservoir is equivocal as marmots are often associated with ground squirrels known to harbor plague (Eskey and Haas 1940). However, plague occurs in the yellow-bellied marmot in Nevada and California. Laboratory evidence repeatedly confirmed the presence of *Y. pestis*, associated with human cases, in California. Marmots appear to be highly susceptible and undergo intense local die-offs (Nelson 1980).

Although plague occurs in prairie dogs (*Cynomys gunnisoni*) in Gunnison County, Colorado (Räyor 1985), within 50 miles of our study sites, we never observed any evidence of plague in our marmot metapopulation and blood samples from 35 marmots tested negative for *Y. pestis* (Van Vuren 1996).

Ectoparasites

From 1982 through 2001, trappers recorded the presence of ectoparasites. Two young had maggots in a head wound and one young had an orange egg mass in the AG area; all survived through their subsequent hibernation. Three adult female, two yearling male, and one yearling female yellow-bellied marmots had mange, a skin disease caused by parasitic mites. Five survived through their next hibernation; two of the adult females weaned litters, and the fate of a male disperser is unknown. No impact on demography was detected.

Ectoparasites, such as fleas and mites, can cause a loss of fitness for their hosts other than by serving as disease vectors. Because the probability of contact with other individuals is greater within social groups than in individuals living singly, the risk of increased parasitism is considered a cost of sociality (Alexander 1974, Hoogland 1979). The major flea in our *M. flaviventris* population is *Oropsylla standfordi*; the number of fleas on a marmot was not correlated with age or sex (Van Vuren 1996). Yearlings with greater flea infestations grew more slowly; of 32 marmots that hibernated, 4 died and 3 of them had five times the number of fleas that survivors had. There was no significant difference in the number of fleas on colonial and non-colonial marmots (the non-colonial actually had more) or on polygynous and monogamous males. Thus, there was no indication that ectoparasitic infection was a cost of sociality.

The flea is probably the most common ectoparasite in marmots; mites and ticks are also widespread (Bassano 1996, Bibikow 1996). Their effects on marmots are understudied. One interesting study of the effects of a single parasite, the mite *Echinomyssus blanchardi*, which accounted for 99.7% of all ectoparasites identified in the study, on mortality of *M. marmota* revealed that infant winter mortality increased with the ectoparasite load of the family. Litters produced by females exposed to a high number of mites during the preceding winter were weaned later and late weaners had a lower chance of surviving hibernation. There was no relationship between ectoparasite load and group size or marmot density; the number of parasites was lower in areas where marmot groups were clumped; i.e., on favorable habitats (Arnold and Lichtenstein 1993). The parasite loads did not differ between territorial individuals and those of lower rank; variation in parasite loads was greater among groups than within groups. The authors concluded that ectoparasites decrease individual fitness but are not a cost of sociality.

Both studies (fleas and mites) demonstrated a fitness cost on individuals but neither study discovered any major demographic impact on their respective populations.

Endoparasites

Endoparasites may have demographic impacts, but there is no clear evidence for such impacts. Numerous species of helminthes have been reported from at least 12 marmot species (Bassano 1996) and it is safe to assume that all marmot species harbor one or more helminthes species. There is some tendency for a helminth species to infect only one (or a few) marmot species. An early comparison of phylogenetic trees of *Marmota* and the helminth *Citellina* indicated parallel patterns between the two groups, which could explain strict host–parasite specificity (Hugot 1980). On the other hand, in North America, the cestode *Diandrya composita* infects all the species of the subgenus *Petromarmota* and *M. broweri* (Rausch and Rausch 1971, Rausch 1980), but not *M. monax*. An additional species, *D. vancouverensis* occurs only in *M. vancouverensis* (Nagorsen 1987). This pattern suggests that *D. composita* spread among the *Petromarmota* as they speciated and infested *M. broweri* through environmental contact, probably between *M. broweri* and *M. caligata* as *M. caligata* is the only marmot species with which *M. broweri* has a parapatric distribution (Gunderson *et al.* 2009). *D. vancouverensis* probably represents a recent evolutionary divergence within one of the *M. vancouverensis* metapopulations, probably the Nanaimo Lakes metapopulation (Bryant 1998).

The best information on the prevalence and abundance of these parasites comes from studies of the alpine marmot. All marmots collected in late summer were infected with helminths (Manfredi *et al.* 1992, Prosl *et al.* 1992). As many as six species of helminths occurred in a population, but any individual had 1–3 helminth species. Similarly, a steppe marmot population was infected by five species (Zhaltsanova and Shalaeva 1990). Three species dominated the helminth population: *Ctenotaenia marmotae*, *Citellina alpina*, and *Ascaris laevis* (Manfredi *et al.* 1992, Callait *et al.* 2000). The marmots generally were reported in good condition (Prosl *et al.* 1992), but with some evidence of low mass gain by some infected marmots (Callait *et al.* 1996). Several times we trapped young yellow-bellied marmots in late summer that were much smaller than other young in their social group. We detected ascarids in one young; the young survived the subsequent hibernation. A young brought to the laboratory for hibernation studies failed to gain mass. We gave it a vermifuge, ascarids were expelled, and mass gain followed. It may be that undersized young in late summer suffer from ascarid infections; these young do not survive hibernation. However, there is no evidence to indicate that helminth infection is a major source of mortality.

A major reason why helminth infections may not cause major mortality in marmots is that at the time of hibernation when the intestine atrophies (Bassano *et al.* 1992b, Callait and Gauthier 2000), the parasite load decreases. *C. marmotae* and *A. laevis* adults are expelled. *C. marmotae* is reinfected from oribatid mites during the active season and *A. laevis* reinfects marmots when they ingest their eggs. Some *C. alpina* remain in the caecum (Callait and Gauthier 2000); infections are low in the adults in May and juvenile marmots are parasite-free (Prosl *et al.* 1992).

Does the intestine atrophy to eliminate parasites or is the expulsion of parasites a secondary effect? It seems most likely that eliminating parasites is a secondary effect of reducing intestinal mass to greatly decrease the metabolic costs of hibernation. It is difficult to conceive that mammals could go through the process of intestinal atrophy and re-growth during normal euthermic activity. There would be a considerable drain on any energy reserves that could not be replaced by foraging. Water balance could not be maintained. Thus, the morphological adjustments for hibernation secondarily provide at least partial control over major endoparasites.

When marmots are introduced into new areas, some of the parasites are left behind. The intracellular *Eimeria*, which is not eliminated during hibernation, maintains a high prevalence. Some new species of unknown significance may be acquired (Gortazar *et al.* 1996). Parasite infestations decreased in reintroduced populations in eastern Austria and Switzerland (Preleuthner *et al.* 1996). We do not know if this loss of parasite species has any major effects on demography, but there may be selection effects. The Pep-1 genotypes differed in their degree of infestation by *C. alpina* and *C. marmotae*; the S/S genotype was over-represented among non-infected individuals for *C. alpina* and the S/F genotype occurred at a higher frequency among non- or weakly-infected individuals for *C. marmotae* (Preleuthner *et al.* 1995).

Summary

The age structure of marmot populations is unstable as a consequence primarily of temporal variation in reproduction. Age of first reproduction and fertility significantly affect both population growth and decline in all local populations, which follow their own trajectories. These two variables are determined mainly by social behavior and provide a direct link between social behavior within the matrilineal structure and population dynamics. Reproductive variables rather than survival drive population growth.

The major causes of mortality with significant demographic consequences are predation and major weather events, especially those that affect the length of the growing season. Parasites and disease cause mortality, but no major demographic effects have been reported.

18 Dispersal and immigration

Dispersal and immigration are two ends of the process of an individual leaving its home site, traveling for some distance across a landscape, and settling in a favorable location. Immigration cannot occur without dispersal, but dispersal can occur without immigration. When a disperser leaves its natal environment, there is no certainty that it will find a new residence. For marmots, avoidance of predators, finding food, and locating a hibernaculum are critical activities. Movement across an unknown landscape increases vulnerability to predation because the marmot likely has no known burrow in which to take refuge (Van Vuren, pers. com.); however, some dispersers may find shelter under boulders, tree roots, etc. (Fig. 18.1). In general, predation on marmots is associated with movement (Van Vuren 1990). Marmots are successful at finding hibernacula; mortality during hibernation does not differ between dispersing and philopatric yearlings (Van Vuren and Armitage 1994a). However, the likelihood of immigrating into a colony is low (18.8% successful) and survival of dispersers is lower than that of philopatric residents during their second summer (Table 18.1); waiting to disperse until age 2 or older did not improve survival (Van Vuren 1990). The higher rate of mortality and difficulty of successful immigration raises the question of why females disperse.

Female dispersal

Several lines of evidence suggest that female dispersal, as with male dispersal, is conditional. When adults were removed from a population, no yearling female dispersed (Brody and Armitage 1985). When an experimental colony of young females was established without adults, none dispersed (Armitage 1986b). Short-term studies identified factors such as density of adult females (Armitage 1962), social behavior (Armitage 1977, 1986b, 1986c), kinship (Armitage and Johns 1982), and patterns of space use (Armitage 1975) as possible proximal factors affecting the probability of dispersal.

For the long-term study, residency or dispersal status was confirmed either by radio-telemetry (10 years) or by trapping/visual identification (37 years) for 231 yearling females (Armitage *et al.* 2011). We examined 24 variables for possible factors causing dispersal, defined as moving away from the natal site (dispersal movement, Dobson 2013). The 24 variables were chosen to test three general explanations for dispersal in mammals: inbreeding avoidance, competition for mates, and competition for environmental resources

Table 18.1 Likelihood of successful immigration into a colony by female yellow-bellied marmots first trapped as a yearling disperser.

Number trapped at a colony	48
Number that were transients	35
Number that settled peripherally	4
Successful immigrant	1
Disappeared	3
Successful immigrant	9
Eventually reproduced	2
Mortality rate of known dispersers	2.4 times that of residents

Figure 18.1   A group of dispersing yearlings at a rock shelter in a meadow.

(Greenwood 1980). We also tested five major hypotheses about the proximate causes of dispersal: inbreeding avoidance, population density, kin competition, social intolerance, and body size of dispersers. There is overlap between proximate and ultimate factors; e.g., inbreeding avoidance. Also, competition for mates is related to kin competition and social intolerance. Body size is related to who should disperse; if some subset of the population disperses and others of dispersal age do not, individuals should stay or leave depending on which process is facilitated by large size.

Univariate analyses revealed that none of the measured variables was consistently associated with inducing dispersal; each variable had some probability of affecting the likelihood of dispersal. A positive estimate of the slope parameter indicates that the variable tended to induce dispersal whereas a negative estimate indicates a tendency toward philopatry. Of the 24 variables, 7 significantly influenced the probability of dispersal ($p < 0.05$).

Table 18.2 Number of occurrences and results of the univariate analyses for the categorical variables used in the analysis of dispersal by female yearlings. The slope parameter is for logistic regression modeling dispersal probability as a function of the listed variable. The p value tests for slope = 0; significant p marked with an asterisk (*). SE = standard error. R^2 is the generalized max-rescaled coefficient of determination.

Variable	Occurrence		Slope parameters			
	Yes	No	Estimate	SE	p	R^2
Father absent	131	100	0.079	0.142	0.577	0.0019
Yearling mass above the mean	133	81	0.068	0.151	0.653	0.0013
Mother absent	65	166	0.340	0.152	0.026*	0.0295
Mother non-reproductive	57	109	−0.265	0.192	0.168	0.0171

Inbreeding avoidance

Inbreeding avoidance was invoked frequently as the ultimate cause of dispersal (Bowler and Benton 2005) and some instances of dispersal patterns were best accounted for by inbreeding avoidance (Pusey 1987, Pusey and Wolff 1996). Inbreeding avoidance is more likely when one sex disperses, but is unlikely if both sexes disperse and inbreeding avoidance can be confounded by kin competition (Perrin and Goudet 2001). However, about half of female yellow-bellied marmot yearlings disperse and the decision to disperse could be to avoid inbreeding. We tested inbreeding avoidance by asking were females more likely to disperse if their father was present. Female yearlings were no more likely to disperse when the father was present (probability of dispersal = 0.31) than when the father was absent (probability of dispersal = 0.34). This variable was statistically insignificant and explained virtually none of the variation in dispersal (Table 18.2).

Population density

Density-dependent population regulation is often considered to be a fundamental population dynamic consequence of dispersal (Hanski 2001, Bowler and Benton 2005), but examples of density-dependent and density-independent dispersal have been reported for mammals (see discussion in Armitage *et al.* 2011). The lack of a consistent relationship between density and the probability of dispersal clearly removes density as a universal proximate cause of dispersal. The patchy distribution of yellow-bellied marmot populations suggests that dispersal could increase as a local population approached saturation.

Mean density of resident adult females varied from 1.43 to 6.35 among sites. Therefore, density was calculated as a ratio of the number of adult females present in a given year divided by the long-term average for that site when occupied. Because social structure and dynamics occur through matrilines, matriline density was also determined; matriline size for a particular year was divided by mean matriline size for that site to

Table 18.3 Mean values and results of the univariate analyses of the continuous variables used in the analysis of dispersal by female yearling yellow-bellied marmots. The slope parameter is for logistic regression modeling dispersal probability as a function of the listed variable. The p value tests for slope = 0; significant p marked with an asterisk (*). N = sample size, SD = standard deviation, R^2 is the generalized max-rescaled coefficient of determination, SE = standard error.

	N	Mean	SD	Slope parameters			
				Estimate	SE	p	R^2
Density variables							
Adult female density	231	1.144	0.799	0.041	0.176	0.816	0.0003
Matriline density	231	0.935	0.689	−0.159	0.210	0.450	0.0035
Number of female yearlings	231	2.953	1.758	0.023	0.080	0.777	0.0005
Social variables (#/animal/h)							
Adult male grapples	102	0.019	0.032	−11.522	9.391	0.220	0.023
Play (yearling females)	102	0.073	0.141	−35.252	11.959	0.003*	0.266
Amicable behavior							
Mother	77	0.039	0.070	−56.046	27.214	0.039*	0.285
Matriline females	56	0.020	0.034	−17.514	12.788	0.171	0.063
Non-matriline females	63	0.0008	0.003	92.665	86.042	0.282	0.03
Agonistic behavior							
Mother	75	0.019	0.083	−39.648	32.737	0.226	0.073
Matriline females	58	0.023	0.047	−6.795	8.859	0.443	0.019
Non-matriline females	63	0.004	0.008	−107.7	69.718	0.123	0.009
Proportion amicable							
Mother	48	0.638	0.441	−1.746	1.019	0.087	0.121
Matriline females	48	0.411	0.445	−0.654	0.847	0.440	0.020
Non-matriline females	29	0.114	0.312	0.914	1.358	0.501	0.024
Space-use overlap (%)							
With mother	77	0.289	0.191	−10.881	2.887	0.0002*	0.448
With matriline females	56	0.217	0.177	−7.928	2.810	0.005*	0.277
With non-matriline females	71	0.035	0.045	−2.431	6.734	0.718	0.003
With yearling females	87	0.321	0.222	−4.904	1.341	0.0003*	0.258
Kinship (average relatedness)							
With matriline females	188	0.421	0.090	−2.739	1.736	0.115	0.018
With yearling females	194	0.420	0.097	3.270	1.730	0.059	0.027

provide an index of matriline density. Mean number of adult females and mean matriline size were not correlated ($r = 0.25$, $p > 0.05$, $N = 11$), thus, these two measures of density are independent. Because a yearling could disperse if the potential competition for residency was great (large number of yearlings) or become a resident if potential competition was low (small number of yearlings), the third measure of density was the number of female yearlings in the matriline.

No density variable significantly affected the probability of dispersal nor explained the variation in dispersal (Table 18.3). Density, when measured over many years in several, independent local populations, is neither a proximal nor an ultimate factor affecting the probability of dispersal.

Table 18.4 Conditions associated with density effects on dispersal of female yearling yellow-bellied marmots.

Site	Year	Conditions
Picnic	1990	Nine adult, 12 yearling females present.
Upper		Recruit had three times the space overlap with the two adult females as the dispersers; recruit had amicable social behavior.
Middle		Yearling dispersed early, no space-sharing with adult.
Lower		Two of nine yearlings were recruits; one moved to Middle by 1991 when adult failed to return. Dispersing yearlings had no space overlap with one or more adults; all social interactions agonistic. Space use of second recruit overlapped those of all resident adults; amicable interactions seven times more frequent than agonistic.
Marmot Meadow	1981	Two unrelated matrilines; three adults at Main Talus, two adults at Aspen Burrow.
		Space overlap of adult females averaged 17% with dispersers, 28% with recruits. Only agonistic interactions between adult females and dispersers. Amicable behavior between adults and the four yearlings that remained until late summer was three times more frequent than agonistic.
River	1986	Three adult females weaned litters, three female yearlings dispersed. Space overlap between adults and yearlings averaged 10%, among adults, 39%.
	1988	Two adults and five yearling females, all yearlings recruited after settling in an area little utilized by the adults.
	1990	Six adult and four yearling females; all yearlings recruited. Yearlings remained in home range of mother, space overlap between yearlings and mother averaged 51.3%; among yearlings, 55%, between yearlings and other five females, 13.6%; between their mother and other females, 17.6%.

When looking at individual sites, some evidence supported a role for either adult female density or matriline density. Density of adult females increased the probability of dispersal at Picnic Colony (p = 0.008) and matriline density increased the probability of dispersal at Marmot Meadow Colony (p = 0.02). These density measures also increased the probability of dispersal at Boulder and River Colonies, but p values were insignificant. Because density so rarely affected the probability of dispersal, it is instructive to examine the conditions associated with density effects at four sites.

At Picnic all major burrow sites were occupied; there was no empty site within the locality to which yearling females could move or where they could forage without potential encounters with adult females. Only 3 of 12 female yearlings were recruited (Table 18.4). Successful yearling recruits adopted a home range pattern (see Fig. 11.1, eighth year, for example of space overlap patterns of the recruit with the adults) coupled with amicable (or lack of agonistic) behavior that enabled them to be philopatric at a time of high density (Table 18.4).

At Marmot Meadow, both major burrow sites were occupied (Table 18.4), as a consequence, there was no available space into which female yearlings could move.

Two of the seven yearling females were recruits; one disappeared in late summer, and two moved to a nearby satellite site in late summer. Space overlap with the adult females was greater than that of the dispersers, but was much less than that of a recruit (43%) who was the only yearling present with two adults 2 years earlier. Thus, when population density was high, the pattern of space use by the yearlings indicates avoidance of the adults by the yearlings. Despite favorable social behavior (Table 18.4), the high density of adults and yearlings resulted in only one recruit. When density decreased 2 years later, one of the yearlings returned from the satellite site and was a resident for 4 years.

Although the effect of density on dispersal was not statistically significant at River, density effects were indicated in certain years (Table 18.4). The yearlings avoided the adults by moving early in the season to nearby sites where they apparently established residency. However, two were predated and the third died over winter. Recruitment occurred in subsequent years when yearlings could avoid adults by moving into little utilized space or by sharing space with their mother and avoiding other female adults who were distributed across the colony site (Table 18.4, see Fig. 16.1 for pattern of recruitment and matriline structures). This pattern indicates that not only the density of adults, but also their dispersion in the colony affects recruitment.

Boulder typically has less than two resident females (Table 8.1). In 1987, when two reproductive adults were present, all five yearling females dispersed even though the reproductive females were the mothers of the dispersers. Even when yearlings remained, such as in 1985 (Fig. 17.2), they often dispersed as 2-year olds. The only instance in which a large number of yearlings was recruited and remained as 2-year olds occurred in 1978 when only the mother was present.

These examples indicate that density may increase the likelihood of dispersal at some colonies in some years and the effect of density is ameliorated by the dispersion of the adults in the colony. Overall density is much less important than the presence of the mother and the pattern of space overlap between adults and yearlings.

Body size

In general, larger individuals are thought to be competitively superior and more likely to achieve residency in the natal site; thus smaller individuals should disperse. On the other hand, the larger individuals may be more likely to be successful immigrants and should disperse (Bowler and Benton 2005). Both patterns are reported among mammals (Armitage et al. 2011). Among yellow-bellied marmots, dispersal of yearling males, but not of yearling females, was delayed, but not eliminated when they were underweight (Downhower and Armitage 1981).

Yearling marmots begin to gain mass soon after emergence from hibernation. There are two problems with evaluating the role of body mass: (1) individuals are captured on different days and no individual at this time of year is captured frequently enough to calculate an individual growth curve; and (2) emergence times vary from year to year and are later at UV (the higher elevation sites). Therefore, we calculated regressions for the population of yearling females for the higher elevation sites (2930–3043 m) and the lower elevation sites (2711–2867 m) for June and recorded for each female whether its body mass was above or below the regression.

There was no significant effect of body size on the probability of dispersal and body size explained virtually none of the variation (Table 18.2). The probability of dispersal was 0.34 for the larger females and 0.31, for the smaller.

Sociality and kinship

Neither kinship variable explained much of the variation in the probability of dispersal. This result is not surprising because kinship is high within the matriline and among yearlings and there is little variation (Table 18.3).

Five of the variables associated with sociality, which includes the social and space-use-overlap variables, were significant in the univariate analyses: play, amicable behavior with mother, and space-use overlap with mother, with matriline females, and with yearling females (Table 18.3). Each variable explained a considerable proportion of the variation in the probability of dispersal; e.g., play, 26.6%, space-use overlap with mother, 44.8%. The latter variable explained more of the variation than any other variable. All slope estimates for these variables are negative, which indicates that they favor residency rather than dispersal.

It seemed likely that some of these variables could be related. For example, play is more likely to occur if yearlings share space. Therefore, all social and space-use variables were combined and a stepwise variable selection was conducted to identify which variables significantly affected dispersal (Armitage et al. 2011). Only space-use overlap with mother was significant. As discussed in Chapter 10, space-sharing is highly related to kinship; space-sharing integrates both close kinship and social tolerance. Individuals are not integrated into a group unless space is shared.

Because agonistic behavior variables were not significantly related to the probability of dispersal (Table 18.3), social intolerance is rejected as a general cause of dispersal. This rejection implies that social cohesiveness leads to philopatry. The social cohesiveness hypotheses (Bekoff 1977) is supported by the significant effects of play, amicable behavior with the mother, space-use overlap with matriline females and with yearlings and, most importantly, with mother. This hypothesis was supported in a social network analysis in which female yellow-bellied marmots that were more socially embedded in their groups were less likely to disperse (Blumstein et al. 2009b). Both the social cohesion and social network hypotheses fail to recognize that dispersal is multifactorial and do not recognize the central role of kinship and the mother in the decision-making process of whether to disperse or remain philopatric.

Role of the mother

A yearling female was significantly more likely to disperse (Table 18.2) when her mother was absent (probability of dispersal = 0.43) than when her mother was present (probability of dispersal = 0.28). Why does mother absent not always cause dispersal? I identified 36 cases in which a yearling female remained philopatric when mother was absent. For 14 (39%) yearlings, no adult female was present. Thus, there was no conflict with unrelated or distantly related adult females that could cause dispersal. Although agonistic behavior with non-matriline females explained virtually none of the variation in dispersal (Table 18.3), it could be important when the mother is absent as the yearling does not have the mother's home range within which she can function buffered from hostility of non-matriline females. Social interactions with non-matriline females are predominantly agonistic (Tables 11.1, 11.3) and occur at a frequency much greater than would be expected from the probability of contact between yearlings and adults (Armitage and Johns 1982). Resident adult females were observed chasing yearling females whose mother was absent and those yearlings dispersed (Armitage 1973, 1986c).

Fifteen (42%) yearling females became resident by moving into an area of the site that was not occupied by adult females in that year (philopatric movement, Dobson 2013). These events occurred in the five largest sites. What was critical was not the number of adult females present, but their dispersion in the site. Two (6%) yearlings remained in their mother's former area where each associated with one adult female from their mother's matriline with whom they had associated as young. This pattern emphasizes the importance of familiarity in establishing social bonds.

The internal movements to an unoccupied location within a site illustrate the complexity of choices available to yearlings. Several yearlings traveled between their natal site and a nearby site. In one instance, two yearlings hibernated in their natal site and then dispersed to the adjoining site at age 2. In another instance, two yearlings hibernated in the nearby site, then moved back to their natal site as 2-year olds when the resident females, unrelated to the yearlings, emigrated, thus creating a vacancy that the yearlings quickly filled. One consistent feature of these movements is that the yearlings settled where they were free from conflict with unrelated adults.

These movement patterns suggest that a female yearling should seek residency wherever she has a high probability of future reproduction and that she should move as little distance as possible. The median dispersal distance of 350 m for females, which includes the internal movements within a site, supports this interpretation. By contrast, the median dispersal distance of males is 1825 m (Van Vuren 1990). As Van Vuren emphasized, males must locate undefended females whereas females need to find the nearest available resources. Female movements may result in philopatry (remaining in the natal site) and, if mother is present, remaining in her natal home range.

Individuality

Why do some female yearlings disperse when their mother is present? One factor is whether the mother is reproductive. The probability of dispersal is 0.31 if the mother is reproductive and 0.21 if she is not. Although not statistically significant, a yearling is more likely to remain if her mother is non-reproductive; this variable explains 17.1% of the variation and is likely of biological importance (Table 18.2).

Another factor influencing the dispersal decision is individuality, the expression of individual behavioral phenotypes (see Chapter 12). Female yearlings with shy phenotypes are more likely to be recruited and those with bold phenotypes are more likely to disperse. Note: shy animals are those that are more aggressive and bold individuals are more social. There was an interaction between mothers and daughters; shy mothers recruited nearly all of the shy yearlings but only one-third of the bold yearlings. By contrast, a bold mother recruited both shy and bold yearlings and none of the yearlings dispersed (Armitage and Van Vuren 2003). These differences in behavioral phenotypes likely account for some of the variability in dispersal, but behavioral phenotypes are lacking for most yearlings which preclude making a quantitative assessment of its overall importance.

Immigration

Successful dispersal requires that a female yearling settle in a habitat patch where she can reproduce. What is the likelihood that a female can immigrate into a site given that, except for unusual climatic or predation events, it likely has residents? I examined the dynamics of immigration and recruitment for five colonial and one satellite site.

Neither recruitment nor immigration occurred every year at any site (also see Figures 16.1, 16.2, 17.2, 17.3). The smallest site (Boulder, a satellite site) and the largest site (North Picnic) had rates of immigration higher than those of recruitment; rates of recruitment were greater than those of immigration at the other five sites (Table 18.5).

Table 18.5 Characteristics of recruitment and immigration in seven populations of yellow-bellied marmots. An immigrant is a marmot born elsewhere and resides at a site for at least one active season. A recruit is a marmot who becomes a resident in the site of her birth. An event means that immigration or recruitment occurred regardless of the number of immigrants or recruits. The rate per year was calculated as the total member of recruits or immigrants divided by N. The sites are listed in order of smallest (Boulder, a satellite site) to largest. R = recruitment, I = immigration, N = number of years observed.

Name of site	N	Number of years event occurred I	R	Rate per year I	R	Total number I	R	Number of years both I and R occurred
Boulder	39	8	6	0.308	0.231	12	9	1[a]
Gothic Townsite	22	5	11	0.214	0.851	6	21	4[b]
Cliff	26	4	5	0.154	0.385	4	10	
Marmot Meadow	38	4	12	0.106	0.473	4	18	
River	40	6	17	0.175	0.925	7	37	
Picnic	40	8	23	0.275	1.25	11	50	4[b]
North Picnic	37	17	6	0.568	0.297	21	11	3[b]
Total	242	52	80	0.269	0.645	65	156	12

[a] Density of adult females at the mean.
[b] Density of adult females below the mean.

Rates of recruitment and immigration were not correlated among sites and neither immigration nor recruitment was correlated with area (Armitage 2003i). The lack of a significant relationship between area and immigration or recruitment suggests that density of adult females is likely more critical than area. Area and adult female density are not correlated. I examined the relationship between density and immigration or recruitment by recording for each year in which immigration and/or recruitment occurred when the density of females was above or below the mean density for that site.

When the density of adult females was below the mean, there was no significant difference in the number of immigration (N = 45) or recruitment (N = 43) events (Armitage 2003i). However, when the density of adult females was above the mean, immigration events (N = 8) were significantly fewer than recruitment events (N = 36). Density of adult females acted differently on immigration and recruitment. Recruitment was equally likely when the density of resident adult females was below or above the mean (G = 0.23, p > 0.5), whereas immigration was significantly more likely when the density of adult females was below the mean (G = 21.4, p < 0.001).

Although both immigration and recruitment are uncommon (most litters do not produce recruits), recruitment occurs 2.4 times more often than immigration (Table 18.5). This difference is highly significant (for 242 colony-years [a colony-year is one colony or site in 1 year], G = 5.1, p = 0.025). Immigration and recruitment rarely occurred in the same colony-year and was most likely when the density of adult females was below the mean.

The low frequency of immigration is consistent with the relative stability of marmot populations (Fig. 17.1). Colonies are essentially closed systems and immigration is successful when a decline in the resident population creates open space available for colonization. This space may be occupied either by philopatric movements of recruits or by immigrants. The closed nature of marmot colonies is more apparent when transient or peripheral adult females are considered. Forty-one of these were trapped and marked; this number is a minimal estimate of the potential immigrants as not all animals that were sighted were subsequently trapped. None of 12 transients became immigrants and only 6 of 29 peripheral females were successful immigrants. Furthermore, when seven adult females were introduced into three marmot colonies, none became resident.

When an immigrant settles in a colony, she is unrelated to the other adult females. Recruits, on the other hand, are highly related to members of their matrilines. Recruitment is an individual fitness strategy; females recruit daughters into the mother's matriline. Because fitness increases as matriline size increases (Armitage and Schwartz 2000), an individual's inclusive fitness increases if she recruits daughters that eventually reproduce. No immigrant female has ever joined a matriline. Toleration of such immigrants would greatly reduce average relatedness within the matriline and decrease the probability that a female will recruit reproductive descendents, hence reducing her inclusive fitness.

Individual fitness and dispersal

Although dispersal is usually approached and modeled as a population parameter, there is increasing realization that dispersal is an individual fitness phenomenon with population consequences (Stenseth and Lidicker 1992). One population consequence in yellow-bellied marmots is that each colony develops its genetic characteristics (Schwartz and Armitage 1981) thus producing heterogeneity among colonies (Schwartz and Armitage 1980). Heterogeneity occurs because mating is restricted to the social group, because recruitment occurs at a higher rate than immigration, and because of the low rate of exchange of individuals between groups. Because of immigration, especially of males, the fixation of genetic variants is prevented. Some of the genetic loci may have obvious fitness consequences; e.g., the transferrin (TRF) locus has been associated with aggressiveness (Schwartz and Armitage 1981).

The proper comparison of fitness would be to compare the fitness of a philopatric female yearling with the fitness she would have as a disperser. Obviously, such a comparison is not possible; fitness consequences must be inferred from other evidence. A yearling female cannot be forced to become a resident in a colony in the presence of unrelated adult females; all introductions of female yearlings into colonies failed to produce residents (Armitage 2003i). Because yearling females do not disperse when adult females are absent and are more likely to disperse when their mothers are absent, I suggest that female yearlings disperse to attempt to gain fitness by finding a place where they can reproduce.

The fitness of the potential disperser must be considered in the context of the fitness of the mother. The mother gains fitness by producing offspring that eventually reproduce. Females do not compensate for a loss of reproduction (direct fitness) by assisting relatives (indirect fitness) to reproduce as indirect fitness does not compensate for the loss of direct fitness (Chapter 16).

How can a female increase the probability that a daughter will live to be reproductive? One possibility is to recruit her daughter into her matriline. One effect of recruitment is reduced mortality of yearlings. Because the age of first reproduction is 2 years, a female cannot produce reproductive descendents unless her daughters live to at least age 2. The number of yearling recruits is significantly related to the number of 2-year-old recruits (Armitage and Van Vuren 2003); thus the first step of recruiting a yearling daughter is a good indication of the likelihood of producing a reproductive daughter.

The recruitment of yearling daughters by adult females suggests the possibility of reproductive competition in the next active season. Potential competition is partially offset by a fitness gain; survivorship increases as matriline size increases (Armitage and Schwartz 2000), in addition to the reproductive gains described in Chapter 16. The older female may also suppress the reproduction of her young daughter thus gaining the advantage of producing young with whom she is related by $r = 0.5$ rather than allowing her daughter to wean young with whom she is related by $r = 0.25$. Thus, the older adults increase the likelihood of producing descendents by weaning more young with whom to share resources while also increasing the likelihood of eventually having grandoffspring by recruiting their daughters. For example, one female recruited six yearling daughters; their average age of first reproduction was 2.8 years and they produced 63 grandoffspring (Armitage 1992).

Why do female yearlings remain in their natal colony in the face of possible reproductive suppression? An essential factor must be that the yearling experiences a habitat that is associated with successful reproduction and survival and the yearling functions in a social group that provides some degree of protection from predators and agonistic conspecifics. Also, the yearling may breed as a 2-year old for several reasons. First, reproductive suppression is not absolute; the yearling has about one chance in seven of reproducing as a 2-year old when the older female is reproductive and that chance increases to about one in three if the older female is non-reproductive (Armitage 2003f). Second, the older female may not survive and the recruit escapes reproductive suppression. Third, the recruit may escape suppression by moving to an unpopulated area within the site.

Nevertheless, on average, a 2-year old has about a 20% chance of weaning a litter. If she fails to reproduce as a 2-year old, she has about a 33% chance of a weaning her first litter as a 3-year old. Because fitness is inversely related to age of first reproduction, a yearling that remains in her natal area has a high probability of losing fitness.

The alternative strategy, for the mother and the yearling, is for the daughter to disperse. But, as we noted, the likelihood that the yearling can successfully immigrate into a colony site is very low with a probability of about 0.19. And no yearling successfully immigrated when philopatric yearlings were present. Thus, dispersers are much more likely to be successful immigrants as adults.

To be a successful adult immigrant, the marmot must survive one or more years as a transient. Virtually nothing is known about how and where dispersers survive. Marmots may occupy shelters under fallen logs, at rock outcrops and by stumps or at the base of a tree (Downhower 1968). Such mini-sites probably enable some dispersers to survive until they can obtain residency in a nearby patch. When adults were removed from North Picnic Colony, two 3-year-old females, four 2-year-old and two older males moved into the site (Brody and Armitage 1985). The males were transients (some from nearby colonies) and the females moved from peripheral sites. Some marmots continue to live at a mini-site and may reproduce. The availability of these sites probably explains why yellow-bellied marmots are not retained in their natal sites to an older age as occurs in those species that form family groups (Table 5.1, Armitage 1996b).

The philopatric female becomes an adult recruit at age 2, but the average age of known-age immigrant adults was 2.8 years. In effect, an immigrant is unlikely to reproduce before age 3 (Armitage 2003i) and only 20 of 77 (26%) immigrant adults reproduced in their year of immigration (Armitage and Schwartz 2000). By contrast, 39.6% of philopatric residents reproduce by age 3. The lower probability of surviving and reproducing by dispersers than by philopatric residents indicates that dispersal and subsequent immigration into a colonial site probably has lower fitness than remaining and coping with reproductive suppression.

To summarize, dispersal by female yearling yellow-bellied marmots is conditional and multifactorial. Dispersal is unlikely when the yearling is socially integrated into her mother's matriline as evidenced by high space-sharing and amicable behavior with mother. Dispersal is an alternative strategy that has lower fitness than philopatry, but provides some opportunities to gain some fitness through successful immigration.

The effects of kin cooperation or competition are likely critical factors underlying dispersal patterns (Dobson 2013). Emphasis traditionally focused on kin competition following the prediction of Hamilton and May (1977) that individuals should disperse to avoid competition with nearby kin. However, in three species of prairie dogs, the probability of natal dispersal by females varied strongly and significantly with the number of close kin within the natal territory (Hoogland 2013). For both prairie dogs and yellow-bellied marmots, cooperation among kin provides more fitness benefits than competition among kin imposes fitness costs. The ultimate factor causing dispersal is the effort by individuals to maximize individual fitness and future research on dispersal should focus on fitness benefits. Dispersal is not a population process but has important population consequences. There is no universal cause of dispersal; dispersal can be understood only when treated as part of the life-history strategy of individuals.

Dispersal in other marmot species

Intensive studies of dispersal in other species of marmots are lacking, but some short-term studies provide insight into the mechanisms promoting dispersal. There is evidence that dispersal in woodchucks is conditional and that population density is an important proximal factor. When woodchuck density was reduced experimentally (Table 18.6), dispersal decreased, philopatry of young increased, and immigration occurred primarily in the area where the greatest number of woodchucks was removed (Davis *et al.* 1964). Dispersal in a high-density population in Maine was delayed or did not occur because

Table 18.6 Demographic characteristics associated with dispersal in other marmot species.

Species	Demographic pattern
Woodchuck (Pennsylvania)	Percentages of young and yearlings increased following removal of several hundred animals. Lack of dispersal; immigration primarily of young restored population numbers.
Woodchuck (Maine)	High proportion of animals philopatric; males more likely to disperse in their second summer; females more likely to be recruits in a high-density population. About 54% of juveniles of both sexes remained within natal home range.
Olympic	Dispersal of >0.5 km extremely rare among 101 radio-tagged marmots; no known female settled >0.4 km from her natal home range. Three-year-olds more likely to disperse; median distance of male dispersal 3.7 times greater than that of females. More males than females dispersed; 80% survived to their second hibernation following dispersal. Reproductive females moved <500 m; most females remained within their original habitat patch.
Alpine	19% of subordinates obtained residency in natal or neighboring territory; 19% of newly established territorials were immigrants. Social behaviors in a French population characterized as aggression, play, avoidance, and amicable. Dominants characterized by aggressive behaviors; subordinates, mainly non-aggressive. Avoidance behavior characterized dispersers.

available places for settlement were few, which suggests an inverse density effect, not in the family, but in the wider population (Maher 2006). About two-thirds of the females remained philopatric and as adults established home ranges within or adjacent to their natal range (Maher 2009a). This pattern suggests that young woodchucks use some cues to assess future reproductive possibilities and decide whether to disperse based on future fitness. For woodchucks, dispersal apparently occurs when an individual has a better opportunity to find resources and reproduce elsewhere, otherwise they remain near their natal area.

Dispersal in Olympic marmots apparently is conditional. Recently, populations declined drastically, local extinctions were widespread, and recolonizations have not occurred (Griffin *et al*. 2008). Dispersal was virtually non-existent; 2- and 3-year olds that did disperse did so before their first reproduction (Griffin *et al*. 2009). This pattern indicates that dispersers were seeking an opportunity to reproduce; three of eight females bred the year after dispersal; all females moved short distances (Table 18.6). Most short-distance movements were into home ranges that recently had become vacant or patches where marmot density had declined (Griffin *et al*. 2009). There were few immigrants into the study sites. The pattern of short-distance movements is similar to that of yellow-bellied marmots; females move only as far as needed to settle in a location where they are likely to reproduce.

Delayed dispersal characterizes the alpine marmot. Lower dispersal rates occurred when immatures lived in the group, which may, along with later dispersal of males, indicate benefits from the subordinates' role in social thermoregulation (Arnold 1990a). Dispersal was unrelated to body mass, but dispersers lost more mass during the previous hibernation than non-dispersers. Dispersal apparently was related to the opportunity to achieve terri-torial status; most animals obtained a territory by evicting the previous tenant of the same sex, an activity never observed in yellow-bellied marmots. Females obtained territorial status at an earlier age than males. Long distance dispersal bore a high mortality risk; few such dispersers obtained territories and many individuals gained residency in their habitat patch (Table 18.6). Similar to immigration patterns in yellow-bellied marmots, some alpine marmots joined a group when the territorial position was vacant or they occupied a vacant site (Arnold 1990a). These patterns are consistent with dispersal being conditional and occurring at a time when an individual can increase its fitness by dispersing.

Social behavior apparently is a proximal factor that induces dispersal by alpine marmots. Chasing of 3-year olds by territorial females occurred in a high altitude colony; none remained (Lenti Boero 1994), and dispersal of 2-year-old males was associated with high levels of aggression by the adult male (Perrin *et al*. 1996). The role of social behavior was examined in France in families with the territorial pair and at least two subordinates (Magnolon *et al*. 2002). Behavioral and dispersal patterns were complex (Table 18.6), but two relationships were similar to those of yellow-bellied marmots. Play was associated with non-dispersers and dispersers frequently received less amicable behavior than expected from the territorials, which suggests that dispersal was related to formation of social bonds. However, aggression by dominants was not necessary to induce dispersal and some individuals dispersed who apparently were integrated into their social group. The overall patterns indicate that individuals made dispersal decisions based on their future fitness. Long-term studies of known individuals plus experimental manipulations of family composition along with the reproductive success of residents and dispersers are needed to determine how dispersal is related to fitness.

Dispersal clearly is sex-biased in social marmots. The possibility that sex-biased dispersal is related to social complexity was examined in 11 species, two *Marmota* two *Cynomys*, and seven spermophiles. In all cases, males dispersed more than females (Devillard *et al.* 2004). When phylogeny was accounted for, male-biased dispersal and male dispersal rates increased with increasing sociality, but female rates were not related to sociality. The authors interpret this pattern as strong support for inbreeding avoidance as a major role in the evolution of dispersal for most social mammalian species. However, the mating system through the level of polygyny could be a confounding factor (Devillard *et al.* 2004). As I argued previously, male sex-biased dispersal is consistent with the need for males to find mating opportunities, which are likely to be few when an older dominant male is present. Females need resources to breed and are more likely to find opportunities in their natal area, but do encounter competition, which causes some females to disperse. Teasing apart the causes and consequences of dispersal requires more long-term research on individually identified animals (Clutton-Brock and Sheldon 2010).

Summary

The available evidence from marmot research supports the interpretation that dispersal by marmots is conditional, that social cohesiveness is important, that proximal factors probably vary with the social system and the life-history strategy of individuals, and that the ultimate factor inducing dispersal is the attempt to maximize individual fitness.

19 Metapopulation dynamics

The term "metapopulation" describes a group of spatially delimited local populations that are to some degree coupled by dispersal from and immigration into the individual, local populations; i.e., they are joined together by the exchange of individuals. Metapopulation ecology especially applies to species that occupy a patchy habitat distributed in a broader landscape matrix that is unsuitable for sustaining a reproductive population. Such spatial structure may arise naturally, as in montane marmot populations, or due to human-caused fragmentation. Two important characteristics of metapopulations are demographic and genetic (Hanski and Gaggiotti 2004). A critical demographic effect is that a patch that goes extinct may be repopulated by immigrants from other patches, the so-called "rescue" effect. The genetic effect occurs when immigrants bring new alleles (or change allelic frequencies) to the patch where they settle. Some genetic consequences of immigration into yellow-bellied marmot populations were described in the previous chapter.

The metapopulation concept has wide acceptance and popularity, but few empirical studies conform to the definition of the classical metapopulation. The classical meta-population describes a set of local populations that are linked by dispersal in a dynamic equilibrium of extinction and recolonization (Fronhofer *et al.* 2012). Four conditions must be fulfilled by a classic metapopulation: (1) any local population must be prone to extinction; (2) recolonization must be possible; (3) each local habitat patch must be capable of supporting a reproductive population; and (4) population dynamics of local populations should be asynchronous. Many mammalian species, usually those of >1 kg body mass, do not meet these criteria. Many studies deduced metapopulation structure if dispersal occurred, but only 5 of 75 species met the classic criteria (Olivier *et al.* 2009).

Many aspects of metapopulation dynamics are species specific (Krohne 1997); such variation among species complicates the application of classic metapopulation criteria to individual species. Classic criteria are narrowly focused; in this discussion a metapopulation is identified when the population is spatially structured; i.e., into discrete, local breeding populations from which dispersal occurs.

Marmot metapopulations

Reviews of mammalian metapopulation structure have not included marmots. Probably most species of marmots form metapopulations, especially those with a mosaic settlement pattern. Little is known about metapopulation structure other than that dispersal occurs in all species and patches with reduced numbers are subject to immigration. When marmots are introduced or reintroduced into an area, vacant habitat patches are colonized by dispersers (Borgo *et al.* 2009, Barrio 2012). Little is known about the frequency and pattern of the exchanges of individuals among patches, although variation in family structure and immigration into occupied patches suggest exchanges occur (e.g., Mikhailuta 1991, Mashkin 2003, Rumyantsev *et al.* 2012).

The historic distribution of the Olympic marmot in Olympic National Park (Barash 1973a) offered an opportunity for modeling metapopulation dynamics (Griffin 2007). Griffin intended to model gene flow, determine movement rates, and identify suitable habitat. She assumed that a stable population existed in the protected national park. However, she discovered that marmots were missing from many areas where they had been recorded and that marmot numbers were declining. The marked decline in dispersal (see Chapter 18) eliminated any rescue effect.

The effect of dispersal on metapopulation dynamics is also illustrated by studies of the Vancouver Island marmot metapopulation (Bryant 1998). Clear-cutting of forest tracts created new habitat; many clear-cuts were colonized and colonization events were limited by the nearness of natural colonies. The association of immigration with nearby colonies suggests that, in general, Vancouver Island marmots disperse only as far as required to locate suitable habitat. During the same time that clear-cuts were colonized, some natural colonies went extinct. Regenerating clear-cuts soon became unsuitable for marmots and those sites went extinct (Bryant 1996a). Vancouver Island marmots typically disperse at age 2 years; most residents at a natural colony were born there (Bryant 1998). The population decline was associated with a decline in dispersal; extinct sites were not re-colonized. Bryant's work strongly suggests that a few sites produced a "surplus" of individuals that provided the dispersers to populate sites that otherwise went extinct.

These studies emphasize that dispersal is conditional and related to the fitness strategies of individuals. When their social/demographic systems permitted individuals to remain philopatric, the local populations no longer formed a metapopulation. The demographic changes in these two marmot species support the necessity of long-term population studies (Clutton-Brock and Sheldon 2010) and that metapopulation modeling must be based on local dynamics and not on assumptions based on a patchy distribution (Krohne 1997, Ozgul *et al.* 2009). The following discussion of the yellow-bellied marmot metapopulation describes its characteristics and emphasizes the significance of the demographic dynamics of local populations.

The yellow-bellied marmot metapopulation

We defined the metapopulation to include all of our study sites in the Upper East River Valley. The "true" metapopulation is unknown, but is larger than our study area. Marmots of both sexes usually dispersed 4 km or less, but males dispersed up to 15 km and females, up to 6 km (Van Vuren 1990). These distances exceed the 5 km range of the distribution of our study sites (Fig. 19.1). For example, one yearling male moved to a site

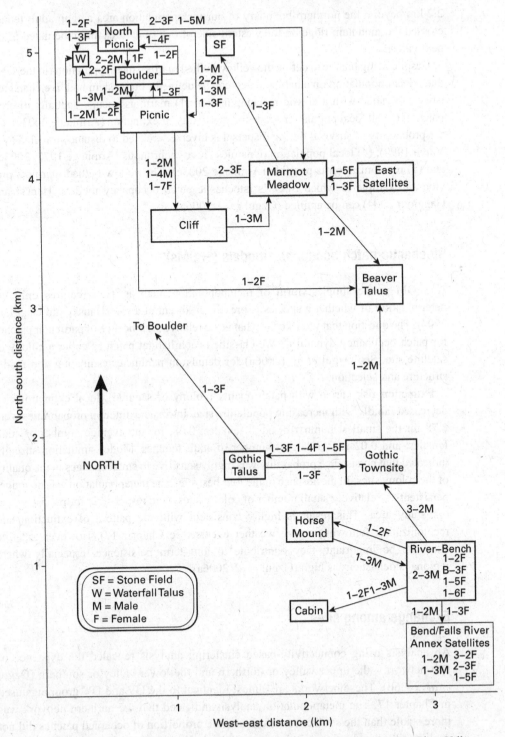

Figure 19.1   Pattern of intersite movements of yellow-bellied marmots in the Upper East River Valley metapopulation. Arrows indicate the direction of the movement. Animals are identified by age and sex and the number of individuals of the same age and sex that moved between sites is indicated; e.g., 1–4F refers to one 4-year-old female; 3–2M refers to three 2-year-old males.

2.2 km beyond the northern boundary of our metapopulation area and an adult female crossed the mountain ridge on the western side of our study area and hibernated in the next valley.

Despite being unable to define the yellow-bellied marmot metapopulation (in the sense that a local population on a habitat patch can be defined), our system had five character-istics associated with a classic metapopulation: (1) marmots occupy spatially discrete patches; (2) all local populations face the possibility of extinction (Armitage 2003c); (3) the probability of survival during dispersal is inversely related to distance traveled (Van Vuren 1990); (4) local population dynamics are asynchronous (Armitage 1977, 2003c), and (5) recolonization is possible (Armitage 2003c). The yellow-bellied marmots pro-vide an example to which SPOMs (stochastic patch occupancy models; Hanski and Gaggiotti 2004) can be applied (Ozgul *et al.* 2006a).

Stochastic patch occupancy models (SPOMs)

A SPOM is the simplest form of metapopulation analysis as it requires only the determination of whether a species is present or absent at a site (Hanski and Gaggiotti 2004). The question that we asked is what is the relative significance of particular patches for patch occupancy dynamics? We classified each habitat patch as either a colony or satellite site. See Ozgul *et al.* (2006a) for details on parameter estimation and model structure and selection.

Extinction risk varied with patch quality (colony or satellite); local extinction risk decreased rapidly with increasing population size. Intrinsic extinction probabilities were 0.78 for the smallest number of adult females; 0.08, for the average number of adult females; and 0.02 for the largest number of adult females. Model simulation strongly indicated that persistence probability was highly sensitive to small changes in the quality of the colony sites. In the absence of the four best sites, the metapopulation was no longer persistent. A relatively small number of colony sites were responsible for persistence in our study area. This result is highly consistent with the pattern of extinction and recolonization following severe weather events (see Chapter 17). However, satellite sites can be important; they contribute to long-term persistence, especially where regional stochasticity is high (Ozgul *et al.* 2006a).

Exchange among sites

An analysis using connectivity-based clustering analysis revealed the existence of two networks: the upper valley or northern and the lower valley or southern (Ozgul *et al.* 2006b). These networks are almost identical to the UV and DV groupings used in Chapter 17. The metapopulation analysis indicated that the northern network was more stable than the southern network and the proportion of occupied patches did not decline either under low or high stochasticity during 100 years of simulations. By contrast, the southern network was highly unstable; the proportion of occupied patches frequently declined to zero, and few of the simulated replicates survived to the end of 100 years.

The indication that the two networks in the marmot metapopulation have very different persistence probabilities and that exchange among sites is critical for long-term persistence leads to an examination of what we know about exchange of marmots among the local sites. I identified 24 males and 45 females that moved from one site to another (Fig. 19.1). In the 40 years over which these data were collected, only one male and one female moved between the two networks. Both moved from the southern to the northern network.

In the southern network, 23 of the 36 movements were within two sites, River/Bench and Bend/Falls/Annex. When we initiated the marmot project, we recorded locations where marmots were trapped and treated those as independent sties. For example, River and Bench were considered separate as were Bend, Falls, and River Annex. However, as data accumulated, we learned that individuals moved between River and Bench as if the two sites formed one large habitat patch with burrow sites and foraging areas distributed across the landscape. Similarly, individuals moved between Bend, Falls, and River Annex; the same resident female could be at Bend one year and Falls the next. From the marmot perspective, the area was utilized as one patch and movements between burrow sites did not differ in principle from movements between burrow sites at Gothic Townsite or Picnic Colonies. Thus, I treat River/Bench and Bend/Falls/River Annex as single sites.

As a consequence of reducing five sites to two, there was very little movement between sites in the southern network (Table 19.1). Most movement was by females in the southern network and both sexes moved equally in the northern network. Fewer movements between sites in the southern network likely account for the differences in the long-term stability between the two networks.

It is of considerable interest to compare intersite movements with overall immigration. Thirty-five (59.3%) successful immigrants were from outside our metapopulation. The high proportion of immigrants from outside the Upper East River metapopulation cannot be attributed to a lack of dispersers (Table 19.1). Using the number of dispersers as an estimate of potential immigrants, over the tenure of the study, only about 12% of the female dispersers and 5% of the male dispersers successfully immigrated into a patch in the metapopulation.

Major factors affecting occupancy

Three major factors affect patch occupancy dynamics. First, and most important, is the social structure. The social system enhances recruitment from within the patch and inhibits immigration. As a consequence, exchange of individuals between patches is minimal. Second, there is no relationship between the number of dispersers from the patches and the number of immigrants settling on a patch. The number of dispersers greatly exceeds the number of immigrants and does not limit the potential for recolonization. Third, successful immigration is highly stochastic. A disperser in

Table 19.1 Some characteristics of yellow-bellied marmot movement between sites. A marmot that was present in its natal site in mid-August was considered a resident; i.e., a non-disperser. A marmot that disappeared during May, June, or July, when most dispersal occurs (Van Vuren 1990), was considered a disperser.

Movement within sites	In southern network, 23 (63.8%) of 36, includes 74.1% of female movements, only three males. None in the northern network.
Movement between sites	Six male movements in southern network. Fourteen males and 17 females in northern network.
Immigrant movement into sites	Only 24 of 59 adult female immigrants into six major population sites were from known sites in our study area.
Number of dispersers	Female yearlings (194), male yearlings (332), 2-year-old males (38), total males (370).

Figure 19.2   A solitary marmot occupying a small habitat patch with rocks that provide a burrow site. Such sites provide shelter for dispersers until possible immigration into a colony.

any given year has little chance of finding unoccupied space in a patch available for settlement. Because the availability of space for colonization varies among years, a disperser must survive, probably in a mini-patch (Fig. 19.2) until it locates available space in a colony. Most immigrants will be long distance dispersers who survived to age 2 or older.

Despite the low probability of success, dispersal prevents extinction of the major sites. For 229 colony-years, extinction was avoided 18 times because an immigrant colonized

Table 19.2 Number of times extinction was avoided because of successful immigration. A colony-year is one colony in 1 year.

Site	Total	Number of recolonization events Within the network	Number of colony-years
River	3	1	41
Cliff	3	2	27
Marmot Meadow	2	1	41
Picnic	0	0	41
Boulder	5	2	40
North Picnic	5	3	39
Total	18	9	229

the site that particular year. In half of the colonizations, the immigrant came from within the network. One colonization occurred when a yearling immigrant the previous year survived to become the resident adult female. The rescue effect did not occur at Picnic Colony as extinction never occurred there (Table 19.2).

Metapopulation theory

Recently developed methods of matrix metapopulation modeling and transient sensitivity analysis were used to investigate the significance of local demography on short-term (transient) versus long-term (asymptotic) dynamics of the yellow-bellied marmot population (Ozgul *et al.* 2009). A major conclusion was that both long-term and short-term dynamics depended primarily on a few colony sites and metapopulation dynamics were highly sensitive to changes in demography at those sites. Also the relative importance of sites differed between long-term and short-term dynamics; e.g., spatial structure and local population sizes were influential on short-term but insignificant for long-term dynamics. The vital rates that were most influential on local dynamics were the most important on short- and long-term metapopulation dynamics. In essence, local demographic factors are important drivers of metapopulation dynamics and "explicit consideration of local demography is essential for a thorough understanding of the dynamics and persistence" of spatially structured populations (Ozgul *et al.* 2009).

Two additional factors are important for the future development of models. First, the function of dispersal was unrealistic. The analysis assumed that after demographic changes took place within each site, dispersal redistributed individuals among sites. Clearly, dispersal does not function in this manner (Fig. 19.1, also see Chapter 18). Second, the social complexity of yellow-bellied marmots was not considered. Social behavior affects dispersal, breeding probability, and survival. The importance of social complexity was demonstrated in a study of the alpine marmot (*M. marmota*). Only a behavioral model predicted population sizes or densities under transient dynamics and predicted dispersal behavior and group-size distributions. By contrast, a model that included spatial structure but ignored social behavior produced unrealistic dispersal events (Stephens *et al.* 2002a). Social behavior can moderate the effect of environmental factors on population dynamics, especially on winter mortality (Grimm *et al.* 2003).

Spatiotemporal variation in survival rates

Because local dynamics are critical for understanding metapopulation dynamics and environmental conditions vary over space and time (e.g., major predation events are unpredictable and may impact one site, whereas other sites are unaffected), a consideration of how survival and reproductive rates vary over space and time is necessary for understanding the effects of local demography on the metapopulation.

Spatiotemporal variation in age-specific annual survival rates was analyzed for 28 years (1976–2003) for colonial and satellite sites (Table 19.3). Seven climatic variables served as temporal covariates. Capture–mark–recapture (CMR) models were used to analyze the data. Several models with different age structures were tested; the most parsimonious model indicated that the survival rate of juveniles (0–1 year) and yearlings (1–2 year) varied significantly among sites whereas there was little support for site effect on adult (>2 year) survival rates; however, adult survival rates were higher in colony sites than in satellite sites (Table 19.4). Adult survival rates were generally higher than yearling and juvenile survival rates in colonies, but no trend was apparent in the satellite groups. The lowest spatial variation in survival rates was observed in adults (0.04); was slightly higher in young (0.08) and was highest in yearlings (0.11). The high variation and generally lower survival rates of yearlings is probably a consequence of dispersal. As noted previously, virtually no yearlings were successful immigrants into our study sites and therefore would be categorized as non-survivors. The variation in timing of dispersal, especially by males (not dispersing until age 2) would produce higher survival rates. In effect, the nature of the social system and the conditional nature of dispersal affect the apparent survival of yearlings. Although spatial variation in survival rates occurred in all three age classes, temporal variation in survival rates (0.15 to 0.89) was apparent only for juveniles (Ozgul *et al.* 2006b).

Table 19.3 Parameters for spatiotemporal variation in annual survival rates.

Locations	Major colonies: River, Gothic Townsite (RMBL), Marmot Meadow, Picnic
	Satellite groups: South, West, East, North
Climatic variables	Annual precipitation
	Annual amount of snow fall
	Length of the growing season: number of days between first bare ground in the spring and the first killing frost in the autumn
	Monthly mean summer temperature
	Duration of permanent snow cover
	Julian date of first bare ground;
	Julian date of first permanent snow pack

Table 19.4 Survival rates and breeding probabilities of female yellow-bellied marmots. All values are mean ± standard error.

	Sites							
	Colonial sites					Satellite sites		
Survival	River	Gothic	Marmot Meadow	Picnic	South	West	East	North
Juvenile	0.67 ± 0.04	0.52 ± 0.05	0.53 ± 0.05	0.54 ± 0.04	0.75 ± 0.07	0.61 ± 0.08	0.78 ± 0.11	0.57 ± 0.05
Yearling	0.53 ± 0.06	0.48 ± 0.07	0.30 ± 0.07	0.54 ± 0.06	0.33 ± 0.07	0.51 ± 0.11	0.78 ± 0.15	0.57 ± 0.07
Adult	0.80 ± 0.03	0.73 ± 0.05	0.72 ± 0.07	0.76 ± 0.04	0.66 ± 0.07	0.65 ± 0.08	0.62 ± 0.09	0.63 ± 0.05
Breeding probability						Mean for all satellites		
Sub-adult	0.27 ± 0.05	0.23 ± 0.07	0.57 ± 0.14	0.57 ± 0.07		0.40 ± 0.06		
Adult	0.61 ± 0.06	0.71 ± 0.06	0.88 ± 0.05	0.55 ± 0.05		0.59 ± 0.05		

Weather and survival

Severe weather events affect marmot populations (see Chapter 17). Do typical weather events or other environmental factors, such as those associated with site, affect either spatial or temporal survival rates? Site-specific variation in juvenile survival was positively influenced by aspect; survival rates were higher on southwest than on northeast facing slopes. Snowmelt occurs earlier, hence the growing season is longer on the southwest facing slopes. Juvenile survival was negatively affected by elevation where snow cover persists for a longer time (Van Vuren and Armitage 1991). Site-specific variation in yearling survival rates was positively influenced by elevation. There is no obvious explanation for this relationship, but it may reflect the timing of dispersal. More yearling males remained in their natal sites and dispersed as 2-year olds at the high-elevation sites (e.g., Picnic, Cliff, North Picnic) than at the lower elevation sites (e.g., River, Gothic), and this pattern increased the apparent survival of yearlings. Site-specific variation in the rates of adult survival was positively influenced by average group size, a result that is consistent with the positive affect of matriline size on survival (Armitage and Schwartz 2000). Temporal variation in survival rates of juveniles was negatively affected by the length of permanent snow cover (survival decreased as length of snow cover increased) and is consistent with the site-specific influences. Snow cover is an important factor in marmot biology and will be discussed more extensively in Chapter 21.

As would be expected, spatiotemporal variation in age-specific survival rates was reflected in spatial and temporal variation in population growth rates. Both population growth rate and adult growth rate were primarily affected by site-specific differences in juvenile survival rates. Spatial variation in adult population growth rate was significantly influenced by juvenile survival, but not by that of yearlings or adults. Adult population growth rate closely co-varied over time with a 1-year time lag with juvenile survival. This pattern is directly related to recruitment. When juvenile survival was determined, juveniles had become yearlings. A 1-year time lag makes them age 2, which is the age of recruitment as an adult (Armitage 1984, 1988) and is the main process by which the number of resident adults increases (Armitage 1996b). It is not surprising that yearling survival did not affect adult population growth rate as yearling survival was confounded by dispersal. This relationship between juvenile survival and adult population growth rate is also consistent with the positive correlations between the number of young weaned and the number of yearling daughters and the number of 2-year-old recruits (Armitage and Van Vuren 2003). In effect, the more young weaned, the greater the likelihood that the number of resident adults will increase. To summarize, survival rates vary spatially and this spatial variation affects metapopulation dynamics.

Spatiotemporal variation in reproduction

Metapopulation dynamics should also be affected by spatial variation in reproductive rates. Spatial and temporal variation in breeding probabilities and litter size were investigated for 2-year-old females, sub-adult females (age of first reproduction > 2), and adult females (age >2 and have previously reproduced). There was no site effect on the breeding probability of 2-year olds (0.25 ± 0.03). This result is consistent with repro-ductive suppression of 2-year olds by older females and is a function of the matrilineal social organization that does not differ among sites. However, the breeding probability of sub-adults and adults did vary among sites (Table 19.4). Marmot Meadow had higher breeding probabilities for adults than all other sites and, for sub-adults, than all other sites except Picnic. The higher breeding probabilities at Marmot Meadow may be related to the very low rate of immigration (Table 18.5). Immigrant sub-adults had a lower breeding probability than philopatric sub-adults and, in general, immigrants are unlikely to reproduce in their year of immigration. In general, breeding probability was higher in adults than sub-adults. The lower breeding probability of sub-adults is also likely a function of reproductive suppression. Sub-adults are usually younger than adults and female marmots are significantly less likely to reproduce when older females are present (Nuckolls 2010).

Environmental factors and breeding probability

Breeding probability of adults was higher on southwest than on northeast facing slopes; the pattern was reversed for sub-adults. This difference probably is related to reproductive suppression. A younger female is less likely to reproduce if the older female is reproductive (Armitage 2003f). The lower likelihood of an adult female reproducing on a northeast facing slope may relax reproductive suppression and permit the sub-adult to breed. By contrast, the higher reproductive probability of adult females on southwest facing slopes should reduce the likelihood of sub-adult reproduction. Elevation acted similarly; adults had a lower breeding probability at higher elevation and sub-adults, at lower elevation. The pattern is consistent with the expected role of reproductive suppression.

The proportion of temporal variation in breeding probability explained (r^2) by climatic factors was very small. Severity of the winter ($r^2 = 0.07$) negatively influenced and rainfall the previous summer ($r^2 = 0.08$) positively influenced breeding in adult females, and a delay in the onset of summer ($r^2 = 0.04$) decreased breeding probability in sub-adults.

Litter size at weaning varied among sites; the largest mean litter size was 5.03 ± 0.19 at Marmot Meadow and the smallest was 3.74 ± 0.14 at satellites (Ozgul *et al.* 2007). There was no year effect and a small age effect; average littler size of 2-year-old females (3.79 ± 0.16) was slightly smaller than that (4.22 ± 0.08) of older females, a finding in agreement with the life table analysis (Schwartz *et al.* 1998).

The annual realized population growth rate varied spatially from 1.00 ± 0.01 (Satellites) to 1.04 ± 0.01 (Gothic) and varied yearly from 0.69 ± 0.04 (1981–82) to 1.51 ± 0.13 (2003–04). Interestingly, 1981 was a year of major badger predation that greatly reduced the number of young and adults that survived to 1982. Also, 1982 was characterized by late snow cover and low June rainfall that reduced the number of animals in the juvenile and yearling age classes (Schwartz and Armitage 2003). Litter size had the strongest influence on site-specific ($r^2 = 0.56$) and temporal ($r^2 = 0.74$) variation in population growth rate. Although temporal variation in the probability of

Table 19.5 The effects of breeding probabilities on spatial and temporal variation on population growth rate of yellow-bellied marmots. r^2 = proportion of variation explained.

Age group	Breeding probabilities	
	Spatial	Temporal
2-year old	$r^2 = 0.35$	$r^2 = 0.14$
Sub-adult	$r^2 = 0.19$	$r^2 = 0.17$
Adult	$r^2 = 0.19$	$r^2 = 0.12$

breeding was little affected by age group, spatial variation in the probability of breeding was more influenced by 2-year olds than by older females (Table 19.5). The greater influence of 2-year-old females exemplifies the role of reproductive competition within the matriline; reproductive suppression reduces the probability of a 2-year old reproducing which increases the age of first reproduction and decreases population growth rate (Oli and Armitage 2004). When 2-year olds reproduce, population growth rate is higher; in large, successful matrilines all age groups reproduce and NRR is high. In large matrilines with low success, young females fail to breed and NRR is low (Armitage and Schwartz 2000, Armitage 2007). Breeding probabilities also influenced temporal variation in growth rate, but their effects were much less than that of litter size.

The analysis of spatiotemporal variation in survival (Ozgul *et al.* 2006b) and reproduction (Ozgul *et al.* 2007) indicate that components of recruitment (litter size, breeding probability, and juvenile survival) are the major demographic factors driving fluctuations in population growth rate.

Comparison of survival rates of other marmot species

Population studies of alpine marmots in France (Farand *et al.* 2002) and the Vancouver Island marmot (Bryant 1998) allow comparisons with the yellow-bellied marmot. These three species are all social, but differ in the degree of sociality (Table 5.2). The alpine marmot study considered one population of 17 contiguous family groups. In a sense, this population is similar to a single colonial patch of yellow-bellied marmots, but habitat variability within the large patch provides some environmental variability. The Vancouver Island marmot study was a metapopulation analysis, but for some comparisons, I use the metapopulation means of all three species.

Sex did not affect survival rates of alpine marmots. However, beyond age 1, survivorship of females was greater than that of males at all ages for yellow-bellied and Vancouver Island marmots. This difference, in part, is related to the social system. Later age of dispersal and movement of individuals between family groups reduces male mortality in alpine marmots, whereas earlier dispersal of yellow-bellied and Vancouver Island males results in their higher mortalities.

The survival rate of young is 0.62 for alpine marmots, 0.53 to 0.54 for yellow-bellied marmots (Schwartz *et al.* 1998) and 0.49 for Vancouver Island marmots. Given the variation in survival rates among sites for yellow-bellied marmots (at four sites the survival rate was 0.61 or higher, Table 19.4), it is unclear whether juvenile survival is actually higher in alpine marmots or whether the study site was particularly favorable. For example, juvenile survival rate was 0.37 in a colony in Italy with the same social system (Lenti Boero 1994). There was no time effect on juvenile survival for alpine marmots in contrast to the highly temporal variation in Vancouver Island marmots (0.0 to 0.74) and yellow-bellied marmots (0.15 to 0.89). Spatial variation in juvenile survival rates of Vancouver Island and yellow-bellied marmots varied considerably.

The average rate of survival of older age classes (0.71) of alpine marmots was similar to the rates at yellow-bellied marmot colony sites. Adult female survival rates for Vancouver Island marmots averaged about 0.68, lower than that of the alpine marmot and of yellow-bellied marmot colonial females. Both spatial and temporal variation characterized the Vancouver Island marmot population, whereas the yellow-bellied marmot survival rates varied among sites but not significantly over time. The survival rates of older alpine marmots also varied over time and that variation was strongly related to the intensity of the fall frost, which may act by reducing soil temperature at the beginning of hibernation that subsequently increases the energetic cost of hibernation.

No environmental effect on juvenile survival of alpine marmots was detected. Quite possibly social thermoregulation in this species buffers juveniles from such effects. However, the range of environmental effects may not have been sufficient to affect survival. This result contrasts markedly with the yellow-bellied marmot research that concluded that juvenile survival rates were mainly influenced by environmental factors (Ozgul *et al.* 2006b). There was some indication that summer rainfall affected survival of juvenile alpine marmots. The area normally receives ample rain, droughts are rare, and when summers have excessive rainfall, juveniles remain in their burrows for long periods of time. Cool, wet environments increase the costs of thermoregulation (Turk and Arnold 1988, Melcher *et al.* 1990), mass gain is reduced and may be insufficient for surviving hibernation.

The difference in survival rates between the Vancouver Island marmot and the other species is a consequence of relatively stable alpine and yellow-bellied marmot populations and a declining Vancouver Island marmot population.

Summary

The analysis of metapopulation dynamics of yellow-bellied marmots provides two major conclusions. First, metapopulation dynamics is driven by the population dynamics of the individual patches, especially the larger colonies. In particular, rates of immigration and dispersal from these sites emphasize the importance of local processes and that metapopulation models should not use assumptions of equal exchange rates among local populations. Second, metapopulation dynamics is strongly influenced by social dynamics. The functions of the social system need to be incorporated into metapopulation models in order to approach a higher degree of biological reality.

20 Population regulation or population limitation

In Chapter 1, I explained that my interest in the yellow-bellied marmot was as a model to investigate the possible role of social behavior as a population-regulation mechanism. Now that we have explored many facets of marmot biology, it is appropriate that we examine rigorously population regulation and determine whether the concept of population regulation applies to yellow-bellied marmot population dynamics. This examination is especially appropriate in that extensive studies of rodent social behavior question whether social behavior impacts population dynamics (Krebs *et al.* 2007). Social behavior can affect population dynamics by limiting the size of the breeding population, by determining the age of sexual maturity, and by controlling the rates of natality, mortality, and dispersal.

The model

The possible role of social behavior in population regulation is related to the argument of whether populations are limited by extrinsic factors (weather, food, parasites, and predators) or by intrinsic factors (social behavior). A debate developed over whether regulation was density-dependent or density independent (see Krebs 1985, for review). It was widely assumed that population control was density-dependent, but arguments developed that the populations could be self-limiting and that a major role could be played by dispersal, which would maintain the population below carrying capacity (Lidicker 1962). Much of the early arguments about population regulation considered members of a population to be essentially identical; however, Lomnicki (1988) emphasized the importance of individual variation and derived models of self-regulation from the properties of individuals.

These considerations can be illustrated in a feedback model (Fig. 20.1) which is based on the idea of self-regulation in which social behavior plays a central role. As population increases, non-regulatory mortality (predation, weather) occurs, but the critical factor is an increase in social competition. Social competition causes natality, recruitment, and immigration to decrease and dispersal and mortality to increase, which in turn causes the population to decrease. During population decline, non-regulatory mortality occurs, but the critical factor is decreased social competition. As a consequence dispersal and behaviorally induced mortality decrease and natality, recruitment, and immigration increase. The subsequent population increase initiates the control mechanism and the feedback loop continues.

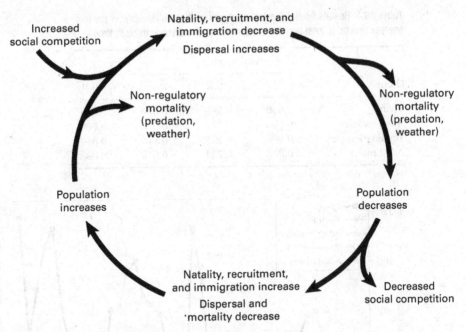

Figure 20.1   A density-dependent feedback model of population regulation.

It is quite difficult to determine which factor(s) regulate population density in natural populations. Although predation is sometimes detected (see Table 17.4), it is usually cryptic. Rather than trying to quantify the contributions of various factors to mortality, I examined population dynamics for evidence of population regulation. First, I examine population structure and changes in numbers in the yellow-bellied marmot metapopulation. Then I repeat the analyses for individual colonies. Finally, I compare the density-dependent and population-limitation hypotheses through a series of density analyses.

Metapopulation analysis

For the metapopulation analysis, I chose four sites (River/Bench, Marmot Meadow, Picnic, and Boulder) that were trapped each year from 1964 through 2002. The numbers were summed over all sites to determine the number of resident young, female yearlings, adult males, adult females, and total number (Fig. 20.2).

Each of the four age/sex groups contributed significantly to total numbers (Table 20.1). The contributions of each age/sex group to the total population are described by the following multiple regression:

$$\text{Total no.} = -0.18 + 1.01 \text{ No. young} + 0.924 \text{ No. female yearlings} + 1.15 \text{ No. adult female} + 0.717 \text{ No. adult males}$$

Because each age/sex group was included in the regression, it is not surprising that the regression adj-$R^2 = 0.994$.

Table 20.1 Results from univariate analyses on the contributions of the four age/sex groups to total population numbers and to change through time.

Group	Total number		Time	
	p	adj-R^2	P	adj-R^2
Young	0.000	0.898	0.040	0.086
Adult females	0.000	0.311	0.002	0.209
Yearling females	0.000	0.535	0.116	0.040
Adult males	0.001	0.234	0.610	0.000

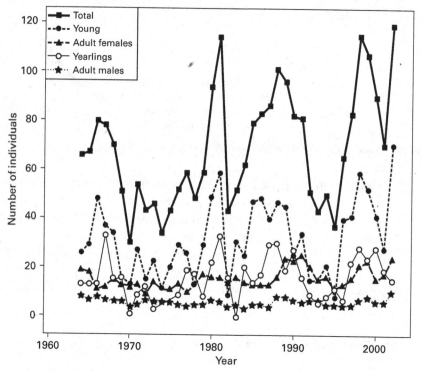

Figure 20.2   The variation in numbers over time for age/sex classes in a yellow-bellied marmot metapopulation.

. Despite the wide fluctuations from year to year, total number significantly increased over time (Fig. 20.3). Among age/sex groups, only the number of young and adult females significantly increased over time (Table 20.1). The effect of year explained only about 15% of the variation in total number and its effect might be a consequence of the strong contribution of the number of adult females. Therefore, the multiple regression was repeated with the four age/sex groups and year. All partial coefficients were statistically significant except year (partial p = 0.19). However, year is probably biologically significant because of the contribution of adult females. Their significant

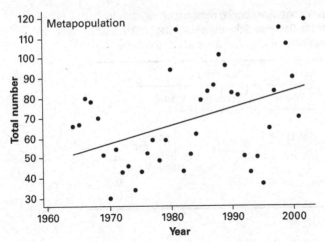

Figure 20.3 The relationship between total numbers in a metapopulation and time. Total no. $= -1655 + 0.89$ year; $p = 0.009$, adj-$R^2 = 0.149$.

contribution to the population increase from 1964 through 2002 is similar to their significant contribution to the population increase since 2002 (Fig. 21.4), an increase primarily caused by their increased survivorship (Ozgul *et al.* 2010). Also, there was a slightly positive λ between 1976 and 2002 prior to the rapid population increase ($\lambda = 1.18$) after 2002 (Ozgul *et al.* 2010). The rapid increase in total number and number of adult females raises the question of whether the yellow-bellied marmot population is regulated. This question will be discussed fully later in this chapter.

Because of their low numbers in a colony, males were considered to make little contribution to population dynamics (Armitage 1991a). However, the number of adult males contributed significantly, albeit the smallest contribution, to the variation in total number of residents (Table 20.1). The major effect of males on population dynamics may be through their fitness effects on females (Rankin and Kokko 2007). Male absence results in no reproduction (Armitage 2003f) and more young are produced when a familiar male compared to when a new male is present (Armitage and Downhower 1974). Male activity has population consequences, but these consequences are not regulatory.

The number of young contributed most to the variation in the total number of residents (Table 20.1). The number of young can be predicted from the number of adult females:

$$\text{No. young} = 7.49 + 1.55 \text{ No. females}, p = 0.008, \text{adj-}R^2 = 0.15$$

According to the feedback model, as the number of adult females increases, the number of young should decrease (decrease in natality), which is contrary to the positive coefficient in the regression analysis. However, the number of females explains only a small part of the variation in the number of young. Much of the variation is explained by reproductive suppression within the matriline, which is independent of the number of

Table 20.2 Univariate correlations among demographic variables for yellow-bellied marmots. Data from Schwartz and Armitage (2002, 2005). R_o = net reproductive rate, * $p < 0.02$, ** $p < 0.01$, *** $p < 0.001$.

| | % survival | | | |
	Young	Yearling	Adult	R_o
Survival				
Total	0.83***		0.54**	
Young		0.81**	0.56*	
Yearling			0.64**	
Density				0.71**
Mean litter size				0.56*

older females in the matriline (Armitage 2003f) and by other factors such as the age structure of the female population (Nuckolls 2010).

The number of yearling females in a year can be predicted by the number of young the previous year:

$$\text{No. yearlings} = 0.26 + 0.503 \text{ No. young, } p = 0.000, \text{ adj-}R^2 = 0.649$$

Thus, the number of young explains most of the variation in the number of female yearlings. The remaining variation can be explained by summer mortality (Table 17.4) and differential survival among colonies due to predation, drought, and prolonged snow cover (Armitage 1994, 2004, Ozgul et al. 2006). Thus mortality of young does not increase at higher population numbers as expected from the feedback model; mortality of young appears to be density independent.

Dispersal is unrelated to density (Armitage et al. 2011, Chapter 18); recruitment is unrelated to density, and immigration is weakly related to density (Armitage 2003i). Both immigration and recruitment are dependent on the social structure of matrilines; not to density. Survival is significantly positively correlated among young, yearlings, and adults (Table 20.2). There is no indication that a particular age class is affected when total survival changes; good or poor conditions for survival affect all marmots. If population regulation was involved in density changes, we would expect a pattern of differential survival with regulatory mechanisms more severely impacting the more vulnerable young. Mean litter size and R_o increase as population density increases (Table 20.2). This density-dependent relationship between population size and R_o is contrary to the expectations of the feedback model and indicates an Allee effect (Gregory et al. 2010).

These interpretations of a lack of evidence for density-dependent regulation are consistent with earlier work that tested for the effect of density dependence on vital rates, including the probability of reproduction (Oli and Armitage 2003) and survival (Ozgul et al. 2006b). Population density did not influence age of first reproduction or individual fitness and survival rates were higher at higher densities. I conclude that the evidence is inconsistent with a population-regulation model when examined at the

Table 20.3 Linear regressions of relationships between population variables for three colonies. No. = number, A = adult, F = females, Yg = young, Ylg = yearlings.

Colony	Regression	p	adj-R^2
A. Boulder	No. Yg = −0.397 + 3.48 No A F	0.000	0.494
River	No. Yg = 3.66 + 0.961 No A F	0.017	0.118
Picnic	No. Yg = 0.18 + 2.29 No A F	0.001	0.255
B. Boulder	Total no. = 1.26 + 1.19 No Yg	0.000	0.940
River	Total no = 3.58 + 1.18 No Yg	0.000	0.780
Picnic	Total no. = 5.43 + 1.17 No Yg	0.000	0.947
C. Boulder	No. age 2 F = 0.053 + 0.863 No Ylg F	0.000	0.794
River	No. age 2 F = 0.070 + 0.832 No Ylg F	0.000	0.865
Picnic	No. age 2 F = 0.032 + 0.926 No Ylg F	0.001	0.793
D. Boulder	Total no. = −108 + 0.058 year	0.455	0.000
River	Total no. = −898 + 0.459 year	0.000	0.458
Picnic	Total no. = −220 + 0.120 year	0.324	0.000

metapopulation level. However, population dynamics is highly dependent on the dynamics of individual colonies and these will now be examined.

Analysis of individual colonies

First, I examined population relationships for three colonies: Picnic, Boulder, and River. These colonies were chosen because they were monitored every year from 1962 through 2002 and they represent the diversity in size and location of marmot colonies.

At all three sites, the number of young is significantly related to the number of adult females (Table 20.3A). The amount of variation in the number of young that is explained by the number of adult females varies among the sites and is highest at Boulder. Boulder has the smallest mean matriline size among these three sites (Table 8.1); as a consequence, reproductive suppression plays a lesser role in determining the number of young produced than at River and Picnic where mean matriline sizes are larger and where multiple matrilines occur in many years. These results are the same as those obtained for the metapopulation analysis except the colony analysis reveals the differences among sites.

The total number of residents is significantly affected by the number of young at all three sites (Table 20.3B). The number of young explains most of the variation in the total number of residents, similar to the metapopulation analysis. However, there is no significant correlation over 41 years between the total numbers at any two sites (Table 20.4). Boulder and Picnic are adjacent; the difference in their population trends is unlikely to be environmental; the two sites are at a similar elevation with similar aspects, rainfall, and snowfall. The contribution of yearlings to the total number varies among the three sites; as a consequence the amount of variation in total number explained by adults is greatest for Boulder (Table 20.4). I conclude the differences in

Table 20.4 Comparison of female age contributions to total number of residents at three sites. Total numbers are not correlated. Boulder vs. River, r = 0.26; Boulder vs. Picnic, r = 0.13; River vs. Picnic, r = 0.02. Values in the table are for adjusted-R^2.

	Contribution to total number	
Site	Yearlings	Adults
Boulder	0.050	0.589
Picnic	0.000	0.350
River	0.122	0.422

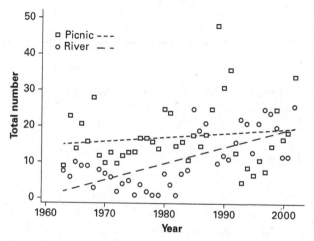

Figure 20.4   The relationship between total numbers and year for River and Picnic Colonies.

population trends are most strongly affected by the demographic processes within each site.

The recruitment of 2-year-old daughters represents the addition of reproductive adults to the resident population. The number of 2-year-old females is significantly and positively related to the number of female yearlings (Table 20.3C). This relationship explains 79–86% of the variation in the number of 2-year-old females at the three sites. Some of the variation results from the death of yearling females during the active season or during hibernation (Van Vuren and Armitage 1994a). This pattern of recruitment is consistent with an adult female's attempt to maximize her individual fitness by recruiting reproductive daughters into her matriline and does not indicate a response to the density of matriline residents.

Finally, the total number of residents was related to year only at River (Table 20.3D, Fig. 20.4). The slight effect of year on population numbers in the metapopulation analysis probably reflects the major influence of River and does not reflect the population trends at all sites. Overall, population numbers and their relationships at the level of individual sites do not support density-dependent population regulation.

Population regulation versus population limitation

It is widely assumed that populations are regulated by density-dependent mechanisms (see discussion in Murray 1999). Most models of population dynamics incorporate the assumption of density dependence on the basis that density-dependent regulation is a logical necessity. Murray (1999) proposed a population-limitation hypothesis in which density-independent processes have a density-dependent effect; i.e., reduce the birth rate or increase the death rate with increasing density. Above some upper critical density, birth rate declines, death rate increases, or both occur. Thus, a density-dependent effect can occur above the upper critical density. Mortality could affect population growth rate and may or may not affect population numbers. What is important is the relationship between density (population numbers) and the limiting resource, not density per se.

Although population processes based on numbers are contrary to predictions from a density-dependent feedback model, an analysis based on rates could be consistent with those predictions. For example, density-dependent models predict either a linear or curvilinear relationship between birth rates and population density across a range of densities. By contrast, the population-limitation hypothesis predicts that birth rate is independent of density over a wide range of densities and may decline only at very high densities (Murray 1999). Therefore, I examined the relationship between density and several demographic variables for the three sites used in the previous analyses.

Birth rates

I calculated birth rates of all young and of female young separately. Birth rate of female young was not significantly related to the number of adult females; only at River Colony was birth rate for all young significantly related to the number of adult females, but the relationship explained only 7.8% of the variation in birth rate (Table 20.5A). Thus, I conclude that there is no density-dependent regulation of birth rate in these populations of yellow-bellied marmots.

Because yearling females are recruited into their natal population, adult females could use the number of female yearlings as a cue for adjusting birth rates. When there is a high probability of recruiting a daughter, a female might reduce her investment in reproduction and direct her energy to recruitment. Therefore, I examined the relationship between the birth rate of young and the number of female yearlings. Several regressions had relatively low p values, but the regressions explained only 6.3% or less of the variation (Table 20.5B). Contrary to expectations from density-dependent hypotheses, the birth rate tended to increase with the number of female yearlings, except at Picnic Colony, but there was considerable variation in birth rate especially at low numbers of female yearlings (Fig. 20.5). Analysis of a larger data set (all marmots at all sites) revealed a significant, positive association between the average number of female yearlings and the lifetime production of young (Nuckolls 2010). This relationship is interpreted to mean that where conditions favor the survival of female yearlings, they also favor the production of young. Again, this relationship contradicts predictions from density-dependent hypotheses.

Table 20.5 Linear regressions describing the relationship between birth rate and population numbers of adult and yearling females for three colonies. BR = birth rate, F = female, A = adult, Yg = young, Ylg = yearling, No. = number.

Colony	Regression	p	adj-R^2
A. River	BRFYg = 1.17 − 0.608 No. AF	0.235	0.012
	BRallYg = 2.92 − 0.224 No. AF	0.045	0.078
Marmot	BRFYg = 1.07 + 0.246 No. AF	0.175	0.024
Meadow	BRallYg = 2.67 + 0.327 No. AF	0.314	0.021
Picnic	BRFYg = 0.914 − 0.0014 No. AF	0.976	0.000
	BRallYg = 1.81 − 0.0086 No. AF	0.916	0.000
B. River	BRFYg = 0.690 + 0.071 No. FYlg	0.075	0.057
	BRallYg = 1.72 + 0.10 No. Ylg	0.262	0.008
Marmot	BRFYg = 1.18 + 0.010 No. Ylg	0.073	0.059
Meadow	BRallYg = 2.61 + 0.182 No. Ylg	0.067	0.063
Picnic	BRFYg = 0.926 + 0.0039 No. Ylg	0.846	0.000
	BRallYg = 1.76 + 0.0054 No. Ylg	0.877	0.000

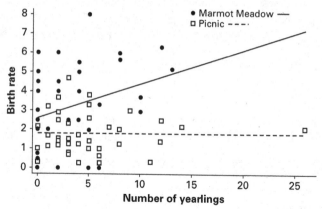

Figure 20.5   The relationship between birth rate for all young and the number of female yearlings for Marmot Meadow and Picnic Colonies.

Survival rates

Survival rate is expected to decrease when density increases in a density-dependent feedback model. The density of adult females had little effect on their survival rate (Table 20.6). Only at Picnic was the relationship significant; however density explained only 14.7% of the variation in the rate of survival. In each colony survival rate varies considerably at any density of adult females; e.g., survival rate at Picnic Colony can be higher at a density of seven females than at a density of five females (Fig. 20.6). At Marmot Meadow the lowest survival rate occurred at the highest density of five females when badger predation killed about 95% of all residents. This incident represents a density-independent factor acting in a density-dependent manner (Murray 1999).

Table 20.6 The relationship between survival rates and population numbers of female adults and young for three colonies. A = adult, S = survival, Yg = young, F = female, No. = number.

Colony	Regression	p	adj-R^2
River	AFS = 0.607 + 0.0163 No. AF	0.450	0.000
	YgFS = 0.772 − 0.0167 No. AF	0.437	0.000
Marmot	AFS = 0.695 − 0.0007 No. AF	0.990	0.000
Meadow	YgFS = 0.799 − 0.101 No. AF	0.102	0.062
Picnic	AFS = 0.991 − 0.0377 No. AF	0.008	0.147
	YgFS = 0.724 − 0.0422 No. AF	0.072	0.064

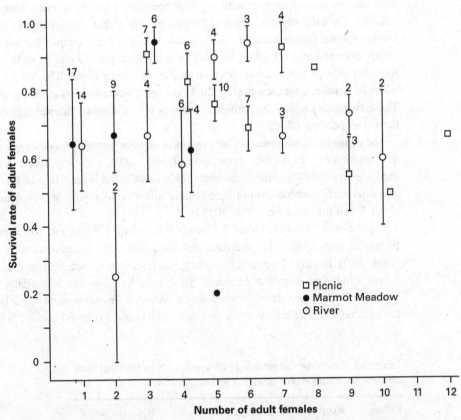

Figure 20.6   Survival rate of adult females as a function of the number of adult females. All values are mean ± SE. The number of adult females is a measure of female density. The numbers by the vertical lines represent the number of years that number of adult females was present. Points without a number are samples of one. Note that Picnic never had fewer than three resident females and Marmot Meadow usually had three or fewer resident females.

The survival rate of young females is little affected by the number of adult females; only about 6% of the variation in survival rates is explained by the regression (Table 20.6). Survival rates of all young also were unrelated to the density of adult females (e.g., River adj-$R^2 = 0$).

Growth rates

Realized per capita growth rate (lambda, λ) could be density dependent. First, I determined whether λ changed significantly over time. There was no significant change in λ over time at either River or Picnic colonies (Table 20.7). The correlation coefficient for λ between the two colonies was insignificant ($r = 0.001$). This result further emphasizes the independence of population processes among colonies.

The density of adult females explains a relatively small (20% or less) amount of the variation in λ of adult females (Table 20.8A). At both River and Picnic, λ varies considerably at any give density of adult females (Fig. 20.7). For example, at both colonies, λ at a density of seven females may be higher than that for four or five females. Environmental factors account for some low λ values. For example, the lowest value at Picnic occurred in 1991 when 10 adult females were present and drought caused high mortality of reproductive females and young (Armitage 1994). The lowest λ at River, when 10 females were present (Fig. 20.7), occurred when badgers killed several females. These two cases of low λ represent instances where density-independent factors have density-dependent effects.

The density of adult females also explains a relatively small (9% or less) proportion of the variation in λ of the total population (Table 20.8B); a slightly higher percentage is explained by total population numbers (Table 20.8C), but is only 16% or less. Birth rate as a univariate variable did not significantly affect population λ and explained at most about 3% of the variation (Table 20.8D).

Population λ is little affected by density; therefore, I tested the contribution of six factors to population λ for the three colonies. The three regressions were significant (Table 20.9). Because the number of young was highly correlated with the other variables, it was eliminated from the equation. No density variable was statistically significant (Table 20.9); therefore, density variables were removed. The subsequent multiple regression for each colony was highly significant ($p < 0.001$) and explained about 25–61% of the

Table 20.7 The relationship between λ and year (N = 40) for River and Picnic colonies. No. = number, A = adult, F = female.

Colony	Regression	p	adj-R^2
River	λ(total No.) = $-11.3 + 0.00572$ year	0.485	0
	λ(AF) = $-7.0 + 0.00356$ year	0.601	0
Picnic	λ(total No.) = $-4.4 + 0.00224$ year	0.661	0
	λ(AF) = $-2.7 + 0.00140$ year	0.788	0

Table 20.8 The relationship between λ and measures of population density at three colonies. No. = number, A = adult, F = female, BR = birth rate, Tot. = total number.

Colony	Regression	P	adj-R^2
A. River	λ(AF) = 0.185 – 0.0475 No. AF	0.092	0.049
Marmot Meadow	λ(AF) = 0.717 – 0.381 No. AF	0.009	0.146
Picnic	λ(AF) = 0.487 – 0.0832 No. AF	0.002	0.200
B. River	λ(Tot.) = 0.083 – 0.0185 No. AF	0.592	0.000
Marmot Meadow	λ(Tot.) = 0.530 – 0.265 No. AF	0.034	0.089
Picnic	λ(Tot.) = 0.287 – 0.0475 No. AF	0.09	0.050
C. River	λ(Tot.) = 0.247 – 0.0159 Tot.	0.117	0.039
Marmot Meadow	λ(Tot.) = 0.563 – 0.0407 Tot.	0.006	0.160
Picnic	λ(Tot.) = 0.291 – 0.0118 Tot.	0.025	0.102
D. River	λ(Tot.) = 0.068 – 0.0263 BR	0.592	0.000
Marmot Meadow	λ(Tot.) = 0.340 – 0.101 BR	0.137	0.033
Picnic	λ(Tot.) = 0.169 – 0.0823 BR	0.146	0.030

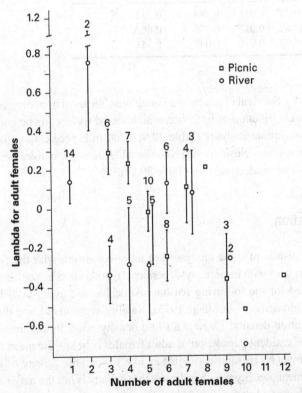

Figure 20.7 λ for adult females as a function of the number of adult females. Values are mean ± SE except for those instances when a particular number of females occurred only once. Numbers adjacent to vertical lines are the number of years that number of resident females occurred.

Table 20.9 Results from multiple regression analyses of the effects of number of adult females, number of female yearlings, total number, birth rate of young and survival rates of young and of adult females on population λ.

Colony	p	adj-R^2
River	0.023	0.291
Marmot Meadow	0.000	0.570
Picnic	0.003	0.334

Number of young removed: partial p for density varied from 0.120 to 0.954.

Table 20.10 Partial p values for a multiple regression of the contribution of three demographic variables (adult and young survival and birth rate) to population λ for three colonies.

Colony	Survival		Birth rate	Regression adj-R^2
	Adult	Young		
River	0.037	0.036	0.488	0.251
Marmot Meadow	0.002	0.016	0.000	0.609
Picnic	0.002	0.036	0.042	0.341

variation (Table 20.10). Survival of adults and young were the most important contributors to population λ. The contribution of birth rate, which varied widely among colonies, was insignificant in the univariate analyses (Table 20.8D) and in the regression that included density variables, but was significant for two colonies when density variables were removed and survival variables were accounted for (Table 20.10).

Population limitation

I conclude that the results of these analyses are more consistent with the population-limitation hypothesis than with the density-dependent hypothesis. The population model (Fig. 20.1) is rejected for the following reasons. Amicable and agonistic behavior are unrelated to population density (Armitage 1975). Natality, recruitment, and dispersal do not vary with population density. There is a slight density effect in that immigration is more likely when the resident population of adult females is below the mean number of females at a site. However, recruitment is just as likely when the density of females is above or below the mean density which indicates that density is not the major factor that determines the number of adult females added to a population.

In an extensive review of population regulation in mammals, Wolff (1997) concluded that intrinsic regulation was unlikely to occur in polygynous species and was most likely to occur when female territoriality, the threat of infanticide, and the presence of male relatives in the natal home range prevailed. Wolff's model predicts that young females exhibit reproductive suppression and that dispersal had limited potential to regulate population density. Yellow-bellied marmots do not fit this model of intrinsic regulation; e.g., male relatives typically are not present and female territoriality is matrilineal, not individual. However, reproductive suppression is consistent with the model, if one considers infanticide to be a form of reproductive suppression.

Intrinsic limitation

Intrinsic mechanisms related to the function of matrilines could limit yellow-bellied marmot populations. Matriline territoriality reduces space available for non-relatives, eliminates immigration into the matrilineal space, but allows for an increase in matriline size by recruiting daughters. When matrilines form, home range area of individual females does not increase, but their space overlap increases (Armitage 1975). The number of matrilines increases as population numbers increase; mean number of female residents is significantly correlated ($r_s = 0.601$, $0.05 > p > 0.01$) with the mean number of matrilines, but not with the mean size of matrilines ($r_s = 0.086$, $p \gg 0.05$). This relationship indicates that the mean size of matrilines is intrinsic to the reproductive and recruitment patterns of individual matrilines. The mean number of matrilines is positively ($r_s = 0.886$, $p = 0.05$) and mean size of matrilines is negatively ($r_s = -0.771$, $p = 0.1$) related to area (Armitage 1988). The mean size of matrilines is limited by competition within the matriline; large matrilines undergo fission that reduces competition and increases the number of matrilines. Thus more matrilines, not larger matrilines occur in larger areas.

These relationships suggest that there is an upper limit to the number of matrilines at a site and that some factor, such as the availability of resources, sets an upper limit on population size. Evidence suggests that the yellow-bellied marmot population is not food limited (Kilgore and Armitage 1978, Frase and Armitage 1989, Woods and Armitage 2003b). However, home range area is related, in part, to vegetation biomass (Armitage 2009b). Marmots need access to sufficient biomass to meet requirements for energy and nutrients such as protein and dietary fatty acids. As discussed in Chapter 18, evidence indicates that density at its upper limits increased dispersal, which effectively limits population size. This pattern of density dependence at the upper limits of population density is predicted by the population-limitation hypothesis. In conclusion, social mechanisms related to matriline size, reproductive suppression of young females by older females (Chapter 16), and density effects on dispersal are intrinsic mechanisms that contribute to limit population numbers.

Extrinsic limitation

Extrinsic factors have a major role in limiting marmot populations. Environmental factors, such as drought and prolonged snow cover, limit reproduction and increase mortality. These factors are not related to density. Possibly both extreme and typical weather patterns reduce the marmot population below carrying capacity. This hypothesis is supported by the recent finding that earlier snow melting resulted in increased survival, especially of adult females, which produced a major increase in population numbers (Fig. 21.4) (Ozgul *et al*. 2010). The significance of snow cover as an extrinsic mechanism was dramatically illustrated in the winter of 2010–11 when prolonged snow cover late into the spring resulted in the mortality of about 50% of the adults and 80% of the young of 2010, and greatly reduced reproduction in 2011 (Blumstein, pers. com.). Predation causes mortality and can significantly decrease a local population. Disease causes mortality, but its significance is unknown.

Carrying capacity

The population-limitation hypothesis explains population fluctuations of yellow-bellied marmots much better than the population-regulation hypothesis. Factors that affect reproductive success and mortality maintain these populations below carrying capacity. Density-dependent regulation occurs rarely when a set of favorable environment factors over several years results in a high population density.

It is unclear what determines carrying capacity. A critical factor appears to be space. The infrequency of matriline formation in the small satellite sites suggests that space is inadequate to support multiple females. A small colony site such as Marmot Meadow never supports the number of adult females characteristic of the larger River and Picnic sites. Within this framework, social factors play a predominant role in the number of females at a major burrow area. For example, at Marmot Meadow three adult females may occupy Main Talus in some years, but only one female in other years; a single female may not recruit daughters even though the site can support a larger number. Shy behavioral phenotypes (more aggressive females) are less likely to recruit daughters. Social behavior prevents unrelated conspecifics from obtaining residency and utilizing some of the resources.

The importance of space is supported by a comparison of reproductive success of colonial and non-colonial females (Van Vuren and Armitage 1994b). Frequency of reproduction and litter size did not differ between the two groups. However, a lower percentage of non-colonial young were recaptured and survival of non-colonial offspring from age 1–2 years was lower, probably because of a higher dispersal rate. These differences in recruitment likely result from less space and possibly lower quality space at the non-colonial (satellite) sites. I conclude that space in yellow-bellied marmots provides the required resources for immediate needs and must be sufficient to provide for future recruitment of daughters and the development of a matriline. Insufficient space ultimately limits population density.

Population limitation characterizes population dynamics of the alpine marmot. Field studies provided no evidence for density-dependent mortality due to food limitation. Strong density dependence occurred only above intermediate population sizes (Stephens *et al*. 2002a). Space also is important; because of a limited number of territories on a site, population growth produces an increase in the proportion of subordinate adults and floaters (marmots without family residency) with subsequent decrease in fecundity and increase in mortality (Stephens *et al.* 2002a). Similar to yellow-bellied marmots, space limits carrying capacity and only at higher population numbers do density mechanisms operate.

During restoration following extermination of a gray marmot population, over 6 years the number of families increased slightly, the family home range decreased from 3.6 ha to 2.1 ha, mean family size increased to a peak of 5.4 individuals by the fourth year and changed slightly thereafter, population density increased throughout the 6-year period, but much less than expected after the fourth year (Pole and Bibikov 1991). At high density only 3% of 2-year-old females reproduced compared to 25% at low density and older females alternated breeding and barren years, whereas some bred in three successive years at low density. This pattern indicates density mechanisms operate at high population numbers probably through reproductive suppression of younger females and poorer physiological condition of adult females (Pole and Bibikov 1991).

Summary

The model of density-dependent population regulation is rejected; density explains little of the variation in birth, survival, and population growth rates. Both intrinsic and extrinsic factors contribute to limit yellow-bellied-marmot numbers. The major intrinsic factor is social behavior which affects the size of the breeding population and the age of first reproduction. The availability of space with its essential resources determines carrying capacity. Population limitation characterizes population dynamics of marmots and social behavior has the major impact in limiting population density.

Part VI

The future of marmots

21 Climate change and conservation

The way climate change may affect marmots is examined over both evolutionary and ecological time scales. The fossil record provides some information on the historic climate space occupied by marmots and on changes in geographic distribution through evolutionary time. Recent shifts in habitat utilization document changes in ecological time.

Marmots may be more susceptible to climate change than many other species of mammals if their evolutionary history is tightly connected to their climate preferences (Davis 2005). Extreme shifts in climate might similarly affect closely related species. The relationship between marmot phylogenetic history and the climate space occupied by 12 extant species was examined by a regression between phylogenetic and climate distance matrices. The relationship was significant; marmot sister species live in more similar environments than would be expected by chance. Marmots do not evolve randomly with respect to climate; however, the correlation explained only about 11.4% of the variance, which suggests that climate may not be the most important factor shaping the evolutionary history of marmots (Davis 2005). An analysis of the successful introduction of the alpine marmot into the Pyrenees Mountains provides some insight into the role of climate. The climate was compared between native and introduction sites; climate (precipitation and temperature) matching was relatively low whereas habitat matching was relatively high (Lopez *et al.* 2010). Possibly habitat characteristics account for some of the dissimilarity in climate space occupied by sister species.

Davis concluded that marmot sister species live in more similar environments than would be expected by chance and that marmots do not evolve randomly with respect to climate. This view is supported by the known changes in marmot distribution since the last glaciation (Chapter 2). The shift in the distribution of *M. monax* suggests that marmots tend to occupy their climate space and shift geographic distribution as climatic distribution changes.

A paleoclimatic reconstruction for Porcupine Cave in Colorado supports this suggestion (Wood and Barnosky 1994). *M. flaviventris* is common and abundant in Late Pleistocene fossil sites of the montane West, but probably is not the species in the Pit sequence (Barnosky and Hopkins 2004), which probably was *M. monax*. The high number of *Marmota* and other rodents parallels the situation today at 2900 m on the southwest edge of South Park, in central Colorado where Porcupine Cave is located. Marmots first appeared coincident with a slight decrease in *Spermophilis* and *Cynomys* in level eight (about 900 000 years ago) during an interglacial (Barnosky 2004a). Probably marmots immigrated into the area during a cool, wet period. Marmot densities were high relative to *Cynomys* and *Spermophilis* (averaging 36%) during the relatively cool and moist glacial and interglacial levels, but decreased to about 9% during the driest and warmest intervals about 780 000 years ago (Barnosky *et al.* 2004). Thus, as the aridity increased, *M. monax* became less abundant in the fossil record and *Cynomys* and *Spermophilis* became more abundant. The functional stability of the ecosystem seems to have been maintained, but species-composition changed (Barnosky 2004b); *M. flaviventris* replaced *M. monax* and is the current marmot species in the area. By the end of the ice age about 12 000–13 000 years ago, *M. monax* was a member of a fauna occupying a landscape of spruce and pine parklands with open patches of grasslands (Tankersley and Redmond 2000). This species is still associated with woodland/grass-land habitats at low elevations.

Ecological niche modeling

Ecological niche modeling (ENM) uses environmental data from a species distribution to build models to predict potential past, current, and future distributional patterns. Two assumptions underlie ENM: (1) the distribution of a species is mainly determined by the environment, especially climate; and (2) a species conserves its niche through time and space (Waltari and Guralnick 2009). ENM was used to model the availability of suitable habitat for 13 montane mammals in the Great Basin, North America in the present and during the Last Glacial Maximum (LGM). For *M. flaviventris*, the average LGM elevation was 1717 m; the average present elevation is 1868m. There has been an upward shift of 151 m (Waltari and Guralnick 2009). The yellow-bellied marmot currently occupies 15 of the 16 ranges predicted by ENM, but only three LGM ranges were predicted under the higher threshold of model agreement and present basin connections with suitable habitat were rated marginal. The lower threshold model predicted 16 LGM ranges and abundant connections (Waltari and Guralnick 2009). The lower threshold model is consistent with genetic studies of Great Basin populations that reported a strong correlation between geographic and genetic distance, which indicates that the populations were linked by dispersal (Floyd *et al.* 2005). Floyd *et al.* suggest that increased lowland aridity resulting from climate changes may isolate the boreal faunas on Great Basin mountaintops.

Sleep-or-hide (SLOH) behavior

Certain life-history traits may buffer a species against climate change by allowing it to escape changes in climate and effectively expand its ecological niche. One possible trait is sleep-or-hide (SLOH) behavior, which includes the use of shelters; e.g., burrows, and/ or hibernation and torpor to avoid unfavorable climatic conditions. An examination of 4536 mammal species for SLOH behavior found that SLOH mammals are underrepresented in the "at high risk" International Union for Conservation of Nature (IUCN) Red List Categories. Mammals with SLOH behavior function at lower metabolism or are buffered from the changing physical environment (Liow *et al.* 2009). When IUCN contrasts were regressed on contrasts of diet, SLOH behavior, body mass, and geographic range, all four contributed to lower extinction risk, but diet and body mass were not significant. The authors conclude that mammals with SLOH behavior have a greater propensity to survive. However, the analysis does not consider that marmots are highly adapted to cool environments and are severely stressed by drought. Nor does this analysis consider possible indirect effects to be discussed later.

One could argue that niche conservation with its attendant physiological adaptations is why marmots are threatened by climate change. For example, the rise in the lower border of snow cover resulted in about 60% of the former habitats of *M. menzbieri* in Uzbekistan becoming drier. The population decreased because the foraging vegetation deteriorated. The upward shift in vertical distribution resulting from decreased snow and the significant reduction and fragmentation of habitats led to some surviving marmots aggregating near sparse springs where they are extensively killed by herdsmen (Bykova and Esipov 2008). Thus, climate change has both direct (loss of habitat) and indirect (intensive hunting) effects on Menzbier's marmot.

Timing of emergence

One consistent effect of climate change is that warming is associated with earlier emergence from hibernation (Inouye *et al.* 2000). Since 1998, woodchucks in Maine have been emerging significantly earlier (calculated from data provided by C. Maher):

$$\text{Emergence date} = 3018 - 1.48 \text{ year}, p = 0.001, \text{adj-}R^2 = 0.584$$

Woodchucks in 2010 were emerging about 17 days earlier then in 1998. Woodchuck emergence is complex because they may emerge initially and then re-immerge, especially after a heavy snowfall. Once final emergence begins, activity persists regardless of snow cover (Maher, pers. com.).

Since 1976, when Billy Barr at the RMBL began recording emergence date of yellow-bellied marmots, the date of sighting of the first marmot was 23 days earlier by 1999 (Inouye *et al.* 2000). The earlier appearances have continued and by 2010 they were emerging 30 days earlier than in 1976. Year accounts for a substantial variation in the day of first sighting (Table 21.1A). Many of the environmental variables and year are highly correlated (Table 21.2); therefore, variables with significant effects were identified by

Table 21.1 Regression equations describing relationships among factors related to seasonal phenology. Day is Julian day.

Regression	p	adj-R^2
A. First sighting = 1853 − 0.87 year	0.000	0.355
B. First sighting = − 4.7 − 0.0182 Total snow + 1.01 Bare ground	0.001	0.349
C. First sighting = 22.1 + 0.672 Day of bare ground	0.000	0.356
D. Bare ground = 427 − 0.169 year + 0.043 total snow + 0.232 April snow	0.000	0.859
E. Mean weaning day = 627 − 0.22 year	0.021	0.133
F. Day of UV weaning = 131 − 0.0172 Total snow + 0.443 Day of bare ground + 0.0318 April snowfall + 0.120 Day of first sighting	0.000	0.532

Table 21.2 Correlation matrix of environmental variables related to time when yellow-bellied marmots were first observed. * p = 0.05, ** p < 0.01. N = 29 years. Year arranged from early to late.

	Year	Total snowfall	April snowfall	Bare ground
First sighting	−0.608**	0.571**	0.249	0.616**
Year		−0.277	−0.086	−0.363*
Total snowfall			0.468*	0.924**
April snowfall				0.501**

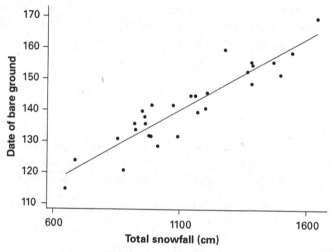

Figure 21.1   The relationship between total snowfall and day of bare ground. Day is reported as the Julian date. Bare Ground = 89.0 + 0.0466 Total Snow, p = 0.000, adj-R^2 = 0.849.

regression analysis. The time of first sighting is related to total snowfall and the date when bare ground occurs (Table 21.1B), but only the effect of bare ground is statistically significant (p = 0.029).

The day of first sighting can be estimated from the day of bare ground (Table 21.1C). The strong relationship between day of first sighting and day of bare ground and the lack of a significant relationship with measures of snowfall indicate that time of snowmelt rather than the amount of snowfall affects the timing of emergence. However, the date of bare ground is significantly correlated with snowfall variables and year (Table 21.2). Although the regression was highly significant (Table 21.1D), only the partial p for total snow was significant. The amount of winter snowfall determines when bare ground appears (Fig. 21.1) which in turn markedly affects when marmots are first sighted above ground.

Timing of weaning

The earlier emergence of marmots could result in earlier reproduction and weaning of young. Therefore, I examined possible factors that could affect the timing of weaning.

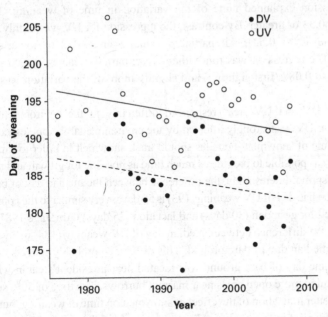

Figure 21.2   The relationship between the Julian day of weaning and year for up-valley (UV) and down-valley (DV) marmot sites.

Because snow cover persists longer at UV than DV (Van Vuren and Armitage 1991), I treated each area separately. Mean weaning day DV (184.39 ± 0.90 SE) was significantly (t = 7.2, p < 0.001, N = 46) earlier than mean weaning day UV (193.32 ± 0.73 SE). Thus, on average, young are weaned 9 days earlier at DV. Weaning times at the two areas were poorly correlated (r = 0.249, ns). The lack of a significant correlation between the two areas suggests that weaning time in the two areas is affected either by different factors or by factors varying in the two areas in any given year.

Both DV and UV weaning days were earlier in 2008 than in 1976 (Fig. 21.2); on average at each area weaning was 5 days earlier in 2008. Over the same time period, the snowmelt date advanced by 4.14 day/decade (Lambert *et al.* 2010). Thus weaning date advanced much less than snowmelt date, which suggests that the year-to-year variation in addition to the long-term trend in snowmelt affects the timing of weaning. This interpretation is consistent with year-to-year variation in snowfall, neither total snowfall nor April snowfall is significantly related to year (Table 21.2).

Because no young were weaned in some years at either the DV or UV areas, weaning days (Julian day) were averaged over both areas for two time periods, 1962–75 and 1976–2008, and regressed against year. Mean weaning day and year were unrelated for the years 1962–75 (p = 0.959); but time of weaning decreased between 1976 and 2008 (Table 21.1E). Year explained only a small percentage of the variation in weaning day, which indicates that the year-to-year variation is the more important factor affecting time of weaning. Therefore, weaning day was regressed against the group of variables (Table 21.2) for DV and UV areas. For DV, the regression was insignificant (p = 0.574) and the regression explained none of the variation in time of weaning. The partial p values were 0.33 or greater. By contrast, the regression for UV was highly significant ($p = 0.000$, adj-$R^2 = 0.52$). All partial p values were 0.078 or lower except for year (0.518). The regression was run without year; partial p values ranged form 0.015 (bare ground) to 0.088 (first sighting) and slightly more of the variation was explained (Table 21.1F).

Clearly, the DV and UV areas responded differently to the environmental factors. Weaning time at UV was strongly affected by the environmental factors, especially those indicative of time of snowmelt. Females should track snowmelt in order to initiate reproduction as early as possible to provide as much time as possible for growth and fattening of mother and offspring. Interestingly, the difference between the median day of bare ground (141) and the median day of UV weaning (193) is 52 days, very similar to the approximately 55 days required for gestation (30 days) and lactation (25 days) (Armitage 1981).

By contrast, the difference between median day of DV weaning (184) on average is 43 days after the median day of bare ground. This difference probably reflects that the site used to determine day of bare ground was located near a resident's cabin where snow melted later than in more open areas near marmot burrows. The time of 50% snow cover probably is a better indication of the effect of snowmelt on time of weaning, but such data are too few to be used in this analysis. The mean day for 50% snow cover was about 13 days later at UV than at DV (Van Vuren and Armitage 1991). Because snowmelt is later UV, the timing of bare ground at the DV location serves as a proxy for predicting snowmelt UV.

Environmental cues

The variation in emergence and weaning dates suggests that marmots use some cue or cues to initiate emergence. We know that hibernation is driven by a circannual rhythm and that arousal cycles become more frequent as the end of the hibernation period nears. Adults spontaneously terminate hibernation, whereas young continue hibernation until they become emaciated or are fed (French 1990). Hibernation can be terminated prior to spontaneous termination during a periodic arousal if the marmots receive an appropriate cue. We do not know the nature of the cue or how a marmot in its burrow detects the cue. Marmots will emerge through the snow (Fig. 21.3), which makes sense if the marmot terminates hibernation. Early emergence of marmots has occurred through the snowpack and the early emergence DV was correlated with mean minimum April air temperature; marmots emerged earlier as air temperature increased (Inouye et al. 2000). This highly significant relationship emphasizes the problem of how marmots detect the correct environmental cue.

Detection of an appropriate environmental cue to allow earlier emergence can have a pronounced effect. The combination of earlier snowmelt and earlier weaning produced a pattern of coupled dynamics of body mass and population growth (Ozgul et al. 2010). The longer growing season resulted in a larger body mass at hibernation. As a consequence of larger size, heavier marmots, especially adult females, survived better after

Figure 21.3 Yellow-bellied marmot snow tunnels.

2000 and the population steadily increased (Fig. 21.4). The population growth rate was 1.02 (approximately stable) from 1976 to 2000, and was 1.18 (rapid increase) after 2000. Juvenile survival did not change substantially, but the small increase was caused by the change (i.e., increase) in mass. This change in mass and its effect on survival and population growth is phenotypic, not genetic; i.e., not controlled by specific gene changes.

The significance of snow cover

In the previous section, I emphasized the importance of snow cover, particularly focusing on the relationship between timing of snowmelt and timing of weaning and that earlier weaning enhances survival and population growth. In this section, I explore more broadly the effects of snow cover on marmots with emphasis on the effect of climate change.

We previously noted that the longer persistence of snow in the UV network is associated with a smaller proportion of females weaning litters and a smaller litter size than in the DV network (Van Vuren and Armitage 1991). In NPB at an elevation of about 3400 m, where snow persists much longer than in the East River Valley, no adult female was known to wean a litter in successive years (Johns and Armitage 1979). Variation in

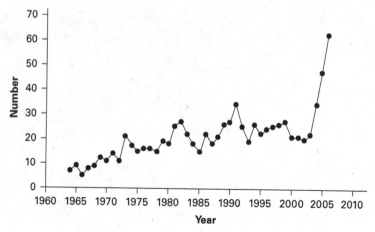

Figure 21.4   The number of known-age adult females at the better-studied sites in the Upper East River Valley. Figure courtesy of A. Ozgul.

the timing of snowmelt affected the frequency of reproduction in the NPB population. In 2 years of early snowmelt 25–37.5% of the adult females weaned litters; in 3 years of late snowmelt, 10–16% of the females weaned litters. Prolonged vernal snow cover accounts for reproductive skipping in several species of marmots (Armitage and Blumstein 2002). These normal effects of snow cover are exacerbated in years of prolonged snow cover. In 1995, snow cover in the UV area persisted well into June (Fig. 17.10). Reproduction and survival decreased; Boulder and North Picnic went extinct and only one animal remained at Cliff (Armitage 2003c, also see Chapter 17). In NPB, in 2008, snow cover lasted until mid-July. No young emerged and none of the young of the previous year were recovered as yearlings (Woods *et al.* 2009).

Reproduction in yellow-bellied marmots in the Central Sierra Nevada at an elevation of 1925 m to 2484 m, where marmots are active for 5 months, was affected by snow cover. After the long winter of 1966–67, when marmots emerged 23 days later than the previous year, no juveniles from 1966 were caught and the sub-adult and adult population was greatly reduced (Nee 1969). Marmots emerged much earlier when snowmelt was earlier. It seems likely that the timing of snowmelt is important to yellow-bellied marmots throughout their range.

Other marmot species

Spring snow cover affects reproductive frequency of Olympic marmots. Considerable snow cover typically persists into May or later (Fig. 21.5). Heavy snow depth was associated with higher winter mortality of young and delayed dispersal by 2-year olds. During the years of study (1967–69), no female was parous in successive years (Barash 1973a). However, in 2004 and 2005, when the snowpack was greatly reduced, an average of half of the females bred in successive years. In the high snowpack year of 2003, as in the high snowpack years of 1968 and 1969, no female weaned successive

Figure 21.5   An Olympic marmot in a snow-covered landscape in mid-June. (See plate section for color version.)

litters (Griffin *et al.* 2007a). Similarly, prolonged snow cover in 2010 (snowmelt 4–8 weeks behind average) resulted in reduced survival of Vancouver Island marmots (Fig. 21.6) that were reared in captivity and released in the previous summer, and in reduced reproduction by wild marmots. Wild marmots survived well, in part because in most areas they moved to lower elevations (a behavior not previously reported) presumably to find food (Doyle, pers. com.). As with the yellow-bellied marmot, these responses to snow cover are phenotypic and not genetic. The Olympic, Vancouver Island, and yellow-bellied marmot studies indicate that marmots have considerable plasticity in their responses to environmental factors.

In a hoary marmot population in Alaska, in years of delayed snowmelt food availability was delayed by about 2 weeks; because of reduced time for foraging, the body mass of females at hibernation was not significantly larger than it was at emergence the previous spring. Females entered hibernation about 1 kg larger in years of early snowmelt than in years of late snowmelt (Holmes 1984a). Possibly the pattern of snowmelt reduced reproductive frequency; females weaned litters on average every 3.3 years.

The time of snowmelt clearly affected survival and reproduction in a hoary marmot population in Canada. When snowmelt was late in 2000, weaning rate was 67% compared to 2001 when snowmelt was early and weaning rate was 79%. Late snowmelt was associated with low juvenile survival (38%) the following winter. However, juvenile survival (74%) was much higher in a year of late snowmelt that followed the year when the juveniles were born early because of early snowmelt (Karels and Hik 2003). The difference in juvenile survival indicates that the length of the growing season is critical

Figure 21.6   A Vancouver Island marmot on the snow near a snow burrow in mid-June. (See plate section for color version.)

for survival during the subsequent hibernation as a longer growing season allows the marmots to hibernate in better condition which provides sufficient energy reserves for coping with later snowmelt.

A late spring snow storm indirectly increased mortality of the tarbagan in Mongolia. Because the late snow caused a large number of cattle deaths, marmot hunting increased (Kolesnikov *et al.* 2009b).

Delayed snowmelt in the spring reduced reproduction in the black-capped marmot (Mosolov and Tokarsky 1994) and increased the mortality of young and yearling alpine marmots (Arnold 1993a). An entire family group can perish on poor sites, characterized as typically having late snow cover. The marmots tended to clump on the good sites, characterized by early snowmelt (Arnold 1990b). Heavy snow at the time of emergence, which covered the area for a month, resulted in reduced survival and reproduction (Sala *et al.* 1993). Snowmelt may have a lag effect on survival. When early snowmelt occurs, an adult female alpine marmot is in better condition and produces young in better condition and juvenile survival increases in the subsequent hibernation. This pattern occurs in the absence of helpers (Allainé, pers. com.).

Winter snowpack

Although early snowmelt or low snowpack in the spring favors increased reproduction and survival, heavy snowpack during winter insulates against cold temperatures and favors survival. In the winter of 2011/12, winter snowfall was the second lowest on record at RMBL, there was only 1.9 m through December 2011 and March 2012 had the lowest snowfall for that month on record. The day of bare ground (113) was the earliest since records were kept (mean day of bare ground is 140). Marmot mortality was severe; the major effect was DV (Blumstein, pers. čom.). Winter mortality of Olympic marmots was high when winter snow was sparse and was lowest during a winter of heavy snow (Barash 1973a). A shallow snowpack negatively impacted survival of hoary marmots (Braun *et al.* 2011). The probability of survival decreases for alpine marmots when the winter snowpack is too thin (Allainé *et al.* 2008). Avalanches can affect snowpack by removing it from sites with hibernacula. The potential effect on marmots is greater if the avalanche occurs relatively early in the winter and no snowfall subsequently insulates the site. In the Western Tatra Mountains marmots in a hibernaculum at an avalanche-exposed site died. Alpine marmots in colonies on avalanche slopes successfully hibernated in burrows several meters above the starting zone for an avalanche (Ballo and Sykora 2005).

Annual variation in the survival of older age classes of alpine marmots was strongly related to the intensity of the autumnal frost. It may act by lowering soil temperature at the beginning of hibernation with subsequent effects on the energetic costs of hibernation (Farand *et al.* 2002). This relationship emphasizes the importance of early snow cover to delay the onset of temperature stress in the hibernaculum.

Snowpack also significantly impacts reproduction of alpine marmots. Over a 20-year period, winter snow cover thinned and litter size and body mass of the mother progressively decreased (Tafani *et al.* 2013). The decline in female body mass likely accounts for the decrease in litter size. This effect occurred despite the positive effect of early snowmelt and suggests that snowpack and snowmelt can produce the same or contrasting effects on marmot biology. Tafani *et al.* (2013) emphasize that effects of climate change cannot be extrapolated to another, even closely related, species. Climate change impacted the alpine and yellow-bellied marmot differently. During the same time period, heavy winter snowpack and early snowmelt enhanced yellow-bellied marmot survival, whereas light winter snowpack and early snowmelt negatively impacted alpine marmot reproduction. Thus winter snowpack was the major factor affecting one species and spring snowmelt was the major factor affecting the other. However, the yellow-bellied marmot suffers from low snowpack and the alpine marmot benefits from early snowmelt. It is likely that in the long-term early snowmelt and low snowpack will be the critical factors impacting marmot species, but responses to the impacts may differ among species. Which environmental factor is critical at any time depends on the local climatic condition.

The burrow and cooling

The burrow could be a conduit for cold air to penetrate to the hibernaculum. Marmots plug their burrows at the onset of hibernation. It has long been assumed that the function of the plug is to deter predators, but Arnold *et al.* (1991) point out that the plug could slow the rate of cooling. The importance of plugging the burrow is illustrated by the decline in the burrow temperature of yellow-bellied marmots from 11.3°C in late August to 9.1°C in early October, which differed little from the burrow temperature of 9.0°C in June

Table 21.3 Hibernation characteristics and burrow temperatures of the Kazakhstan subspecies *Marmota bobak schaganensis* and European subspecies *M. b. bobak.*

Hibernation immergence	*M. b. bobak* plugs burrow in September
	M. b. schaganensis plugs burrow in early August
Air temperature	Lower for *M. b. schaganensis*
Hibernation temperature	*M. b. bobak* 14–15°C
	M. b. schaganensis 10°C
Burrow depth	140–450 cm for both subspecies
Hibernation emergence	March and April over entire range
Elevation	Rarely above 300 m

(Kilgore and Armitage 1978). Thus, most of the cooling of hibernacula occurs as a result of the cooling of the surrounding soil.

This process was followed for the steppe marmot, *M. bobak* (Nikol'skii 2009). Although two subspecies live at the same elevation, they hibernate at different times and burrow temperatures (Table 21.3). The air temperature is lower for the Kazakhstan subspecies; by plugging its burrow earlier it avoids a burrow temperature that would be 2–4°C lower because of convection associated with a rapid decline in air temperature at the time of hibernation. This study illustrates the capacity of a marmot species to adjust its annual cycle to prevailing environmental conditions. Because the annual cycle of marmots is phenotypically plastic (Armitage 2005), the hibernation differences (immergence times) in the steppe marmot populations are likely phenotypic rather than genetic.

Global warming and burrow temperatures

An analysis of the possible effects of global warming on the temperature at which marmots hibernate demonstrates that marmot response to warming depends on local conditions. Burrow temperatures were modeled for *M. bobak* and *M. sibirica* based on a 3°C increase in mean monthly temperatures. Three scenarios were used: (1) the increase occurs in all months (equal); (2) the increase occurs only from October to March (winter); (3) the increase occurs only from April to September (summer). In the equal and winter models, temperature during hibernation was considerably higher than at present (Belovezhets 2008). For *M. bobak*, the temperature increases to above the optimal range at burrow depths of 1.5–4.5 m. For *M. sibirica*, the temperature remains optimal at 3 m depth and the period of negative (= stressful) temperatures at 1.5 m depth decreases from 6 to 4 months.

One important assumption in the analysis is that burrow temperatures above the optimal range (reported as 0–5°C for *M. bobak*; Belovezhets 2008) will increase metabolism and the costs of hibernation. This potential harmful effect of global warming depends on where the TNZ is during torpor and to the degree that temperature acclimation can minimize the effects of higher temperature. Stable metabolic rates may occur at temperatures as high as 15°C during torpor in some species (Armitage 2008). Just how global warming will affect the physiology of hibernators is largely unknown, but effects on the frequency of torpor bouts and energy use are possible (Mitchell *et al.* 2008).

Indirect effects of climate change

One anticipated indirect effect is loss of habitat. With global warming, trees are expected to invade montane and alpine meadows (Armitage 2013). The habitat of the Alaska marmot is shrinking because of an increase in the abundance and upslope migration of trees and shrubs (Gunderson *et al.* 2012). This species occupies the highest elevations in the northernmost mountains in North America; there is no place to which populations can disperse and they face ultimate extinction if they cannot adapt to their changing environment (Gunderson *et al.* 2009, 2012). Currently tree cover is increasing in sub-alpine meadows utilized by Olympic marmots (Griffin *et al.* 2009). Trees make habitat unsuitable because snowmelt typically occurs later under trees, and may make marmots more susceptible to predation as obstructions block viewlines (Bednekoff and Blumstein 2009). However, reduced snow cover may increase the habitat available to some species of marmots, at least in the short run. Large areas in central Tadzhikistan are not occupied by *M. caudata* because of prolonged snow cover (Davydov 1991). In Cascade Canyon, in the Teton Mountains, early in the summer, yellow-bellied marmots occupied the slopes where the snow had melted; marmots were not active on the opposite snow-covered slopes. However, by late summer marmot activity expanded to the previously snow-covered slope. In the range of the Olympic marmot meadows are observed with unoccupied marmot burrows; obviously marmots had lived there. A survey of those meadows over several years revealed that in years with heavy snowpack and late snowmelt, the meadows were completely snow covered well beyond the time marmots would need to forage to remain alive (Griffin, pers. com.). Presumably in these cases, and doubtless in other areas, early snowmelt as a consequence of global warming will allow marmots to permanently occupy previously unsuitable habitats.

Predictions, such as the one above, do not consider other possible ecosystem responses. Based on early snowmelt and the ability to reproduce in consecutive years, one would predict that Olympic marmot populations would thrive and possibly expand. However, Olympic marmots recently declined significantly and became increasingly fragmented, local extinctions were widespread (Griffin *et al.* 2008, 2009). Adult female survival was considerably lower than it was when Barash did his studies (1967–69). Mortality of adult females in June is consistently high, which strongly suggests predation is the major cause (Griffin *et al.* 2008). Anecdotal evidence suggests that early snowmelt allowed coyotes to access the sites where marmots live and increase predation pressure on vulnerable reproductive females. [Heavy coyote predation has recently been documented (Witczuk *et al.* 2013).] About 30% of the adult females were taken each year from 2002 to 2006 when snowpack was very low. But since then, snowpacks have been above average and loss of adult females dropped to 0 to 10% each year. Sightings and tracks of coyotes were nearly zero (Griffin, pers. com.). The Olympic marmot story not only illustrates the importance of snowpack but also demonstrates the importance of weather variability associated with climate change.

Another possible effect of global warming and early snowmelt is the increased frequency of EPP where a female mates with a male other than the territorial male member of the family group. In alpine marmots, EPP in the absence of subordinate group males occurred at sites already free of snow during the mating season. Normally movement between territories is rare because heavy snowpack restricts movements and polyandry occurred only among group members (Arnold 1990a). Early snowmelt may allow more movement by adult males and EPP by transients (Cohas *et al.* 2007b) could become more common. Possibly multiple mating involving transients or the presence of multiple males in a colony occurred in East River Valley yellow-bellied marmots during the recent period of early snowmelt and rapid population growth (Olson and Blumstein 2010). However, the late snowmelt may cause males to move widely to find food and thus not remain close to their burrows to guard against intruders (Allainé, pers. com.). In conclusion, climate change will produce changes in marmot biology, but the exact nature of those changes is variable and often unpredictable.

Conservation

Problems in the conservation of species of marmots are a consequence of human impacts (Armitage 2013). Conservation of marmots involves three major activities: (1) regulating current exploitation; (2) restoration of an endangered species; and (3) reintroduction of a species into its former range.

Regulating current exploitation

The long-term relationship between the human community and the tarbagan (or Mongolian marmot) in Mongolia illustrates how local people utilize marmots and the conservation problems that emerge as a consequence of their exploitation. Since ancient times the Mongolian people hunted marmots for food, but did not use the skins. However, early in the twentieth century a market for skins developed in Europe. Between 1906 and 1994, an estimated 115 million skins were prepared (Batbold 2002b). There was a slow decline in the trend of skin production to only a few thousand by 1995. The decline was in large part caused by closing many areas, especially those where plague occurred, beginning in 1973. In the next 20 years, the number and total size of closed areas increased so that over 3 million marmots were exempt from hunting (Batbold 2002b).

The situation changed markedly when new markets for marmot skins developed. Estimates indicate that the tarbagan population declined from about 20–24 million in 1990 to about 8 million in 2007. There was no evidence that the decline was caused by predation, plague, or climate change (Kolesnikov *et al.* 2009b). The high price for a pelt (as much as US$7.64) increased the number of hunters and hunting pressure increased. Hunting occurred during the reproductive season, which reduced recruitment. In 2005, hunting was banned throughout the country. The effectiveness of the ban varies: it is

Table 21.4 Effects of hunting on populations of the Mongolian marmot (*M. sibirica*).

Population characteristics	Hunting area	Non-hunting area
Population density	17.1 families/km^2	51.4 families/km^2
Colony sites occupied	39%	100%
Reproductive family groups	16%	29%

ignored in some areas where the decline continues, while in other areas the population appears to be stable but at reduced numbers (Kolesnikov *et al.* 2009b).

There is some indication that marmot populations can recover. Surveys in the Eastern Steppe in 2006 and 2007 indicated an increase in the density of burrow clusters and in percent of active burrow clusters. These results indicate that an effective ban leads to a recovery of the marmot population (Townsend 2009).

A factor that complicates the imposition of a hunting ban is that local populations exploit the Mongolian marmot to provide food, medicinal products, and fur. This subsistence hunting can reduce marmot populations, whereas marmots thrive in areas where hunting is banned or does not occur for religious reasons (Table 21.4) (Kolesnikov *et al.* 2009b). Clearly, conservation of the Mongolian marmot must account for its role in the subsistence of the Mongolian people. Plague in many local populations complicates management. Because of plague, tarbagan were overexploited in non-plague areas to the extent that marmots became extinct in some areas. Batbold (2002b) suggested that a low harvest in plague-free areas and a much higher harvest in plague zones would both decrease the spread of plague and maintain overall population levels and allow local human populations to continue the use of marmots as part of their long-term culture.

One goal of marmot exploitation is that the harvests should maintain viable populations. Extinction probabilities for the alpine marmot were calculated for three harvesting systems: constant-effort harvesting, constant-yield harvesting, and threshold harvesting. Overall, threshold harvesting (exploitation occurs only in those years in which the population exceeds a given threshold and individuals are removed until the population reaches the threshold) provided the highest mean yields in relation to extinction risk (Stephens *et al.* 2002b). This study suggests that indices of sustainability are markedly reduced for species dependent on sociality.

Restoration

The Vancouver Island marmot (*M. vancouverensis*) is one of the most endangered mammal species in the world. The wild population declined to about 27 adults in 2000 (Bryant 2000). The decline was so extensive that source populations no longer provided emigrants and sites went extinct. It became apparent that a captive-breeding program and introduction of captive animals into natural habitat provided the best chance for this species to avoid extinction (Bryant 1996b, Bryant and Page 2005). Marmots were brought into captivity and breeding programs were eventually established at four sites.

Captive animals emerged and reproduced about 5 weeks earlier than wild animals (Bryant 2005), which suggests a phenotypic shift in the phasing of the circannual cycle. The most important factors affecting reproduction in the captive animals (Casimer *et al.* 2007) were age, production of young in any previous year, and the amount of time housed with the current mate; factors that also affect reproduction of wild yellow-bellied marmots (Armitage 2004c, Nuckolls 2010). The importance of pair-bonding was dramatic; of females housed with a mate ≥1 year before the breeding season, 63% produced young but only 25% produced young if paired for <1 year. Subsequently, a release program was initiated and the number of marmots in the wild increased to 300–400 by 2012 (Doyle, pers. com.).

Survival of captive-born marmots released into the wild was significantly lower than survival of wild-born marmots. Predation was the most important cause of mortality; golden eagles were the most important predator of captive-born marmots, whereas wolf and cougar predation was more important for wild-born marmots (Aaltonen *et al.* 2009). Captive-born marmots have higher hibernation mortality which seems to be associated with the inability of the captive-born marmots to cope with delayed snowmelt in the spring (Doyle, pers. com.). It is not clear why captive-born marmots are more vulnerable to predation. When wild-captured and captive-born marmots were presented with taxidermic mounts of cougar (*Felis concolor*) and wolf (*Canis lupus*) plus control stimuli, the marmots responded most to the wolf and cougar. The wild-captured and captive-born marmots did not differ significantly in the amount of time spent vigilant or foraging in response to the stimulus (Blumstein *et al.* 2006a). Because golden eagles are the most important predator of captive-born marmots, there may be a learning component at an early age in the detection of eagles, which is not acquired in captivity.

Population decline: the steppe marmot story

The decrease in the number of marmots most often occurs because of hunting and habitat loss. The steppe marmot in Ukraine occupied the steppe and forest-steppe from the Carpathians as far as the west border of Ukraine and from the southern border of the Ukrainian forested lowland to the Black Sea and Sea of Azov and probably numbered in the millions (Bibikow 1996). The reduction in numbers began in the eleventh and twelfth centuries with the population growth of Kiev Rus (Tokarski *et al.* 1991). By the early twentieth century, only small colonies remained. Marmots survived in large virgin areas occupied by horse-breeding farms and in gullies covered with diverse steppe vegetation. Plowing of the virgin lands that eliminated habitat and hunting contributed to marmot decline. The decline due to hunting was especially severe during the two world wars (1914–18, 1939–45) and during years of famine. Marmots increased considerably after 1960 by naturally occupying gullies, small flat-bottom valleys, pastures, and forest edges. From these habitats, the marmots spread into agricultural fields (Tokarski *et al.* 1991). I observed steppe marmot burrows in a wheat field and in haystacks in the Valikii Burluk district in Ukraine. Steppe marmots are protected from hunting, but poachers probably kill thousands of marmots annually.

Steppe marmot abundance is affected by grazing; numbers are higher under intensive grazing than under moderate grazing; their numbers increased as cattle grazing increased and decreased as cattle grazing decreased (Nikol'skii and Savchenko 1999). Livestock and marmots have separate effects on the vegetation; when both are present their ecological roles are complementary. Their combined effects decrease perennial grasses and increases forbs (Yoshihara *et al.* 2010b). The removal of grazing from an area in the West Tatra Mountains increased the prevalence of grasses and decreased the abundance of forbs preferred by alpine marmots, creating unstable conditions for marmots (Ballo 2006). Grazing enables marmots to utilize the young parts of forage plants and is associated with the highest abundance of legumes. Grazing affects plant succession; marmots prefer species that are in the early stage of succession, species with a long-lasting vegetative period, and with a high ability to re-grow. This relationship with grazing developed historically as marmots occupied steppes where nomads grazed their herds (Ronkin *et al.* 2009). Bibikow (1996) extensively discussed the co-evolution of some Eurasian marmots with ungulates, marmot movements into anthropogenic environments, and programs of reintroductions of marmots into unoccupied areas. In the following section I describe a few current examples from different areas in Europe.

Reintroductions: alpine marmots

Alpine marmots were introduced into the French Pyrenees between 1948 and 1988 to provide food for endangered golden eagles and brown bears (*Ursus arctos*). Marmot preference for southern sunny slopes led to rapid expansion into the Southern Pyrenees where competition and natural predators were lacking (Lopez *et al.* 2009). The marmots occupy lower elevations because higher meadows (>2500 m) are dominated by talus with poor vegetation. At lower elevations, traditional management lowered the tree line which extended the distribution of subalpine meadows that were populated by marmots. Most remaining sites have a low probability of marmot occupancy, but marmots probably can expand throughout the Pyrenees into those alpine and subalpine meadows that provide adequate habitat (Lopez *et al.* 2009).

In similar fashion, marmots introduced into the agricultural landscape in the Central Massif in France first settled in habitats at high elevation where screes and rock outcrops provided natural shelters. The marmots then settled in lower elevations in more open hay meadows where vegetation was more abundant and dry stonewalls provided shelter (Ramousse *et al.* 2009). The reintroduction was necessary because the marmots were eliminated from many areas by hunting as the marmots were classified as game prior to 1939. When the introductions occurred in 1980, the marmots were protected. Marmot losses were significant during releases, 103 animals were introduced to produce a population of 36 in 1988. Despite poor reproduction in some years, the population grew to 492 in 2007. The growth rate in this agricultural area indicated a doubling time of 5.92 years. As in the Southern Pyrenees, the marmots were introduced to provide diverse food for large predators and increased their numbers by utilizing anthropogenic landscapes. An additional motive was to stimulate tourism and this was successful in some areas of the Massif (Ramousse *et al.* 2009). Two principles of reintroductions were proposed: (1) family groups rather than individuals should be released; and (2) genetic stock should be released as close to the original as possible (Ramousse and Le Berre 1993).

In the southern Dolomites in Italy, alpine marmots were hunted from prehistoric times and have become extinct. Reintroductions occurred in 2006 and 2007. Habitat suitability analysis considered elevation, slope, exposure, and availability of meadows to identify two suitable areas for release (Borgo *et al.* 2009). Introductions were successful. Most marmots formed pairs which greatly increased survival during the first winter (52.9% of solitary and 93.6% of group members). More than half of the marmots died by the end of the second summer. Mortality was high when a fox moved into a burrow after killing its occupants and when golden eagle activity increased. Only one pair bred successfully in the year of release, but over 72% of the surviving pairs bred in the following years and by late summer of 2008 more than half of the individuals were wild-born. This high number of recruits occurred despite the high mortality of young born in 2007 when high and persistent snow cover prevailed in the spring of 2008 (Borgo *et al.* 2009).

Summary

These examples indicate that marmots thrive and spread when good habitat is available and hunting is regulated. Climate change that decreases snowpack in the spring resulting in earlier snowmelt increases survival and reproduction in the short term. But if early snowmelt is associated with a thin snowpack or with less moisture during the active season and drier vegetation during the summer, higher mortality and decreased reproduction most likely follow. Indirect effects, such as changes in predation pressure, are difficult to predict, but could have detrimental consequences.

Widespread species, such as *M. flaviventris* and *M. monax*, are in no danger of extinction in the foreseeable future. Probably some local populations will disappear resulting in the loss of genotypic and phenotypic diversity. Species with a narrow distribution, such as *M. broweri*, *M. olympus*, and *M. vancouverensis*, are more threatened. There is little space where these species can find refuge and stochastic population decrease could lead to extinction. Similarly, in Eurasia, *M. menzbieri* appears to be the most threatened, but widespread species, such as *M. bobak* and *M. sibirica*, should persist provided human population pressure does not destroy habitat or eliminate marmots through increased hunting. Both exploitation and management guidelines emphasize the importance of considering the social system in conservation procedures.

22 Major life-history traits

The central adaptation that forms the framework from which marmot life history is derived is hibernation. We envision an evolutionary trajectory beginning with evolution in a harsh environment and culminating with sociality and its attendant life-history traits (Fig. 22.1). Marmots adapted to an open landscape with herbaceous vegetation that provided an abundant seasonal food source; they coped with seasonality by hibernating and evolving large size. Large size facilitated the processing of the herbaceous diet and increased the energetic efficiency of hibernation. The long hibernation period imposed time constraints on growth and reproduction such that most marmot species need at least two growing season to reach adult body size and reproductive maturity. The short active season led to the retention of young in their natal home range and delayed dispersal. Competition for reproductive success, especially among adult females, required that maternal investment in young continue beyond their first hibernation. Retention of young for two or more growing seasons in their natal environment led to sociality with its kin-related behaviors and patterns of reproductive cooperation (e.g., cooperative breeding) and competition (e.g., reproductive suppression).

Matriline structure in yellow-bellied marmots is typical of most species of social mammals; the social groups are highly stable and consist of females of high kinship (Clutton-Brock and Lukas 2011). Social group composition is more complex in marmots with a family structure. In those species with a restricted family organization, social groups may be similar to those of the yellow-bellied marmot except dispersal occurs at an older age. Males are more likely to disperse to a different habitat patch and females are more likely to remain philopatric.

In species with extended family structure younger adults and yearlings of both sexes are philopatric. Among older adults, both sexes may be philopatric or may be unrelated immigrants, but average relatedness within the family is higher that that of the population at large.

The influence of hibernation extends beyond sociality. The physiology of hibernation limits physiological adaptations during the active season. Because marmot physiology is focused on coping with low environmental temperatures, marmots are stressed by high temperatures and solar radiation. The avoidance of heat stress reduces the amount of diurnal activity and establishes the daily activity cycle. The necessity of having adequate energy reserves for hibernation limits the amount of energy available for reproduction and thus reduces the frequency of weaning. Because energy is limiting, marmots evolved

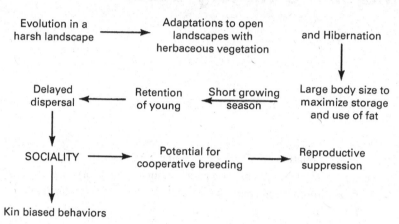

AN EVOLUTIONARY TRAJECTORY

Evolution in a harsh landscape → Adaptations to open landscapes with herbaceous vegetation   and Hibernation

Delayed dispersal ← Retention of young ← Short growing season   Large body size to maximize storage and use of fat

SOCIALITY → Potential for cooperative breeding → Reproductive suppression

Kin biased behaviors

Figure 22.1   The evolutionary trajectory begins with life in a harsh environment and ends with the evolution of sociality and its attendant features. The figure is not meant to indicate sequential steps in the process, but to indicate the critical environmental features and their consequences.

various mechanisms to conserve energy during both the heterothermal and homeothermal periods. Finally, all of these traits impact population dynamics.

Phenotypic plasticity

Phenotypic plasticity is a response to conditional alternatives, especially those due to environmental heterogeneity, although some alternatives may be competition dependent (West-Eberhard 1989). One predictable feature of the marmot environment is a season without food.

The response to seasonal food shortage is hibernation, which is controlled by the endogenous circannual rhythm. The circannual rhythm is under genetic control, but the timing of immergence and emergence is plastic, which is highly adaptable for coping with seasonal unpredictability. Phenotypic plasticity allows marmots to shift the timing of emergence and reproduction according to weather and snowmelt patterns. Plasticity depends on the increasing frequency of arousal cycles as the heterothermal period nears its end. This system is adaptive and probably arose from natural selection (Gotthard and Nylin 1995). Plasticity is limited; the circannual cycle eventually compels marmots to euthermic activity. This spontaneous emergence seems adaptive; it is necessary to provide an active season of sufficient length for successful reproduction.

Dispersal is conditional in both sexes and is determined by the social environment. If a male yearling encounters adult male agonistic behavior, it disperses, if not, it remains philopatric. A female yearling becomes philopatric in a cohesive social environment, otherwise she disperses if unoccupied space is unavailable. Behavioral phenotypes, which are plastic, contribute to the dispersal decision.

Figure 22.2   A Vancouver Island marmot perched on a tree stump. (See plate section for color version.)

Phenotypic plasticity is evident in many other activities. It is unclear to what degree the following examples are adaptive; that is, are expressions of genetic change, or represent phenotypic variations associated with a set of environmental constraints. Whether they are adaptations; i.e., a consequence of natural selection, remains to be determined. In any event, they represent flexibility in reacting to environmental variability and enable marmots to function and persist under varying circumstances.

Habitat use requires some open space, herbaceous vegetation, a place to dig a burrow, and objects on which marmots can sit or lie. Marmots may settle in crevices in cliffs, dig under tree or bush roots, utilize tree stumps (Fig. 22.2), fence posts, or brick walls, and use buildings as burrow and perch sites (Fig. 22.3). Home ranges may be expanded or contracted depending on the distribution of resources, the presence and relatedness of conspecifics, and the proximity and frequency of human or predator activity. That plasticity is the response to these activities is demonstrated by the variability of vigilance behavior, which increases when marmots are confronted by intruders and decreases when intruders depart. Food choice is not species specific and readily adjusts to plant phenology taking advantage of whatever plants are present. The major factor known to limit food choice is the presence of defensive compounds, such as those in fireweed. Hence fireweed may become abundant in the disturbed soil around marmot burrows (Fig. 22.4). Marmot food choice is flexible; marmots feed on the flowers of plants such as columbine, tall larkspur, and fireweed, whose shoots are rejected. There is a limit to this plasticity; marmots do not thrive in the absence of an adequate population of forbs.

Behavioral phenotypes seem not to be genetically determined, but reflect experiences during development. This plasticity adjusts behavior to the current social composition of a matriline and allows females to produce young of varied phenotypes that increase the

Figure 22.3   A yellow-bellied marmot perched on a window sill in a building at Capitol Reef National Park, Utah, USA. (See plate section for color version.)

probability that over the long term some of her descendents will survive in unpredictable and varied environments (Armitage 1986a). The mating system may be monogamous or polygynous depending on the number of adult males present and the number and dispersion of adult females. There is no fixed matriline size; it varies with matriline dynamics. Sex ratio within a litter appears to be plastic and allows females to produce more female young when recruitment of daughters is more likely.

Temperature acclimation provides some plasticity in adjusting metabolic rates to high or low environmental temperatures. Finally, there is some plasticity in social organization and dynamics. Yellow-bellied marmots may live in a group that is cohesive (amicable behavior predominates) or dispersive (agonistic behavior predominates). Juvenile wood-chucks may disperse or spend their first hibernation with their mother; group composition of all the social species varies, primarily as a function of demographic processes. Although reproductive success may vary with social organization, no evidence suggests that a species requires a particular age/sex composition in its social organization to survive and reproduce.

Figure 22.4   Fireweed growing at a yellow-bellied marmot burrow area in Marmot Meadow. Although the foliage is unpalatable, marmots may eat the flowers.

Cooperation and competition

Cooperation generally refers to actions that benefit both participants. The role of co-operation in the evolution of sociality is typically modeled on its function in cooperative breeding species; however, group living may have different functions such as providing shelter, access to resources, or protection or food for offspring (Whitehouse and Lubin 2005). In social marmots, all of these functions operate. Social marmots share burrows, forage in a common home range, and defend or warn against predators. All of these support offspring.

In practice, cooperative actions are often implied rather than directly observed. In yellow-bellied marmots cooperation is indicated by the following: amicable behavior and resource sharing among matrilineal members, increased reproductive success in matri-lines >1, enhancement of an older female's reproductive success when a younger adult female is present, group defense against conspecific intruders, and increased survivorship as matriline size increases.

Competition between matrilines is evident in the high levels of agonistic behavior. Within matrilines competition is sometimes observed (e.g., high level of agonistic behavior between sisters) but usually inferred from demography such as the low rate of reproduction by younger females when older females are present. The mechanism of reproductive suppression in yellow-bellied marmots is unknown; dominance rank did not affect annual female reproductive success and dominance may be unlikely to drive reproductive suppression (Huang *et al.* 2011). However, reproductive suppression probably occurs soon after emergence and could involve scent-gland secretions that are not evident to the human observer. There would be no benefit to impose further dominance as it could cause the loss of the benefits of cooperation and in the long term reduce inclusive fitness.

Are marmots cooperative breeders?

Although yellow-bellied marmot females are highly individualistic, the evidence for cooperation; e.g., reproductive success increases when matrilines are >1; indicates that cooperative breeding occurs. Cooperative breeding is facultative; 72.6% of the matrilines consist of one female (see Table 5.3). By contrast, cooperative breeding in the alpine marmot appears to be obligate if successful reproduction is to occur; non-breeding helpers increase survivorship of the young. In both social systems cooperation and kin selection are possible (Clutton-Brock 2002).

Cooperative breeding is associated with delayed dispersal, reproductive suppression, and helping (care for the offspring of others) (Solomon and French 1997b). Probably all marmot species, with the exception of the woodchuck, are characterized by delayed dispersal and reproductive suppression; the nature and prevalence of helping requires further study (Blumstein and Armitage 1999).

Helping among mammals is strongly associated with providing food for offspring; hunting and food-sharing occur in canids (Moehlman and Hofer 1997). In addition to providing food, non-reproductive naked mole rats groom, huddle, protect, and engage in allocoprophagy with the young of colony members (Lacey and Sherman 1997). The ability to provide food by herbivores constrains helping, although alloparental care in primates is expressed by carrying young (Tardif 1997). In alpine marmots, alloparental care occurs during hibernation when closely related adults huddle with juveniles; huddling actively warms juveniles and increases their survival (Ruf and Arnold 2000). If one includes predator detection and defense of the matrilineal territory as a form of helping, yellow-bellied marmots qualify as cooperative breeders. What distinguishes yellow-bellied marmots from cooperative breeders such as dwarf mongooses (Creel and Waser 1997) is the degree or intensity of competition and the complex roles that helpers may assume. In this context, it is interesting to compare the social system of the facultatively cooperative yellow-bellied marmot with that of the obligatorily cooperative meerkats, *Suricata suricatta* (Clutton-Brock 2007).

For meerkats, group size is critical for maintaining a territory of sufficient size to provide food and burrow resources. Group size is not as critical in yellow-bellied marmots, but coalitions of females may expel a group member, repel a conspecific intruder, and increase reproductive success as group size increases up to a limit at which point reproductive success declines.

In meerkats, helping includes guard duty. Guarding involves baby-sitting at the breeding burrow and assuming sentinel duty to watch for predators. Sentinel duty is associated with distinct alarm calls for terrestrial, aerial, and snake predators; the alarm call informs group members of the nature of the threat and allows them to immediately make the appropriate response. In yellow-bellied marmots, the alarm call indicates the degree of danger. The difference in alarm-calling is associated with foraging behavior. Meerkats forage in larger groups and over more open space than yellow-bellied marmots. When time is of essence and information must be communicated to many individuals over a large area, a more informative alarm call may be critical to survival. In yellow-bellied marmots, each individual is vigilant; no individual assumes the role of a sentinel. Also, no individual remains at the home burrow as a baby-sitter; all individuals forage, although in the early days after weaning the mother may spend more time watching when her young are foraging.

Although reproductive suppression occurs in both species, it is much more severe in meerkats. The dominant female meerkat controls breeding and kills the young of any subordinate female that reproduces. We have numerous examples of failure to wean young in yellow-bellied marmots, but reasons for the failure are unknown. The failure may occur among burrowmates or among matrilineal non-burrowmates and could be caused by infanticide. We detected infanticide between matrilines, but not within matrilines. No obvious cues to the commission of infanticide have been detected and this topic needs further study.

Suppressed meerkat females may initiate lactation and nurse the young of the dominant female. We detected communal nursing in yellow-bellied marmots (Armitage and Gurri-Glass 1994); all participating females weaned litters and lived in the same burrow system. Sometimes, when two females were lactating and young were in one group, the number of weaned young was much less than would be expected from two litters. Possibly one litter was lost and both females nursed the remaining young. Thus, the potential for allonursing exists, but whether and how frequently it occurs, is unknown. There is no indication that a non-reproductive female marmot initiates lactation to help the dominant, reproductive female, as occurs in meerkats.

Why do meerkats and marmots accept reproductive suppression instead of moving to another group or dispersing to establish a new group? For meerkats, immigration into another group is highly unlikely. A female may join a band of males to form a new group, but such a small group is highly susceptible to attacks from established larger groups and is driven from the area. Because the area is saturated with meerkat groups, there is no place for a small group to establish a territory. Female meerkats may evict daughters, sisters, or other kin; yellow-bellied marmots may evict non-littermate sisters, nieces, and cousins, but typically not daughters.

Most dispersing meerkats are probably killed by predators. Dispersing yellow-bellied marmots are more likely to locate a habitat patch, either in unoccupied space in a colony or in an empty mini-habitat patch, than meerkats, but their probability of reproducing is less than that of a philopatric female. For both species environmental constraints reduce the LRS of dispersers and the better alternative is to remain at home and cope with reproductive suppression and, at the minimum, gain indirect fitness benefits.

Qualitatively, both species express the three characteristics associated with cooperative breeding. The major differences are quantitative, especially in the amount and variety of helping and the degree of reproductive suppression. Cooperative and competitive behavior among close kin are essential features of both social systems. Finally, members of each social system attempt to maximize direct or individual fitness. Indirect fitness gains occur as a consequence of direct fitness strategies. Because a female meerkat that fails to reproduce by age 2 is unlikely to ever reproduce, indirect fitness probably is a much larger component of inclusive fitness in meerkats than in yellow-bellied marmots. Indirect fitness is the only component of inclusive fitness for those females that fail to reproduce; in both species indirect fitness does not compensate for the loss of direct fitness. Both subordinate meerkat and yellow-bellied marmot females do the best they can under the circumstances in which they find themselves.

Kin selection and the social system

We noted above that there is some plasticity in marmot social systems. Although a marmot species can live in groups that vary in age and sex structure, this does not mean that sociality is not an essential life-history trait for all the social species. In an extensive review of mammalian sociality, Silk (2007) concluded that sociality influences female fitness in diverse ways; e.g., group size can have a positive effect on juvenile survival and the reproductive success of breeders and that the presence of kin generally enhances female reproductive performance. These effects clearly operate in yellow-bellied marmot social groups and probably in all social marmots. As argued from the biology of cooperatively breeding mammalian species (Clutton-Brock 2009), sociality is the fundamental strategy for maximizing fitness and for many such groups fitness is gained only by living in the group. The fundamental nature of sociality is demonstrated in yellow-bellied marmots that underwent a threefold increase in population size after 2000. During this major demographic event, essential life-history traits were maintained: the central role of kinship in establishing social cohesion, the persistence of social networks (matrilines?), and the proximal causes of dispersal were fundamentally unchanged (Blumstein 2013). Although demography affects the mean relatedness of matriline members and the sex ratio of a colony, it does not change the essential nature of the components of the social system or the tactics used by males and females to gain reproductive success.

Sociality in mammalian societies can be explained by individual selection; group selection is not required (Clutton-Brock 2009). Although sociality makes kin selection possible, we can ask the question: to what degree is kin selection required? To rephrase the question, do marmot social systems require kin selection; i.e., assisting or providing benefits to collateral kin (Wilson 1975)? I suggest that kin selection is not necessary to understand fitness strategies in marmots. The relatively low average relatedness in species such as the alpine marmot occurs because of relatively high rates of dominant takeover by immigrant adults and relatively low levels of ascendancy by young marmots to dominance in their natal family. Average relatedness in this extended family seems to represent the outcome of the demographic consequence of competition for dominant reproductive status. Because some adults and/or sub-adults were born in the family, average relatedness is greater than a random association of individuals from the population at large. The greater involvement of parents and subordinate adults that are full sibs of the young in social thermoregulation and sex-ratio adjustment (producing more male offspring when subordinate male helpers are absent) indicates that alpine marmots follow a direct fitness strategy in which some individuals gain indirect fitness benefits.

Average relatedness is much higher in yellow-bellied marmot social groups because daughters are recruited by their mothers. This pattern is a consequence of individual or direct fitness and does not require kin selection. Although kin recognition may occur, it is not required. Social bonds are readily formed by familiarity, which in many cases seems more important than kinship, per se. Because no immigrant joins a matriline, average relatedness remains high. When demography reduces relatedness, matriline fission reestablishes a high degree of relatedness within each daughter matriline ($r = 0.5$).

Kin selection is not required for understanding patterns of space overlap, social behavior, recruitment (and dispersal), and alarm-calling which are readily understood in terms of classical individual fitness. Competition for direct fitness benefits reduces kin-selected benefits of cooperation (West *et al.* 2002). However, because individuals in groups gain some indirect fitness from assisting kin, inclusive fitness theory is necessary for calculating total fitness of individual marmots. In other words, social and reproductive strategies are directed toward maximizing direct fitness and indirect fitness benefits are a secondary consequence of direct fitness strategies.

Population dynamics

Sociality is the major driver of population dynamics. Demographic change is initiated at the level of the matriline. As the number of adult females in a matriline increases, NRR increases, but declines in the largest matrilines. The increase, decline, and variation in NRR are strongly affected by fertility and age of first reproduction, both of which vary because of reproductive suppression. Both fertility and the age of first reproduction are influenced by the age structure of the matriline and are determined by the reproductive strategies of individual females. The interaction between competition and cooperation that underlies changes in fertility and the age of first reproduction generates population increase or decline. Because these interactions vary from year to year and are not correlated between matrilines or among colonies, the marmot population does not form a stable age distribution.

There is no evidence for density dependence at the basic, functional level of the matriline. Reproductive suppression of a 2-year-old female is independent of the number of older, matrilineal females present. Depressed birth rate when a new territorial male is present is independent of the number of colony females present. Birth rate is low when reproduction by young females is suppressed by one older female. Stochastic events, such as drought, late snowmelt, predation, and length of male residency, are independent of population density and prevent the population from reaching carrying capacity by reducing birth rate, increasing mortality of young and reproductive females, and lowering reproductive rates. Thus both intrinsic and extrinsic factors limit population growth in yellow-belled marmots.

Time

A shortage of time when the rate of increase was positive was considered a very important factor limiting animal populations. The shortage of time kept population numbers below those that would fully exploit the available resources (Andrewartha and Birch 1954). For example, the favorable season was too short for grasshoppers to increase to the point where food supply became limiting (Birch 1957). Similarly, yellow-bellied marmots are not limited by food abundance, but are limited by the time food is available. Plant senescence in late summer affects all individual marmots alike, regardless of population density.

Time is a major factor limiting population size in marmots. Reproductive skipping occurs because marmots have insufficient time to accumulate energy resources needed for both hibernation and reproduction. Time determines the age of yellow-bellied marmot dispersal. At high-elevation sites (>1400 m) marmots disperse as yearlings, but some non-colonial young disperse (Lenihan and Van Vuren 1996b). Marmots disperse in late summer as young at low elevation sites (<1000 m) where a longer growing season allowed marmots to disperse early (Webb 1981). I observed young dispersal at the low-elevation Yellowstone location where the growing season is longer than at the high-elevation Colorado location. This variability in the age of dispersal demonstrates that phenotypic plasticity facilitates its timing. It is the lack of time during the active season for growth to the highly adaptive large body size that delays the age of maturity in those marmots that do not reproduce before age 2 or 3. Both factors limit potential population growth. Mortality of young and reproductive females during hibernation is closely associated with late weaning because time is insufficient for building the fat reserves that provide the energy for metabolism during hibernation. The importance of time was vividly demonstrated when the increased length of the growing season since 2002 at the Colorado location resulted in higher survivorship, especially of adult females, that produced an abrupt increase in population numbers. This increase indicates that marmot populations had not reached their carrying capacity. Also, the added time enabled a yearling female to reproduce.

To sum it up in one sentence, time constraints underlie the evolution of marmot sociality, affect reproductive strategies and the age of maturity, and truncate population growth.

Time had an additional role in the yellow-bellied marmot research project. Only long-term research on the same populations in the same area made the discovery of the relationships among sociality, demography, kinship, and reproductive strategies possible (Armitage 2010, 2012). Long-term research during which the populations were followed in each year revealed that population changes were not correlated among sites and that stochastic events could have profound effects on survival and reproduction and did not affect all populations equally. If one or more years had been missed, critical stochastic events could have been undetected and their importance in population limitation undervalued. And time continuity made possible the detection of the effects of climate change.

References

Aaltonen, K., A. A. Bryant, J. A. Hostetler, and M. K. Oli 2009. Reintroducing endangered Vancouver Island marmots: Survival and cause-specific mortality rates of captive-born versus wild-born individuals. *Biol. Cons.* **142**:2181–2190.

Ageev, V. S., and S. B. Pole 1996. Fleas of the marmots in the plague enzootic areas of the Tien Shan and Pamiro-Alai Mountains. In *Biodiversity in Marmots*, ed. M. Le Berre, R. Ramousse, and L. Le Guelte. Lyon: International Marmot Network, pp. 89–94.

Ageev, V. S., S. B. Pole, V. S. Arakelyanz, and V. I. Sapozhnikov 1997. On the history of a discovery and biocenosis structure of the Kokpak mesofocus of plague. *Abstracts of the 3rd International Marmot Conference*, 114–115.

Aikimbaev, A. M., and S. B. Pole 1996. Use of a test of marmot sensitivity to plague for epidemiological supervision. In *Biodiversity in Marmots*, ed. M. Le Berre, R. Ramousse, and L. Le Guelte. Lyon: International Marmot Network, pp. 95–96.

Aikimbayev, A. M., V. S. Ageev, and S. B. Pole 1997. Additional information on plague in the central Tien Shan. *Abstracts of the 3rd International Marmot Conference*, 115–116.

Aimar, A. 1992. A morphometric analysis of Pleistocene marmots. In *Proceedings of the 1st International Symposium on Alpine Marmot (Marmota marmota) and Genus Marmota*, ed. B. Bassano, P. Durio, U. Gallo Orsi, and E. Macchi. Torino, Italy: Dipartimento di Produzioni animali, Epidemiologia ed Ecologia, pp. 179–184.

Alcock, J. 1989. *Animal Behavior*, 4th edition. Sunderland, MA: Sinauer Associates.

Alexander, R. D. 1974. The evolution of social behavior. *Annu. Rev. Ecol. Syst.* **5**:325–383.

Allainé, D. 2000. Sociality, mating system and reproductive skew in marmots: evidence and hypotheses. *Behav. Processes* **51**:21–34.

Allainé, D. 2004. Sex ratio variation in the cooperatively breeding alpine marmot, *Marmota marmota*. *Behav. Ecol.* **15**:997–1002.

Allainé, D., F. Brondex, L. Graziani, J. Coulon, and I. Till-Bottraud 2000. Male-biased sex ratio in litters of alpine marmots supports the helper repayment hypothesis. *Behav. Ecol.* **11**:507–514.

Allainé, D., A. Cohas, and C. Bonenfane 2008. Demographic effects of climate fluctuations on an alpine marmot *(Marmota marmota)* population. *Abstracts of the 6th International Marmot Conference*, 12–13.

Allainé, D., L. Graziani, and J. Coulon 1998. Postweaning mass gain in juvenile alpine marmots, *Marmota marmota*. *Oecologia* **113**:370–376.

Allainé, D., I. Rodrigue, M. Le Berre, and R. Ramousse 1994. Habitat preferences of alpine marmots, *Marmota marmota*. *Can. J. Zool.* **72**:2193–2198.

Allainé, D., and F. Theuriau 2004. Is there an optimal number of helpers in alpine marmot family groups? *Behav. Ecol.* **15**:916–924.

Allen, K. 1995. *The Legend of the Whistle Pig Wrangler*. Del Mar, CA: Kumquat Press.

Altmann, S. A. 1979. Altruistic behaviour: the fallacy of kin deployment. *Anim. Behav.* 27:958–959.

Andersen, D. C., K. B. Armitage, and R. S. Hoffmann 1976. Socioecology of marmots: female reproductive strategies. *Ecology* 57:552–560.

Andersen, D. C., and D. W. Johns 1977. Predation by badger on yellow-bellied marmot in Colorado. *Southwest. Nat.* 22:283–284.

Anderson, R. M. 1934. Notes on the distribution of the hoary marmots. *Can. Field-Nat.* 48:61–63.

Andrewartha, H. G., and L. C. Birch 1954. *The Distribution and Abundance of Animals.* Chicago, IL: University of Chicago Press.

Anonymous (1963–97). *Climatological Data Annual Summary, Colorado.* Washington DC: National Oceanic and Atmospheric Administration.

Anthony, A. W. 1923. Periodical emigration of mammals. *J. Mammal.* 4:60–61.

Anthony, M. 1962. Activity and behavior of the woodchuck in southern Illinois. *Occas. Papers Adams Ctr. Ecol. Studies* 6:1–25.

Armitage, K. B. 1962. Social behaviour of a colony of the yellow-bellied marmot (*Marmota flaviventris*). *Anim. Behav.* 10:319–331.

Armitage, K. B. 1965. Vernal behaviour of the yellow-bellied marmot (*Marmota flaviventris*). *Anim. Behav.* 13:59–68.

Armitage, K. B. 1973. Population changes and social behavior following colonization by the yellow-bellied marmot. *J. Mammal.* 54:842–854.

Armitage, K. B. 1974. Male behaviour and territoriality in the yellow-bellied marmot. *J. Zool., Lond.* 172:233–265.

Armitage, K. B. 1975. Social behavior and population dynamics of marmots. *Oikos* 26:341–354.

Armitage, K. B. 1976. Scent-marking by yellow-bellied marmots. *J. Mammal.* 57:583–584.

Armitage, K. B. 1977. Social variety in the yellow-bellied marmot: a population-behavioural system. *Anim. Behav.* 25:585–593.

Armitage, K. B. 1979. Food selectivity by yellow-bellied marmots. *J. Mammal.* 60:626–629.

Armitage, K. B. 1981. Sociality as a life-history tactic of ground squirrels. *Oecologia* 48:36–49.

Armitage, K. B. 1982a. Yellow-bellied marmot. In *CRC Handbook of Census Methods for Terrestrial Vertebrates*, ed. D. E. Davis. Boca Raton, FL: CRC Press, pp. 148–149.

Armitage, K. B. 1982b. Marmots and coyotes: behavior of prey and predator. *J. Mammal.* 63:503–505.

Armitage, K. B. 1982c. Social dynamics of juvenile marmots: role of kinship and individual variability. *Behav. Ecol. Sociobiol.* 11:33–36.

Armitage, K. B. 1983. Hematological values for free-ranging yellow-bellied marmots. *Comp. Biochem. Physiol.* 74A:89–93.

Armitage, K. B. 1984. Recruitment in yellow-bellied marmot populations: kinship, philopatry, and individual variability. In *Biology of Ground-dwelling Squirrels*, ed. J. O. Murie and G. R. Michener. Lincoln, NE: University of Nebraska Press, pp. 377–403.

Armitage, K. B. 1986a. Individuality, social behavior, and reproductive success in yellow-bellied marmots. *Ecology* 67:1186–1193.

Armitage, K. B. 1986b. Individual differences in the behavior of juvenile yellow-bellied marmots. *Behav. Ecol. Sociobiol.* 18:419–424.

Armitage, K. B. 1986c. Marmot polygyny revisited: determinants of male and female reproductive success. In *Ecological Aspects of Social Evolution*, ed. D. S. Rubenstein and R. W. Wrangham. Princeton, NJ: Princeton University Press, pp. 303–331.

Armitage, K. B. 1987a. Social dynamics of mammals: reproductive success, kinship, and individual fitness. *Trends Ecol. Evol.* 2:279–284.

Armitage, K. B. 1987b. Do female yellow-bellied marmots adjust the sex ratios of their offspring? *Am. Nat.* **129**:501–519.

Armitage, K. B. 1988. Resources and social organization of ground-dwelling squirrels. In *Ecology of Social Behavior*, ed. C. N. Slobodchikoff. New York: Academic Press, pp. 131–155.

Armitage, K. B. 1989. The function of kin discrimination. *Ethol., Ecol., Evol.* **1**:111–121.

Armitage, K. B. 1991a. Social and population dynamics of yellow-bellied marmots: results from long-term research. *Annu. Rev. Ecol. Syst.* **22**:379–407.

Armitage, K. B. 1991b. Factors affecting corticosteroid concentrations in yellow-bellied marmots. *Comp. Biochem. Physiol.* **98**A:47–54.

Armitage, K. B. 1992. Social organization and fitness strategies of marmots. In *Proceedings of the 1st International Symposium on Alpine Marmot (*Marmota marmota*) and Genus* Marmota, ed. B. Bassano, P. Durio, U. Gallo Orsi, and E. Macchi. Torino, Italy: Dipartimento di Produzioni animali, Epidemiologia ed Ecologia, pp. 89–94.

Armitage, K. B. 1994. Unusual mortality in a yellow-bellied marmot population. In *Actual Problems of Marmots Investigation*, ed. V. Rumiantsev. Moscow: ABF Publishing House, pp. 5–13.

Armitage, K. B. 1996a. Seasonal mass gain in yellow-bellied marmots. In *Biodiversity in Marmots*, ed. M. Le Berre, R. Ramousse, and L. Le Guelte. Lyon, France: International Marmot Network, pp. 223–226.

Armitage, K. B. 1996b. Social dynamics, kinship, and population dynamics of marmots. In *Biodiversity in Marmots*, ed. M. Le Berre, R. Ramousse, and L. Le Guelte. Lyon, France: International Marmot Network, pp. 113–128.

Armitage, K. B. 1996c. Resource sharing and kinship in yellow-bellied marmots. In *Biodiversity in Marmots*, ed. M. Le Berre, R. Ramousse, and L. Le Guelte. Lyon, France: International Marmot Network, pp. 129–134.

Armitage, K. B. 1998. Reproductive strategies of yellow-bellied marmots: energy conservation and differences between the sexes. *J. Mammal.* **79**:385–393.

Armitage, K. B. 1999. Evolution of sociality in marmots. *J. Mammal.* **80**:1–10.

Armitage, K. B. 2000. The evolution, ecology, and systematics of marmots. *Oecologia Montana* **9**:1–18.

Armitage, K. B. 2002. Social dynamics of yellow-bellied marmots: strategies for evolutionary success. In *Holarctic Marmots as a Factor of Biodiversity*, ed. K. B. Armitage and V. Yu. Rumiantsev. Moscow: ABF Publishing House, pp. 9–16.

Armitage, K. B. 2003a. Marmots *Marmota monax* and allies. In *Wild Mammals of North America, Biology, Management, and Conservation*, 2nd edition, ed. G. A. Feldhamer, B. C. Thompson, and J. A. Chapman. Baltimore, MD: The Johns Hopkins University Press, pp. 188–210.

Armitage, K. B. 2003b. Heart rates of free-ranging yellow-bellied marmots. In *Adaptive Strategies and Diversity in Marmots*, ed. R. Ramousse, D. Allainé, and M. Le Berre. Lyon, France: International Marmot Network, pp. 89–96.

Armitage, K. B. 2003c. Recovery of a yellow-bellied marmot population following a weather-induced decline. In *Adaptive Strategies and Diversity in Marmots*, ed. R. Ramousse, D. Allainé, and M. Le Berre. Lyon: International Marmot Network, pp. 217–224.

Armitage, K. B. 2003d. Nesting activities of yellow-bellied marmots. In *Adaptive Strategies and Biodiversity in Marmots*, ed. R. Ramousse, D. Allainé, and M. Le Berre. Lyon, France: International Marmot Network, pp. 27–32.

Armitage, K. B. 2003e. Observations on plant choice by foraging yellow-bellied marmots. *Oecologia Montana* **12**:25–28.

Armitage, K. B. 2003f. Reproductive competition in female yellow-bellied marmots. In *Adaptive Strategies and Diversity in Marmots*, ed. R. Ramousse, D. Allainé, and M. Le Berre. Lyon, France: International Marmot Network, pp. 133–142.

Armitage, K. B. 2003g. Behavioral responses of yellow-bellied marmots to birds and mammals. *Oecologia Montana* 12:15–20.

Armitage, K. B. 2003h. Is the hepatitis virus absent from yellow-bellied marmots? *Oecologia Montana* 12:37–38.

Armitage, K. B. 2003i. Dynamics of immigration into yellow-bellied marmot colonies. *Oecologia Montana* 12:21–24.

Armitage, K. B. 2004a. Badger predation on yellow-bellied marmots. *Am. Midl. Nat.* 151:378–387.

Armitage, K. B. 2004b. Metabolic diversity in yellow-bellied marmots. In *Life in the Cold: Evolution, Mechanisms, Adaptation, and Application*, ed. B. M. Barnes and H. V. Carey. Fairbanks, Alaska: Institute of Arctic Biology, University of Alaska, pp. 162–173.

Armitage, K. B. 2004c. Lifetime reproductive success of territorial male yellow-bellied marmots. *Oecologia Montana* 13:28–34.

Armitage, K. B. 2005. Intraspecific variation in marmots. In *Contribuciones, Mastozoologicas en Homenaje a Bernardo Villa*, ed. V. Sanchez-Cordero and R. A. Medellin. Mexico: Instituto de Biologia, UNAM; Instituto de Ecologia, UNAM, CONABIO, pp. 39–48.

Armitage, K. B. 2007. Evolution of sociality in marmots: it begins with hibernation. In *Rodent Societies: An Ecological and Evolutionary Perspective*, ed. J. O. Wolff and P. W. Sherman. Chicago, IL: University of Chicago Press, pp. 356–367.

Armitage, K. B. 2008. Phylogeny and patterns of energy conservation in marmots. In *Molecules to Migration: The Pressures of Life*, ed. S. Morris and A. Vosloo. Bologna, Italy: Medimond, pp. 591–602.

Armitage, K. B. 2009a. Fur color diversity in marmots. *Ethol., Ecol., Evol.* 21:183–194.

Armitage, K. B. 2009b. Home range area and shape of yellow-bellied marmots. *Ethol., Ecol., Evol.* 21:195–207.

Armitage, K. B. 2010. Individual fitness, social behavior, and population dynamics of yellow-bellied marmots. In *The Ecology of Place*, ed. I. Billick and M. V. Price. Chicago, IL: University of Chicago Press, pp. 134–154.

Armitage, K. B. 2012. Sociality, individual fitness and population dynamics of yellow-bellied marmots. *Mol. Ecol.* 21:532–540.

Armitage, K. B. 2013. Climate change and the conservation of marmots. *Nat. Sci.* 5(5A):36–43.

Armitage, K. B., and D. T. Blumstein 2002. Body-mass diversity in marmots. In *Holarctic Marmots as a Factor of Biodiversity*, ed. K. B. Armitage and V. Yu. Rumiantsev. Moscow: ABF Publishing House, pp. 22–32.

Armitage, K. B., D. T. Blumstein, and B. C. Woods 2003. Energetics of hibernating yellow-bellied marmots (*Marmota flaviventris*). *Comp. Biochem. Physiol. A* 134:101–114.

Armitage, K. B., and M. Chiesura Corona 1994. Time and wariness in yellow-bellied marmots. *IBEX J. M. E.* 2:1–8.

Armitage, K. B., and J. F. Downhower 1970. Interment behavior in the yellow-bellied marmot (*Marmota flaviventris*). *J. Mammal.* 51:177–178.

Armitage, K. B., and J. F. Downhower 1974. Demography of yellow-bellied marmot populations. *Ecology* 55:1233–1245.

Armitage, K. B., J. F. Downhower, and G. E. Svendsen 1976. Seasonal changes in weights of marmots. *Am. Midl. Nat.* 96:36–51.

Armitage, K. B., and G. E. Gurri-Glass 1994. Communal nesting in yellow-bellied marmots. In *Actual Problems of Marmots Investigation*, ed. V. Rumiantsev. Moscow: ABF Publishing House, pp. 14–26.

Armitage, K. B., and D. W. Johns 1982. Kinship, reproductive strategies and social dynamics of marmots. *Behav. Ecol. Sociobiol.* **11**:55–63.

Armitage, K. B., D. Johns, and D. C. Andersen 1979. Cannibalism among yellow-bellied marmots. *J. Mammal.* **60**:205–207.

Armitage, K. B., J. C. Melcher, and J. M. Ward, Jr 1990. Oxygen consumption and body temperature in yellow-bellied marmot populations from montane-mesic and lowland-xeric environments. *J. Comp. Physiol. B.* **160**:491–502.

Armitage, K. B., and C. M. Salsbury 1992. Factors affecting oxygen consumption in wild-caught yellow-bellied marmots (*Marmota flaviventris*). *Comp. Biochem. Physiol.* **103**A:729–737.

Armitage, K. B., and C. M. Salsbury 1993. The effect of molt on oxygen consumption of yellow-blelied marmots (*Marmota flaviventris*). *Comp. Biochem. Physiol.* **106**A:667–670.

Armitage, K. B., C. M. Salsbury, E. L. Barthelmess, R. C. Gray, and A. Kovach 1996. Population time budget for the yellow-bellied marmot. *Ethol., Ecol., Evol.* **8**:67–95.

Armitage, K. B., and O. A. Schwartz 2000. Social enhancement of fitness in yellow-bellied marmots. *Proc. Nat. Acad. Sci.* **97**:12149–12152.

Armitage, K. B., and D. H. Van Vuren 2003. Individual differences and reproductive success in yellow-bellied marmots. *Ethol., Ecol., Evol.* **15**:207–233.

Armitage, K. B., D. H. Van Vuren, A. Ozgul, and M. D. Oli 2011. Proximate causes of natal dispersal in female yellow-bellied marmots. *Ecology* **92**:218–227.

Armitage, K. B., and B. C. Woods 2003. Group hibernation does not reduce energetic costs of young yellow-bellied marmots. *Physiol. Biochem. Zool.* **76**:888–898.

Armitage, K. B., B. C. Woods, and C. M. Salsbury 2000. Energetics of hibernation in woodchucks (*Marmota monax*). In *Life in the Cold*, ed. G. Heldmaier and M. Klingenspor. Berlin: Springer, pp. 73–80.

Armitage, K. B., and K. E. Wynne-Edwards 2002. Progesterone concentrations in wild-caught yellow-bellied marmots. In *Holarctic Marmots as a Factor of Biodiversity*, ed. K. B. Armitage and V. Rumiantsev. Moscow: ABF Publishing House, pp. 41–47.

Armstrong, D. M. 1972. Distribution of mammals in Colorado. *Monograph Mus. Nat. Hist., Univ. Kansas* **3**:1–415.

Arnold, W. 1988. Social thermoregulation during hibernation in alpine marmots (*Marmota marmota*). *J. Comp. Physiol. B* **158**:151–156.

Arnold, W. 1990a. The evolution of marmot sociality: I. Why disperse late? *Behav. Ecol. Sociobiol.* **27**:229–237.

Arnold, W. 1990b. The evolution of marmot sociality: II. Costs and benefits of joint hibernation. *Behav. Ecol. Sociobiol.* **27**:239–246.

Arnold, W. 1993a. Energetics of social hibernation. In *Life in the Cold, Biological, Physiological, and Molecular Mechanisms*, ed. C. Carey, G. L. Florant, B. A. Wunder, and B. Horwitz. Boulder, CO: Westview Press, pp. 65–80.

Arnold, W. 1993b. Social evolution in marmots and the adaptive value of joint hibernation. *Verh. Dtsch. Zool. Ges.* **86**:79–93.

Arnold, W., U. Bruns, F. Frey-Roos, and T. Ruf 2003. Dietary fatty acids and natural hibernation in alpine marmots. In *Adaptive Strategies and Biodiversity in Marmots*, ed. R. Ramousse, D. Allainé, and M. Le Berre. Lyon, France: International Marmot Network, pp. 95–96.

Arnold, W., and J. Dittami 1997. Reproductive suppression in male alpine marmots. *Anim. Behav.* **53**:53–66.

Arnold, W., G. Heldmaier, S. Ortmann, H. Pohl, T. Ruf, and S. Steinlechner 1991. Ambient temperatures in hibernacula and their energetic consequences for alpine marmots (*Marmota marmota*). *J. Therm. Biol.* **16**:223–226.

Arnold, W., M. Klinkicht, K. Rassmann, and D. Tautz 1994. Molecular analysis of the mating system of alpine marmots (*Marmota marmota*). *Verh. Dtsch. Zool. Ges.* **87**:27.

Arnold, W., and V. Lichtenstein 1993. Ectoparasite loads decrease the fitness of alpine marmots (*Marmota marmota*) but are not a cost of sociality. *Behav. Ecol.* **4**:36–39.

Arnold, W., T. Ruf, F. Frey-Roos, and U. Bruns 2011. Diet-independent remodeling of cellular membranes precedes seasonally changing body temperature in a hibernator *Plos ONE* **6**(4): e18641.

Arsenault, J. R., and R. F. Romig 1985. Plants eaten by woodchucks in three northeast Pennsylvania counties. *Proc. Penn. Acad. Sci.* **59**:131–134.

Aschemeier, L. M., and C. R. Maher 2011. Eavesdropping of woodchucks (*Marmota monax*) and eastern chipmunks (*Tamias striatus*) on heterospecific alarm calls. *J. Mammal.* **92**:493–499.

Badmaev, B. B. 1996. The religious traditions of the Transbaikalian nations in marmot protection. In *Biodiversity in Marmots*, ed. R. Ramousse, D. Allainé, and M. Le Berre. Lyon, France: International Marmot Network, pp. 63–64.

Bailey, E. D. 1965. Seasonal changes in metabolic activity of nonhibernating woodchucks. *Can. J. Zool.* **43**:905–909.

Ballo, P. 2006. Monitoring alpine marmot (*Marmota marmota latirostris*) colonies in the West Tatra Mountains – III. *Oecologia Montana* **16**:15–20.

Ballo, P., and J. Sykora 2005. Monitoring of colonies of *Marmota marmota latirostris* in the Western Tatra Mts. – II. Section (2005), Benikov-Ostry Rohac. *Oecologia Montana* **14**:25–32.

Barash, D. P. 1973a. The social biology of the Olympic marmot. *Anim. Behav. Monogra.* **6**:173–245.

Barash, D. P. 1973b. Habitat utilization in three species of subalpine mammals. *J. Mammal.* **54**:247–250.

Barash, D. P. 1974a. The evolution of marmot societies: a general theory. *Science* **185**:415–420.

Barash, D. P. 1974b. The social behaviour of the hoary marmot (*Marmota caligata*). *Anim. Behav.* **22**:256–261.

Barash, D. P. 1974c. Mother–infant relations in captive woodchucks (*Marmota monax*). *Anim. Behav.* **22**:446–448.

Barash, D. P. 1975a. Ecology of paternal behavior in the hoary marmot (*Marmota caligata*): an evolutionary interpretation. *J. Mammal.* **56**:613–618.

Baarash, D. P. 1975b. Marmot alarm-calling and the question of altruistic behavior. *Am. Midl. Nat.* **94**:468–470.

Barash, D. P. 1976a. Pre-hibernation behavior of free-living hoary marmots, *Marmota caligata*. *J. Mammal.* **57**:182–185.

Barash, D. P. 1976b. Social behaviour and individual differences in free living alpine marmots (*Marmota marmota*). *Anim. Behav.* **24**:27–35.

Barash, D. P. 1980. The influence of reproductive status on foraging by hoary marmots (*Marmota caligata*). *Behav. Ecol. Sociobiol.* **7**:201–205.

Barash, D. P. 1981. Mate guarding and gallivanting by male hoary marmots (*Marmota caligata*). *Behav. Ecol. Sociobiol.* **9**:187–193.

Barash, D. P. 1989. *Marmots: Social Behavior and Ecology*. Stanford, CA: Stanford University Press.

Barnes, B. M. 1996. Relationships between hibernation and reproduction in male ground squirrels In *Adaptations to the Cold: Tenth International Hibernation Symposium*, ed. F. Geiser, A. J. Hulbert, and S. C. Nicol. Armidale, Australia: University of New England Press, pp. 71–80.

Barnosky, A. D. 2004a. Faunal dynamics of small mammals through the Pit sequence. In *Biodiversity Response to Climate Change in the Middle Pleistocene*, ed. A. D. Barnosky. Berkeley, CA: University of California Press, pp. 318–326.

Barnosky, A. D. 2004b. Effect of climate change on terrestrial vertebrate biodiversity. In *Biodiversity Response to Climate Change in the Middle Pleistocene*, ed. A. D. Barnosky. Berkeley, CA: University of California Press, pp. 341–345.

Barnosky, A. D., and S. B. Hopkins 2004. Identification of miscellaneous mammals from the pit locality. In *Biodiversity Response to Climate Change in the Middle Pleistocene*, ed. A. D. Barnosky. Berkeley, CA: University of California Press, pp. 169–192.

Barnosky, A. D., M. H. Kaplan, and M. A. Carrasco 2004. *Assessing the Effect of Middle Pleistocene Climate Change on Marmota Populations from the Pit Locality*, ed. A. D. Barnosky. Berkeley, CA: University of California Press, pp. 332–340.

Barrio, I. C., J. Herrero, C. G. Bueno, B. C. Lopex, A. Aldezabel, A. Campos-Arciz, and R. Garcia-Gonzalez 2012. The successful introduction of the alpine marmot *Marmota marmota* in the Pyrenees, Iberian Peninsula, Western Europe. *Mammal Review* **43**:142–155.

Bassano, B. 1996. Sanitary problems related to marmot-other animals cohabitation in mountain areas. In *Biodiversity in Marmots*, ed. M. Le Berre, R. Ramousee, and L. Le Guelte. Lyon, France: International Marmot Network, pp. 75–88.

Bassano, B., I. Grimod, and V. Peracino 1992a. Distribution of alpine marmot (*Marmota marmota*) in the Aosta Valley and suitability analysis. In *Proceedings of the 1st International Symposium on Alpine Marmot (*Marmota marmota*) and Genus Marmota*, ed. B. Bassano, P. Durio, U. Gallo Orsi, and E. Macchi. Torino, Italy: Dipartimento di Produzioni animali, Epidemiologia ed Ecologia, pp. 111–116.

Bassano, B., V. Peracino, and F. Montacchini 1996. Diet composition and feeding habits in a family group of alpine marmot (*Marmota marmota*). In *Biodiversity in Marmots*, ed. M. Le Berre, R. Ramousse, and L. Le Guelte. Lyon, France: International Marmot Network, pp. 135–140.

Bassano, B., B. Sabatier, L. Rossi, and E. Macchi 1992b. Parasitic fauna of the digestive tract of *Marmota marmota* in the western Alps. In *Proceedings of the 1st International Symposium on Alpine Marmot (*Marmota marmota*) and Genus Marmota*, ed. B. Bassano, P. Durio, U. Gallo Orsi, and E. Macchi. Torino, Italy: Dipartimento di Produzioni animali, Epidemiologia ed Ecologia, pp. 13–24.

Batbold, J. 2002a. A study on population genetic structure of the Mongolian marmot and some problems of plague. In *Holarctic Marmots as a Factor of Biodiversity*, ed. K. B. Armitage and V. Yu. Rumiantsev. Moscow: ABF Publishing House, pp. 54–67.

Batbold, J. 2002b. The problem of management of marmots in Mongolia. In *Holarctic Marmots as a Factor of Biodiversity*, ed. K. B. Armitage and V. Yu. Rumiantsev. Moscow: ABF Publishing House, pp. 68–75.

Beckerman, A. P., S. P. Sharp, and B. J. Hatchwell 2011. Predation and kin-structured populations: an empirical perspective on the evolution of cooperation. *Behav. Ecol.* **22**:1294–1303.

Bednekoff, P. A., and D. T. Blumstein 2009. Peripheral obstructions influence marmot vigilance: interesting observational and experimental results. *Behav. Ecol.* **20**:1111–1117.

Bekoff, M. 1977. Mammalian dispersal and the ontogeny of individual behavioral phenotypes. *Am. Natur.* **111**:715–732.

Bekoff, M., and C. Allen 1998. Intentional communication and social play: how and why animals negotiate and agree to play. In *Animal Play: Evolutionary, Comparative, and Ecological Perspectives*, ed. M. Bekoff and J. A. Byers. Cambridge: Cambridge University Press, pp. 97–114.

Bel, M.-C., C. Portoret, and J. Coulon 1995. Scent deposition by cheek rubbing in the alpine marmot (*Marmota marmota*) in the French Alps. *Can. J. Zool.* **73**:2065–2071.

Belovezhets, K. I. 2008. Global warming and marmot's hibernation: Mathematical modeling of burrow temperature in changing climate. *Sixth International Marmot Conference Abstracts*, 18.

Bezuidenhout, A. J., and H. E. Evans 2005. Anatomy of the woodchuck (*Marmota marmota*). *Am. Soc. of Mammalogists Special Publication* **13**:1–180.

Bibikov, D. I. 1992. Marmots are zoonosis provoking carriers. In *Proceedings of the 1st International Symposium on Alpine Marmot (*Marmota marmota*) and Genus* Marmota, ed. B. Bassano, P. Durio, U. Gallo Orsi, and E. Macchi. Torino, Italy: Dipartimento di Produzioni animali, Epidemiologia ed Ecologia, pp. 25–29.

Bibikov, D. I., and V. Yu. Rumiantsev 1996. Current and past relations between people and marmots in the countries of the former Soviet Union. In *Biodiversity in Marmots*, ed. M. Le Berre, R. Ramousse, and L. Le Guelte. Lyon, France: International Marmot Network, pp. 9–22.

Bibikow, D. I. 1996. *Die Murmeltiere der Welt*. Magdeburg, Germany: Westarp Wissenschaften.

Birch, L. C. 1957. The role of weather in determining the distribution and abundance of animals. *Cold Spring Harbor Symp. Quant. Biol.* **22**:203–215.

Black, C. C. 1972. Holarctic evolution and dispersal of squirrels (Rodential: Sciuridae). *Evol. Biol.* **6**:305–322.

Blumstein, D. T. 1992. Multivariate analysis of golden marmot maximum running speed: a new method to study MRS in the field. *Ecology* **73**:1757–1767.

Blumstein, D. T. 1995. Golden-marmot alarm calls. II. Asymmetrical production and perception of situationally specific vocalizations. *Ethology* **101**:25–32.

Blumstein, D. T. 1996. How much does social group size influence golden marmot vigilance? *Behaviour* **133**:1133–1151.

Blumstein, D. T. 1997. Infanticide among golden marmots (*Marmota caudata aurea*). *Ethol., Ecol., Evol.* **9**:169–173.

Blumstein, D. T. 1998. Quantifying predation risk for refuging animals: a case study with golden marmots. *Ethology* **104**:501–516.

Blumstein, D. T. 1999. Alarm calling in three species of marmots. *Behaviour* **136**:731–757.

Blumstein, D. T. 2003. Social complexity but not the acoustic environment is responsible for the evolution of complex alarm communication. In *Adaptive Strategies and Diversity in Marmots*, ed. R. Ramousse, D. Allainé, and M. Le Berre. Lyon, France: International Marmot Network, pp. 31–38.

Blumstein, D. T. 2007. The evolution, function, and meaning of marmot alarm communication. *Adv. in the Study of Behav.* **37**:371–400.

Blumstein, D. T. 2009. Social effects on emergence from hibernation in yellow-bellied marmots. *J. Mammal.* **90**:1184–1187.

Blumstein, D. T. 2013. Yellow-bellied marmots: insights from an emergent view of sociality. *Phil. Trans. R. Soc. B* **368**:20120349.

Blumstein, D. T., and K. B. Armitage 1997a. Does social complexity drive the evolution of communicative complexity? A comparative test with ground-dwelling sciurid alarm calls. *Am. Nat.* **150**:179–200.

Blumstein, D. T., and K. B. Armitage 1997b. Alarm calling in yellow-bellied marmots: I. The meaning of situationally variable alarm calls. *Anim. Behav.* **53**:143–171.

Blumstein, D. T., and K. B. Armitage 1998a. Life-history consequences of social complexity: a comparative study of ground-dwelling sciurids. *Behav. Ecol.* **9**:8–19.

Blumstein, D. T., and K. B. Armitage 1998b. Why do yellow-bellied marmots call? *Anim. Behav.* **56**:1053–1055.

Blumstein, D. T., and K. B. Armitage 1999. Cooperative breeding in marmots. *Oikos* **84**:369–382.

Blumstein, D. T., and W. Arnold 1995. Situational specificity in alpine-marmot alarm communication. *Ethology* **100**:1–13.

Blumstein, D. T., and W. Arnold 1998. Ecology and social behavior of golden marmots (*Marmota caudata aurea*). *J. Mammal.* **79**:873–886.

Blumstein, D. T., L. K. Chung, and J. E. Smith 2013. Early play may predict later dominace relationships in yellow-bellied marmots (*Marmota flaviventris*). *Proc. Royal Soc. B* **280**:20130485.

Blumstein, D. T., L. Cooley, J. Winternitz, and J. C. Daniel 2008a. Do yellow-bellied marmots respond to predator vocalizations? *Behav. Ecol. Sociobiol.* **62**:457–468.

Blumstein, D. T., and J. C. Daniel 1997. Inter- and intraspecific variation in the acoustic habitats of three marmot species. *Ethology* **103**:325–338.

Blumstein, D. T., and J. C. Daniel 2004. Yellow-bellied marmots discriminate between the alarm calls of individuals and are more responsive to calls from juveniles. *Anim. Behav.* **68**:1257–1265.

Blumstein, D. T., J. C. Daniel, and W. Arnold 2002. Survivorship of golden marmots (*Marmota caudata aurea*) in Pakistan. In *Holarctic Marmots as a Factor of Biodiversity*, ed. K. B. Armitage and V. Yu. Rumiantsev. Moscow: ABF Publishing House, pp. 82–85.

Blumstein, D. T., E. Ferando, and T. Stankowich 2009a. A test of the multipredator hypothesis: yellow-bellied marmots respond fearfully to the sight of novel and extinct predators. *Anim. Behav.* **78**:873–878.

Blumstein, D. T., and J. M. Foggin 1997. Effects of vegetative variation on weaning success, overwinter survival, and social group density in golden marmots (*Marmota caudata aurea*). *J. Zool., Lond.* **243**:57–69.

Blumstein, D. T., and S. J. Henderson 1996. Cheek-rubbing in golden marmots (*Marmota caudata aurea*). *J. Zool., Lond.* **238**:113–123.

Blumstein, D. T., B.-D. Holland, and J. C. Daniel 2006a. Predator discrimination and "personality" in captive Vancouver Island marmots (*Marmota vancouverensis*). *An. Conser.* **9**:274–282.

Blumstein, D. T., S. Im, A. Nicodemus, and C. Zugmeyer 2004a. Yellow-bellied marmots (*Marmota flaviventris*) hibernate socially. *J. Mammal.* **85**:25–29.

Blumstein, D. T., A. J. Lea, L. E. Olson, and G. A. Martin 2010. Heritability of anti-predator traits: vigilance and locomotor performance in marmots. *J. Evol. Biol.* **23**:879–887.

Blumstein, D. T., and O. Munos 2005. Individual, age and sex-specific information is contained in yellow-bellied marmot alarm calls. *Anim. Behav.* **69**:353–361.

Blumstein, D. T., A. Ozgul, V. Yovovich, D. H. Van Vuren, and K. B. Armitage 2006b. Effect of predation risk on the presence and persistence of yellow-bellied marmot (*Marmota flaviventris*) colonies. *J. Zool., Lond.* **270**:132–138.

Blumstein, D. T., M. L. Patton, and W. Saltzman 2006c. Faecal glucocorticoid metabolites and alarm calling in free-living yellow-bellied marmots. *Biol. Lett.* **2**:29–32.

Blumstein, D. T., D. T. Richardson, L. Cooley, J. Winternitz, and J. C. Daniel 2008b. The structure, meaning and function of yellow-bellied marmot pup screams. *Anim. Behav.* **76**:1055–1064.

Blumstein, D. T., J. Steinmetz, K. B. Armitage, and J. C. Daniel 1997. Alarm calling in yellow-bellied marmots: II. The importance of direct fitness. *Anim. Behav.* **53**:173–184.

Blumstein, D. T., L. Verneyre, and J. C. Daniel 2004b. Reliability and the adaptive utility of discrimination among alarm callers. *Proc. R. Soc. Lond. B* **271**:1851–1857.

Blumstein, D. T., T. W. Wey, and K. Tang 2009b. A test of the social cohesion hypothesis: interactive female marmots remain at home. *Proc. R. Soc. B* **276**:3007–3012.

Boggs, D. F., and G. F. Birchard 1989. Cardiorespiratory responses of the woodchuck and porcupine to CO_2 and hypoxia. *J. Comp. Physiol. B* **159**:641–648.

Boggs, D. F., D. L. Kilgore, Jr., and G. F. Birchard 1984. Respiratory physiology of burrowing mammals and birds. *Comp. Biochem. Physiol.* **77A**:1–7.

Bonesi, L., L. Lepini, and G. Gregori 1996. Temporal analysis of activities in alpine marmots (*Marmota marmota* L.). In *Biodiversity in Marmots*, ed. M. Le Berre, R. Ramousse, and L. Le Guelte. Lyon, France: International Marmot Network, pp. 157–168.

Bopp, P. 1956. Zur Topographie eines Kolonialterritoriums bei Murmeltieren. *Rev. suisse Zool.* **63**:255–261.

Borgo, A. 2003. Habitat requirements of the alpine marmot (*Marmota marmota*) in re-introduction areas of the Eastern Italian Alps. Formulation and validation of habitat suitability models. *Acta Theriol.* **48**:557–569.

Borgo, A., E. Vettorazzo, and N. Martino 2009. Dynamics of the colonization process in reintroduced populations of the alpine marmot. *Ethol., Ecol., Evol.* **21**:317–323.

Bourke, A. F. G. 2011. The validity and value of inclusive fitness theory. *Proc. R. Soc. B* **278**:3313–3320.

Bowler, D. E., and T. G. Benton 2005. Causes and consequences of animal dispersal strategies: relating individual behavior to spatial dynamics. *Biol. Rev.* **80**:205–225.

Boyeskorov, G. G., M. V. Shchelchkova, and V. N. Vasiliev 1996. Divergence in *Marmota camtschatica*. In *Biodiversity in Marmots*, ed M. Le Berre, R. Ramousse, and L. Le Guelte. Lyon, France: International Marmot Network, pp. 229–230.

Brady, K. M., and K. B. Armitage 1999. Scent-marking in the yellow-bellied marmot (*Marmota flaviventris*). *Ethol., Ecol., Evol.* **11**:35–47.

Brandler, O. V. 2003. Chromosomal speciation and polymorphism in gray marmots (Marmota, Sciuridae, Rodentia). In *Adaptive Strategies and Diversity in Marmots*, ed. R. Ramousse, D. Allainé, and M. Le Berre. Lyon, France: International Marmot Network, pp. 57–62.

Brandler, O. V. 2007. Phylogenetic links in the genus *Marmota* and history of the establishment of its range. In *The Marmots of Eurasia: Origin and Current Status*, ed. A. V. Esipov. Tashkent: International Marmot Network, pp. 16–30.

Brandler, O. V., and E. A. Lyapunova 2009. Molecular phylogenies of the genus *Marmota* (Rodentia: Sciuridae): comparative analyses. *Ethol., Ecol., Evol.* **21**:289–298.

Brandler, O. V., E. A. Lyapunova, A. A. Bannikova, and D. A. Kramerov 2010. Phylogeny and systematics of marmots (Marmota, Sciuridae, Rodentia) inferred from Inter-SINE PCR data. *Rus. J. Gen.* **46**:283–292.

Brandler, O. V., E. A. Lyapunova, and G. G. Boeskorov 2008. Comparative karyology of Palearctic marmots (Marmota, Sciuridae, Rodentia). *Mammalia* **72**:24–34.

Brandler, O. V., A. A. Nikol'sky, and V. V. Kolesnikov 2010. Spatial distribution of *Marmota baibacina* and *M. sibirica* (Marmota, Sciuridae, Rodentia) in a zone of sympatry in Mongolian Altai: Bioacoustic analysis. *Biol. Bull.* **37**:321–325.

Braun, J. K., T. S. Eaton, Jr., and M. A. Mares 2011. *Marmota caligata* (Rodentia: Sciuridae). *Mammalian Species* **43**(884):155–171.

Brody, A. K., and K. B. Armitage 1985. The effects of adult removal on dispersal of yearling yellow-bellied marmots. *Can. J. Zool.* **63**:2560–2564.

Brody, A. K., and J. C. Melcher 1985. Infanticide in yellow-bellied marmots. *Anim. Behav.* **33**:673–674.

Bronson, F. H. 1962. Daily and seasonal activity patterns in woodchucks. *J. Mammal.* **43**:425–427.

Bronson, F. H. 1963. Some correlates of interaction rate in natural populations of woodchucks. *Ecology* **44**:637–643.

Bronson, F. H. 1964. Agonistic behaviour in woodchucks. *Anim Behav.* **12**:470–478.

Brown, J. L. 1980. Fitness in complex avian social systems. In *Evolution of Social Behavior: Hypotheses and Empirical Tests*, ed. H. Markl. Weinhelm, Germany: Verlag Chemie, pp. 115–128.

Bruns, U., F. Frey-Roos, S. Pudritz, F. Tataruch, T. Ruf, and W. Arnold 2000. Essential fatty acids: their impact on free-living alpine marmots (*Marmota marmota*). In *Life in the Cold*, ed. G. Heldmaier and M. Klingenspor. New York: Springer, pp. 215–222.

Bryant, A. A. 1990. Genetic variability and minimum viable populations in the Vancouver Island marmot (*Marmota vancouverensis*). MS Thesis, University of Calgary, Alberta, Canada.

Bryant, A. A. 1996a. Demography of Vancouver Island marmots (*Marmota vancouverensis*) in natural and clearcut habitats. In *Biodiversity in Marmots*, ed. M. Le Berre, R. Ramousse, and L. Le Guelte Lyon: International Marmot Network, pp. 157–168.

Bryant, A. A. 1996b. Reproduction and persistence of Vancouver Island marmots (*Marmota vancouverensis*) in natural and logged habitats. *Can. J. Zool.* **74**:678–687.

Bryant, A. A. 1998. Metapopulation ecology of Vancouver Island marmots (*Marmota vancouverensis*). PhD Dissertation, University of Victoria, Victoria, Canada.

Bryant, A. A. 2000. Progress note. Vancouver Island Marmot Recovery Project, Vancouver Island.

Bryant, A. A. 2005. Reproductive rates of wild and captive Vancouver Island marmots (*Marmota vancouverensis*). *Can. J. Zool.* **83**:664–673.

Bryant, A. A., and D. W. Janz 1996. Distribution and abundance of Vancouver Island marmots (*Marmota vancouverensis*). *Can. J. Zool.* **74**:667–677.

Bryant, A. A., and M. McAdie 2003. Hibernation ecology of wild and captive Vancouver Island marmots (*Marmota vancouverensis*). In *Adaptive Strategies and Diversity in Marmots*, ed. R. Ramousse, D. Allainé, and M. Le Berre. Lyon, France: International Marmot Network, pp. 149–156.

Bryant, A. A., and R. E. Page 2005. Timing and causes of mortality in the endangered Vancouver Island marmot (*Marmota vancouverensis*). *Can. J. Zool.* **83**:674–682.

Bryant, A. A., H. M. Schwantje, and N. I. de With 2002. Disease and unsuccessful reintroduction of Vancouver Island marmots (*Marmota vancouverensis*). In *Holarctic Marmots as a Factor of Biodiversity*, ed. K. B. Armitage and V. Yu. Rumiantsev. Moscow: ABF Publishing House, pp. 101–107.

Bullard, R. W., C. Broumand, and F. R. Meyer 1966. Blood characteristics and volume in two rodents native to high altitude. *J. Appl. Physiol.* **21**:994–998.

Burghardt, G. 1998. The evolutionary orgins of play revisited. In *Animal Play: Evolutionary, Comparative, and Ecological Perspectives*, ed. M. Bekoff and J. A. Byers. Cambridge: Cambridge University Press, pp. 1–26.

Bykova, E., and A. Esipov 2008. Climate change impacts on population of Menzbier's marmot (*Marmota menzbieri*) in Uzbekistan. *Sixth International Marmot Conference Abstracts*, 13.

Calder, W. A. 1984. *Size, Function, and Life History*. Cambridge, MA: Harvard University Press.

Calhoun, J. B. 1952. The social aspects of population dynamics. *J. Mammal.* **33**:139–159.

Callait, M.-P., and D. Gauthier 2000. Parasite adaptations to hibernation in alpine marmots (*Marmota marmota*). In *Life in the Cold*, ed. G. Heldmaier and M. Klingenspor. New York: Springer, pp. 139–146.

Callait, M.-P., D. Gauthier, and C. Prud'homme 2000. Alpine marmots and their digestive parasites: infection kinetic and parasitic strategy. In *Holarctic Marmots as a Factor of Biodiversity*, ed. K. B. Armitage and V. Yu. Rumiantsev. Moscow: ABF Publishing House, pp. 113–122.

Callait, M.-P., D. Gauthier, C. Prud'homme, and B. Sabatier 1996. Impact of the parasitic fauna of digestive tract of alpine marmots (*Marmota marmota*) on their population dynamic. In

Biodiversity in Marmots, ed. M. Le Berre, R. Ramousse, and L. Le Guelte. Lyon, France: International Marmot Network, pp. 97–104.

Cardini, A. 2003. The geometry of the marmot (Rodentia: Sciuridae) mandible: phylogeny and patterns of morphological evolution. *Syst. Biol.* **52**:186–205.

Cardini, A. 2004. Evolution of marmots (Rodentia: Sciuridae): combining information on labial and lingual sides of the mandible. *Acta Theriol.* **49**:301–318.

Cardini A., R. S. Hoffmann, and R. W. Thorington 2005. Morphological evolution in marmots (Rodentia: Sciuridae): size and shape of the dorsal and lateral surfaces of the cranium. *J. Zool. Syst. Evol. Res.* **43**:258–268.

Cardini, A., and P. O'Higgins 2004. Patterns of morphological evolution in Marmota (Rodentia: Sciuridae): geometric morphometrics of the cranium in the context of marmot phylogeny, ecology and conservation. *Biol. J. Linn. Soc.* **82**:385–407.

Cardini, A., and P. O'Higgins 2005. Post-natal ontogeny of the mandible and ventral cranium in *Marmota* species (Rodentia, Sciuridae): allometry and phylogeny. *Zoomorphology* **124**:189–203.

Cardini, A., and R. W. Thorington, Jr. 2006. Postnatal ontogeny of marmot (Rodentia: Sciuridae) crania: allometric trajectories and species divergence. *J. Mammal.* **87**:201–215.

Cardini, A., R. W. Thorington, Jr., and P. D. Polly 2007. Evoultionary acceleration in the most endangered mammal of Canada: speciation and divergence in the Vancouver Island marmot (Rodentia: Sciuridae). *J. Evol. Biol.* **20**:1833–1846.

Cardini, A., and P. Tongiorgi 2003. Yellow-bellied marmots (*Marmota flaviventris*) "in the shape space" (Rodentia: Sciuridae): sexual dimorphism, growth and allometry of the mandible. *Zoomorphology* **122**:11–23.

Carey, H. V. 1985a. Nutritional ecology of yellow-bellied marmots in the White Mountains of California. *Holarctic Ecol.* **8**:259–264.

Carey, H. V. 1985b. The use of foraging areas by yellow-bellied marmots. *Oikos* **44**:273–299.

Carey, H. V., and H. J. Cooke 1991. Effect of hibernation and jejunal bypass on mucosal structure and function. *Am. J. Physiol.* **261**:G37–G44.

Carey, H. V., and P. Moore 1986. Foraging and predation risk in yellow-bellied marmots. *Am. Midl. Nat.* **116**:267–275.

Casimir, D. L., A. Moehrenschlager, and R. M. R. Barclay 2007. Factors influencing reproduction in captive Vancouver Island marmots: implications for captive breeding and reintroduction programs. *J. Mammal.* **88**:1412–1419.

Caumul, R., and P. D. Polly 2005. Phylogenetic and environmental components of morphological variation: skull, mandible, and molar shape in marmots (*Marmota*, Rodentia). *Evolution* **59**:2460–2472.

Chappell, M. A. 1992. Ventilatory accommodation of changing oxygen demand in sciurid rodents. *J. Comp. Physiol. B* **162**:722–730.

Chiesura Corona, M. 1992. Observations on distribution and abundance of the alpine marmot (*Marmota marmota* L.) in the territory of Belluno (South-eastern Alps). In *Proceedings of the 1st International Symposium on Alpine Marmot (*Marmota marmota*) and Genus* Marmota, ed. B. Bassano, P. Durio, U. Gallo Orsi, and E. Macchi. Torino, Italy: Dipartimento di Produzioni animali, Epidemiologia ed Ecologia, pp. 117–121.

Christian, J. J., E. Steinberger, and T. D. McKinney 1972. Annual cycle of spermatogenesis and testis morphology in woodchucks. *J. Mammal.* **53**:708–716.

Clark, A. B., and T. J. Ehlinger 1987. Pattern and adaptation in individual behavioral differences. In *Perspectives in Ethology 7*, ed. P. P. G. Bateson and P. H. Klopfer. New York: Plenum Press, pp. 1–47.

Clutton-Brock, T. H. 1985. Birth sex ratios and the reproductive success of sons and daughters. In *Evolution. Essays in Honour of John Maynard Smith*, ed. P. J. Greenwood, P. H. Harvey, and M. Slatkin. Cambridge: Cambridge University Press, pp. 221–235.

Clutton-Brock, T. 2002. Breeding together: kin selection and mutualism in cooperative vertebrates. *Science* **296**:69–72.

Clutton-Brock, T. 2007. *Meerkat Manor: Flower of the Kalahari*. London: Weidenfeld and Nicolson.

Clutton-Brock, T. 2009. Structure and function in mammalian societies. *Phil. Trans. R. Soc. B* **364**:3229–3242.

Clutton-Brock, T. H., and S. D. Albon 1982. Parental investment in male and female offspring in mammals. In *Current Problems in Sociobiology*, ed. King's College Sociobiology Group. Cambridge: Cambridge University Press, pp. 223–247.

Clutton-Brock, T. H., and G. R. Iason 1986. Sex ratio variation in mammals. *Q. Rev. Biol.* **61**:339–374.

Clutton-Brock, T. H., and D. Lukas 2011. The evolution of social philopatry and dispersal in female mammals. *Mol. Ecol.* **21**:472–492.

Clutton-Brock, T., and B. C. Sheldon 2010. Individuals and populations: the role of long-term, individual-based studies of animals in ecology and evolutionary biology. *Trends Ecol. Evol.* **25**:562–573.

Cochet, N., M. Le Berre, and L. Le Guelte 1992. Rhythm and behaviour in a group of three hibernating alpine marmots. In *Proceedings of the 1st International Symposium on Alpine Marmot (Marmota marmota) and Genus Marmota*, ed. B. Bassano, P. Durio, U. Gallo Orsi, and E. Macchi. Torino, Italy: Dipartimento di Produzioni animali, Epidemiologia ed Ecologia, pp. 55–61.

Cohas, A., C. Bonenfant, J.-M. Gaillard, and D. Allainé 2007b. Are extra-pair young better than within-pair young? A comparison of survival and dominance in alpine marmots *J. Anim. Ecol.* **76**:771–781.

Cohas, A., C. Bonenfant, C. B. Kempenaers, and D. Allainé 2009. Age-specific effect of heterozygosity on survival in alpine marmots, *Marmota marmota. Mol. Ecol.* **18**:1491–1503.

Cohas, A., N. G. Yoccoz, and D. Alliané 2007a. Extra-pair paternity in alpine marmots, *Marmota marmota*: genetic quality and genetic diversity effects. *Behav. Ecol. Sociobiol.* **61**:1081–1092.

Cohas, A., N. G. Yoccoz, C. Bonenfant, *et al.* 2008. The genetic similarity between pair members influences the frequency of extrapair paternity in alpine marmots. *Anim. Behav.* **76**:87–95.

Cohas, A., N. G. Yoccoz, A. Da Silva, B. Goossens, and D. Allainé 2006. Extra-pair paternity in the monogamous alpine marmot (*Marmota marmota*): the roles of social setting and female mate choice. *Behav. Ecol. Sociobiol.* **59**:597–605.

Collier, T. C., D. T. Blumstein, L. Girod, and C. E. Taylor 2010. Is alarm calling risky? Marmots avoid calling from risky places. *Ethology* **116**:1171–1178.

Concannon, P. W., B. Baldwin, P. Roberts, and B. Tennant 1990. Endocrine correlates of hibernation-independent gonadal recrudescence and the late-winter breeding season in woodchucks, *Marmota monax. J. Exp. Zool. Suppl.* **4**:203–206.

Concannon, P. W., V. D. Castracane, R. E. Rawson, and B. C. Tennant 1999. Circannual changes in free thyroxine, prolactin, testes, and relative food intake in woodchucks, *Marmota monax. Am. J. Physiol.* **277**:R1401–R1409.

Concannon, P. W., J. E. Parks, P. J. Roberts, and B. C. Tennant 1992. Persistent free-running circannual reproductive cycles during prolonged exposure to constant 12L:12D photoperiod in laboratory woodchucks (*Marmota monax*). *La. Anim. Sci.* **42**:382–391.

Concannon, P., P. Roberts, B. Baldwin, and B. Tennant 1997. Long-term entrainment of circannual reproductive and metabolic cycles by northern and southern hemisphere photoperiods in woodchucks (*Marmota monax*). *Biol. Reprod.* **57**:1008–1015.

Concannon, P., P. Roberts, B. Baldwin, H. Erb, and B. Tennant 1993. Alteration of growth, advancement of puberty, and season-appropriate circannual breeding during 28 months of photoperiod reversal in woodchucks (*Marmota monax*). *Biol. Reprod.* **48**:1057–1070.

Cooper, W. E., Jr., and T. Stankowich 2010. Prey or predator? Body size of an approaching animal affects decisions to attack or escape. *Behav. Ecol.* **21**:1278–1284.

Coss, R. G., and J. E. Biardi 1997. Individual variation in the antisnake behavior of California ground squirrels (*Spermophilus beecheyi*). *J. Mammal.* **73**:294–310.

Couch, L. K. 1930. Notes on the pallid yellow-bellied marmot. *Murrelet* **11**:3–7.

Coulon, J., D. Graziani, D. Allainé, M. C. Bel, and S. Pouderoux 1995. Infanticide in the alpine marmot (*Marmota marmota*). *Ethol., Ecol., Evol.* **7**:191–194.

Cova, L., C. Jamard, O. Hantz, F. Zoulim, and C. Trepo 2003. Woodchuck (*Marmota monax*) model for the study of chronic hepatitis B virus infection. In *Adaptive Strategies and Diversity in Marmots*, ed. R. Ramousse, D. Allainé, and M. Le Berre. Lyon, France: International Marmot Network, pp. 11–12.

Craig, H., ed. 1880. *Johnson's Household Book of Nature: Containing Full and Interesting Descriptions of the Animal Kingdom*. New York: H. J. Johnson.

Creel, S. R., and P. M. Waser 1997. Variation in reproductive suppression among dwarf mongooses: interplay between mechanisms and evolution. In *Cooperative Breeding in Mammals*, ed. N. G. Solomon and J. A. French. Cambridge: Cambridge University Press, pp. 150–170.

Cross, H. B., D. T. Blumstein, and F. Rosell 2013. Do marmots display a 'dear enemy phenomenon' in response to anal gland secretions? *J. Zool., Lond.* **289**:189–196.

Cushing, J. E., Jr., 1945. Quaternary rodents and lagomorphs of San Josecita Cave, Nueva Leon, Mexico. *J. Mammal.* **26**:182–185.

Daniel, J. C., and D. T. Blumstein 1998. A test of the acoustic adaptation hypothesis in four species of marmots. *Anim. Behav.* **56**:1517–1528.

Davis, D. E. 1967a. The role of environmental factors in hibernation of woodchucks (*Marmota monax*). *Ecology* **48**:683–689.

Davis, D. E. 1967b. The annual rhythm of fat deposition in woodchucks (*Marmota monax*). *Physiol. Zool.* **40**:391–402.

Davis, D. E. 1976. Hibernation and circannual rhythms of food consumption in marmots and ground squirrels. *Quart. Rev. Biol.* **51**:479–514.

Davis, D. E. 1977. Role of ambient temperature in emergence of woodchucks (*Marmota monax*) from hibernation. *Am. Midl. Nat.* **97**:224–229.

Davis, D. E., J. J. Christian, and F. Bronson 1964. Effect of exploitation on birth, mortality, and movement rates in a woodchuck population. *J. Wildl. Manage.* **28**:1–9.

Davis, D. E., and E. P. Finnie 1975. Entrainment of circannual rhythm in weight of woodchucks. *J. Mammal.* **56**:199–203.

Davis, D. E., and J. Ludwig 1981. Mechanism for decline in a woodchuck population. *J. Wildl. Manage.* **45**:658–668.

Davis, E. B. 2005. Comparison of climate space and phylogeny of *Marmota* (Mammalia: Rodentia) indicates a connection between evolutionary history and climate preference *Proc. R. Soc. B.* **272**:519–526.

Davydov, G. S. 1991. Some characters of two populations of the long-tailed marmot. In *Population Structure of the Marmot*, ed. D. I. Bibikov, A. A. Nikolski, V. Yu. Rumiantsev, and T. A. Seredneva. Moscow: USSR Theriological Society, pp. 188–216.

Dawson, T., and S. Maloney. 2008. The significance of fur characteristics in the thermal challenges posed by solar radiation: an underestimated role and source of adaptability. In *Molecules to Migration: The Pressures of Life*, ed. S. Morris and A. Vosloo. Bologna: Medamond, pp. 361–374.

Del Moral, R. 1984. The impact of the Olymic marmot on subalpine vegetation. *Am. J. Bot.* **71**:1228–1236.

Devillard, S., D. Allainé, J.-M. Gaillard, and D. Pontier 2004. Does social complexity lead to sex-biased dispersal in polygynous mammals? A test on ground-dwelling sciurids. *Behav. Ecol.* **15**:83–87.

de Vos, A., and D. I. Gillespie 1960. A study of woodchucks on an Ontario farm. *Can. Field-Nat.* **74**:130–145.

Dimitriev, A. V., Zh. Butbold, B. B. Badmaev, A. S. Oliger, and S. O. Pavlova 2003. Preliminary analysis of one of the Mongolian legends about Tarbagan. In *Adaptive Strategies and Diversity in Marmots*, ed. R. Ramousse, D. Allainé, and M. Le Berre. Lyon, France: International Marmot Network, pp. 227–229.

Dobson, F. S. 2013. The enduring question of sex-biased dispersal: Paul J. Greenwood's (1980) seminal contribution. *Anim. Behav.* **85**:299–304.

Dobson, F. S., V. A. Viblanc, C. M. Arnaud, and J. O. Murie 2012. Kin selection in Columbian ground squirrels: direct and indirect fitness benefits. *Mol. Ecol.* **21**:524–531.

Dousset, L. A. 1996. A structural transformation: the evolution of the relationship between man and marmot in the French Alps. In *Biodiversity in Marmots*, ed. M. Le Berre, R. Ramousse, and L. Le Guelte. Lyon, France: International Marmot Network, pp. 29–36.

Downhower, J. F. 1968. Factors affecting the dispersal of yearling yellow-bellied marmots (*Marmota flaviventris*). PhD dissertation, University of Kansas, Lawrence.

Downhower, J. F., and K. B. Armitage 1981. Dispersal of yearling yellow-bellied marmots (*Marmota flaviventris*). *Anim. Behav.* **29**:1064–1069.

Drucker, P. 1950. Northwest coast. *Anthropological Records* **9**:157–294.

Dugatkin, L. A. 1997. *Cooperation among Animals. An Evolutionary Perspective*. New York: Oxford University Press.

Ebensperger, L. A. 2001. A review of the evolutionary causes of rodent group-living. *Acta Theriol.* **46**:115–144.

Edelman, A. J. 2003. *Marmota olympus. Mammalian Species* **736**:1–5.

English, E. I., and M. A. Bowers 1994. Vegetational gradients and proximity to woodchuck (*Marmota monax*) burrows in an old field. *J. Mammal.* **75**:775–780.

Erbaeva, M. A. 2003. History, evolutionary development and systematics of marmots (Rodentia, Sciuridae) in Transbaikalia. *Russian J. Theriol.* **2**:33–42.

Erbajeva, M. A., and N. V. Alexeeva 2009a. Pliocene-Recent Holarctic marmots: overview. *Ethol., Ecol., Evol.* **21**:339–348.

Erbajeva, M. A., and N. V. Alexeeva 2009b. A new look at Pleistocene marmot diversity in Transbaikalia. *Ethol., Ecol., Evol.* **21**:237–241.

Errington, P. L. 1956. Factors limiting higher vertebrate populations. *Science* **124**:304–307.

Eskey, C. R., and V. H. Haas 1940. Plague in the western part of the United States. *Pub. Health Bull.* **254**:1–83.

Fall, M. W. 1971. Seasonal variations in the food consumption of woodchucks (*Marmota monax*). *J. Mammal.* **52**:370–375.

Farand, E., D. Allainé, and J. Coulon 2002. Variation in survival rates for the alpine marmot (*Marmota marmota*): effects of sex, age, year, and climatic factors. *Can. J. Zool.* **80**:342–349.

Fenn, A. M., S. M. Zervanos, and G. L. Florant 2009. Energetic relationships between field and laboratory woodchucks (*Marmota monax*) along a latitudinal gradient. *Ethol., Ecol., Evol.* **21**:299–315.

Ferrari, C., G. Bogliani, and A. von Hardenberg 2009. Alpine marmots (*Marmota marmota*) adjust vigilance behaviour according to environmental characteristics of their surrounding. *Ethol., Ecol., Evol.* **21**:355–365.

Ferrari, C., C. Pasquaretta, A. von Hardenberg, and B. Bassano 2012. Intraspecific killing and cannibalism in adult alpine marmots *Marmota marmota*. *Ethol., Ecol., Evol.* **24**:388–394.

Ferroglio, E., and P. Durio 1992. Burrows use in marmot colony in Val di Viu' (W. Alps). In *Proceedings of the 1st International Symposium on Alpine Marmot (*Marmota marmota*) and Genus* Marmota, ed. B. Bassano, P. Durio, U. Gallo Orsi, and E. Macchi. Torino, Italy: Dipartimento di Produzioni animali, Epidemiologia ed Ecologia, pp. 247–278.

Ferron, J. 1996. How do woodchucks (*Marmota monax*) cope with harsh winter conditions? *J. Mammal.* **77**:412–416.

Ferron, J., and J.-P. Ouellet 1989. Temporal and intersexual variations in the use of space with regard to social organization in the woodchuck (*Marmota monax*). *Can. J. Zool.* **67**:1642–1649.

Ferron, J., and J.-P. Ouellet. 1991. Physical and behavioral postnatal development of woodchucks (*Marmota monax*). *Can. J. Zool.* **69**:1040–1047.

Fishbein, D. B., A. J. Belotto, R. E. Pacer, *et al.* 1986. Rabies in rodents and lagomorphs in the United States, 1971–1984: increased cases in the woodchucks (*Marmota monax*) in Mid-Atlantic States. *J. Wildl. Dis.* **22**:151–155.

Fleming, W. K., W. M. Haschek, W. H. Gestenmann, J. W. Caslick, and D. J. Lisk 1977. Selenium and white muscle disease in woodchucks. *J. Wildl. Dis.* **13**:265–268.

Fleming, W. J., S. R. Nusbaum, and J. W. Caslick. 1979. Serologic evidence of leptospirosis in woodchucks (*Marmota monax*) in central New York State. *J. Wildl. Dis.* **15**:245–251.

Florant, G. L. 1998. Lipid metabolism in hibernators: the importance of essential fatty acids. *Am. Zool.* **38**:331–340.

Florant, G. L., L. Hester, S. Ameenuddin, and D. A. Rintoul 1993. The effect of a low essential fatty acid diet on hibernation in marmots. *Am. J. Physiol.* **264**:R747–R753.

Florant, G. L., V. Hill, and M. D. Ogilvie 2000. Circadian rhythms of body temperature in laboratory and field marmots (*Marmota flaviventris*). In *Life in the Cold*, ed. G. Heldmaier and M. Klingenspor. New York: Springer, pp. 223–231.

Floyd, C. H. 2004. Marmot distribution and habitat associations in the Great Basin. *West. N.A. Nat.* **64**:471–481.

Floyd, C. H., D. H. Van Vuren, and B. May 2005. Marmots on Great Basin mountaintops: using genetics to test a biogeographic paradigm. *Ecology* **86**:2145–2153.

Ford, A., ed. 1951. *Audubon's Animals. The Quadrupeds of North America*. New York: Thomas Y. Crowell Company.

Formozov, A. N. 1966. Adaptive modifications of behavior in mammals of the Eurasian steppes. *J. Mammal.* **47**:208–223.

Formozov, N. A., A. Yu. Yendukin, and D. I. Bibikov 1996. Coadaptations of marmots (*Marmota sibirica*) and Mongolian hunters. In *Biodiversity in Marmots*, ed. M. Le Berre, R. Ramousse, and L. Le Guelte. Lyon, France: International Marmot Network, pp. 37–42.

Fourcade, P. 1996. Traditional aspects and cultural changes about current practices concerning alpine marmot (*Marmota marmota*) in France. In *Biodiversity in Marmots*, ed. M. Le Berre, R. Ramousse, and L. Le Guelte. Lyon: International Marmot Network, pp. 43–48.

Franceschina-Zimmerli, R., and P. Ingold 1996. The behaviour of alpine marmots (*Marmota m. marmota*) under different hiking pressures. In *Biodiversity in Marmots*, ed. M. Le Berre, R. Ramousse, and L. Le Guelte. Lyon: International Marmot Network, pp. 73–74.

Frank, C. L., E. S. Dierenfeld, and K. B. Storey 1998. The relationship between lipid peroxidation, hibernation, and food selection in mammals. *Am. Zool.* **38**:341–349.

Frase, B. A., and K. B. Armitage 1984. Foraging patterns of yellow-bellied marmots: role of kinship and individual variability. *Behav. Ecol. Sociobiol.* **16**:1–10.

Frase, B. A., and K. B. Armitage 1989. Yellow-bellied marmots are generalist herbivores. *Ethol., Ecol., Evol.* **1**:353–366.

Fraser, D. 1979. Aquatic feeding by a woodchuck. *Can. Field-Nat.* **93**:309–310.

French, A. R. 1986. Patterns of thermoregulation during hibernation. In *Living in the Cold: Physiological and Biochemical Adaptations*, ed. H. C. Heller, X. J. Musacchia, and L. C. H. Wang. Amsterdam: Elsevier Science Publishing Company, pp. 393–402.

French, A. R. 1990. Age-class differences in the pattern of hibernation in yellow-bellied marmots, *Marmota flaviventris*. *Oecologia* **82**:93–96.

Fretwell, S. D. 1972. *Populations in a Seasonal Environment*. Princeton, NJ: Princeton University Press.

Frey-Roos, F. 2005. Long- and short-distance dispersal in subordinate alpine marmots (*Marmota marmota*). In *Adaptive Strategies and Diversity in Marmots*, ed. R. Ramousse, D. Allainé, and M. Le Berre. Lyon: International Marmot Network, pp. 45–46.

Frigerio, D., M. Panseri, and E. Ferrario 1996. Alpine marmot (*Marmota marmota*) in the Orobic Alps: analysis of ecological parameters. In *Biodiversity in Marmots*, ed. M. Le Berre, R. Ramousse, and L. Le Guelte. Lyon: International Marmot Network, pp. 169–174.

Fronhofer, E. A., A. Kubisch, F. M. Hilker, T. Hovestadt, and H. J. Poethke 2012. Why are metapopulations so rare? *Ecology* **93**:1967–1978.

Fryxell, F. M. 1926. An observation on the hunting methods of the timber wolf. *J. Mammal.* **7**:226–227.

Gardner, A., A. S. Griffin, and S. A. West 2010. Altruism and cooperation. In *Evolutionary Behavioral Ecology*, ed. De. Westneat and C. W. Fox. New York: Oxford University Press, pp. 308–326.

Garrott, R. A., and D. A. Jenni. 1978. Arboreal behavior of yellow-bellied marmots. *J. Mammal.* **59**:433–434.

Gasienica Byrcyn, W. 1997. The marmot (*Marmota marmota latirostris* Kratochvil, 1961) population in the Polish Tatra Mountains. *J. Wildl. Res.* **2**:69–81.

Geiser, F., and T. Ruf. 1995. Hibernation versus daily torpor in mammals and birds: physiological variables and classification of torpor patterns. *Physiol. Zool.* **68**:935–966.

Gianini, C. A. 1925. Tree-climbing and insect-eating woodchucks. *J. Mammal.* **6**:281–282.

Gibault, C., R. Ramousse, and M. Le Berre 1996. Hiking influence on feeding behaviour in alpine marmot. In *Biodiversity in Marmots*, ed. M. Le Berre, R. Ramousse, and L. Le Guelte. Lyon, France: International Marmot Network, pp. 233–234.

Giboulet, O., R. Ramousse, and M. Le Berre 2002. Evolution of life history traits and molecular phylogenies: sociality in ground dwelling squirrels as an example. In *Holarctic Marmots as a Factor of Biodiversity*, ed. K. B. Armitage and V. Yu. Rumiantsev. Moscow: ABF Publishing House, pp. 171–175.

Giovannetti, L. 1954. *Max. Das Murmeltier uber das die Welt schmunzelt*. Munich: Wilhelm Heyne Verlag.

Golley, F. B. 1968. Secondary productivity in terrestrial communities. *Am. Zool.* **8**:53–59.

Golombek, D. A., and R. E. Rosenstein 2010. Physiology of circadian entrainment. *Physiol. Rev.* **90**:1063–1102.

Goodwin, H. T. 1989. *Marmota flaviventris* from the Central Mojave Desert of California: biogeographic implications. *Southwest Nat.* **34**:284–287.

Goossens, B., J. Coulon, D. Allainé, *et al.* 1996. Immigration of a pregnant female in an alpine marmot family group: behavioural and genetic data. *C.R. Acad. Sci. Paris, Life Sci.* **319**:241–246.

Goosens, B., L. Graziani, L. P. Waits, *et al.* 1998. Extra-pair paternity in the monogamous alpine marmot revealed by nuclear DNA microsatellite analysis. *Behav. Ecol. Sociobiol.* **43**:281–288.

Gortazar, C., J. Herrero, A. Garcia-Serrano, J. Lucientes, and D. F. Luco 1996. Preliminary data on the parasitic fauna of the digestive system of *Marmota marmota* in the western Pyrenees. In *Biodiversity in Marmots*, ed. M. Le Berre, R. Ramousse, and L. Le Guelte. Lyon, France: International Marmot Network, pp. 105–108.

Gotthard, K., and S. Nylin 1995. Adaptive plasticity and plasticity as an adaptation: a selective review of plasticity in animal morphology and life history. *Oikos* **74**:3–17.

Gray, D. R. 1967. Behaviour and activity in a colony of hoary marmots (*Marmota caligata*) in Manning Park, BC and a comparison of behaviour with other marmot species. BSc(Honours) Thesis, University of Victoria, BC.

Greenwood, P. J. 1980. Mating systems, philopatry and dispersal in birds and mammals. *Anim. Behav.* **28**:1140–1162.

Gregory, S. D., C. J. A. Bradshaw, B. W. Brook, and F. F. Courchamp 2010. Limited evidence for the demographic Allee effect from numerous species across taxa. *Ecology* **91**:2151–2161.

Griffin, S. C. 2007. Demography and ecology of a declining endemic: the Olympic marmot. PhD dissertation, University of Montana, Missoula.

Griffin, S. C., P. C. Griffin, M. L. Taper, and L. S. Mills 2009. Marmots on the move? Dispersal in a declining montane mammal. *J. Mammal.* **90**:686–695.

Griffin, S. C., M. L. Taper, R. Hoffman, and L. S. Mills 2008. The case of the missing marmots: are metapopulation dynamics or range-wide declines responsible? *Biol. Conser.* **141**:1293–1309.

Griffin, S. C., M. L. Taper, and L. S. Mills 2007a. Female Olympic marmots (*Marmota olympus*) reproduce in consecutive years. *Am. Midl. Nat.* **158**:221–225.

Griffin, S. C., T. Valois, M. L. Taper, and L. S. Mills 2007b. Effects of tourists on behavior and demography of Olympic marmots. *Conser. Biol.* **21**:1070–1081.

Grimm, V., N. Dorndorf, F. Frey-Roos, *et al.* 2003. Modeling the role of social behavior in the persistence of the alpine marmot *Marmota marmota*. *Oikos* **102**:124–136.

Grizzell, R. A. 1955. A study of the southern woodchuck, *Marmota monax monax*. *Am. Midl. Nat.* **53**:257–293.

Gudger, E. W. 1935. Animal carts. How marmots, badgers and beavers serve as sleds or wagons. *Sci. Monthly* **40**:153–157.

Gunderson, A. M., B. K. Jacobsen, and L. E. Olson 2009. Revised distribution of the Alaska marmot, *Marmota broweri*, and confirmation of parapatry with hoary marmots. *J. Mammal.* **90**:859–869.

Gunderson, A. M., H. C. Lanier, and L. E. Olson 2012. Limited phylogeographic structure and genetic variation in Alaska's arctic and alpine endemic, the Alaska marmot. *J. Mammal.* **93**:66–75.

Hackländer, K., and W. Arnold 1999. Male-caused reproductive failure of female reproduction and its adaptive value in alpine marmots (*Marmota marmota*). *Behav. Ecol.* **10**:592–597.

Hackländer, K., and W. Arnold 2012. Litter sex ratio affects lifetime reproductive success of free-living alpine marmots, *Marmota marmota*. *Mammal Review* **42**:310–313.

Hackländer, K., E. Möstl, and W. Arnold 2003. Reproductive suppression in female alpine marmots, *Marmota marmota*. *Anim. Behav.* **65**:1133–1140.

Hafner, D. J. 1984. Evolutionary relationships of the Nearctic Sciuridae. In *The Biology of Ground-dwelling Squirrels*, ed. J. O. Murie and G. R. Michener. Lincoln, NE: University of Nebraska Press, pp. 3–23.

Hager, R., and C. B. Jones, eds. 2009. *Reproductive Skew in Vertebrates*. Cambridge: Cambridge University Press.

Hall, F. G. 1965. Hemoglobin and oxygen affinities in seven species of Sciuridae. *Science* 148:1350–1351.

Hamilton, W. 1964. The genetical theory of social behaviour. I. II. *J. Theor. Biol.* 7:1–52.

Hamilton, W. D., and R. M. May 1977. Dispersal in stable habitats. *Nature* 269:578–581.

Hamilton, W. J., Jr. 1934. The life history of the rufescent woodchuck *Marmota monax rufescens* Howell. *Ann. Carnegie Mus.* 23:85–178.

Hansen, R. M. 1975. Foods of the hoary marmot on Kenai Peninsula, Alaska. *Am. Midl. Nat.* 94:348–353.

Hanski, I. 2001. Population dynamic consequences of dispersal in local populations and in metapopulations. In *Dispersal*, ed. J. Clobert, E. Danchin, A. A. Dhondt, and J. D. Nichols. Oxford: Oxford Universsity Press, pp. 283–298.

Hanski, I., and O. E. Gaggiotti 2004. Metapopulation biology: past, present, and future. In *Ecology, Genetics, and Evolution of Metapopulations*, ed. I. Hanski and O. E. Gaggiotti. London: Elsevier Academic Press, pp. 3–22.

Harkness, D. R., S. Roth, and P. Goldman 1974. Studies on the red blood cell oxygen affinity and 2,3-diphosphoglyceric acid in the hibernating woodchuck (*Marmota monax*). *Comp. Biochem. Physiol.* 48A:591–599.

Harrison, R. G., S. M. Bogdanowicz, R. S. Hoffmann, E. Yensen, and P. Sherman. 2003. Phylogeny and evolutionary history of the ground squirrels (Rodentia: Marmotinae). *J. Mamm. Evol.* 10:249–276.

Hayes, S. R. 1976. Daily activity and body temperature of the southern woodchuck, *Marmota monax monax*, in northwestern Arkansas. *J. Mammal.* 57:291–299.

Hayssen, V. 2008a. Patterns of body and tail length and body mass in Sciuridae. *J. Mammal.* 89:852–873.

Hayssen, V. 2008b. Reproductive effort in squirrels: ecological, phylogenetic, allometric, and latitudinal patterns. *J. Mammal.* 89:582–606.

Hayssen, V. 2008c. Reproduction within Marmotine ground squirrels (Sciuridae, Xerinae, Marmotini): patterns among genera. *J. Mammal.* 89:607–616.

Heard, D. C. 1977. The behaviour of Vancouver Island marmots, *Marmota vancouverensis*. MS. Thesis, University of British Columbia.

Hébert, P., and C. Barrette 1989. Experimental demonstration that scent marking can predict dominance in the woodchuck, *Marmota monax*. *Can. J. Zool.* 67:575–578.

Hébert, P., and J. Prescott 1983. Etude de marquage olfactif chez la marmotte commune (*Marmota monax*) en captivite. *Can. J. Zool.* 61:1720–1725.

Heldmaier, G., S. Ortmann, and G. Körtner 1993a. Energy requirements of hibernating alpine marmots. In *Life in the Cold*, ed. C. Carey, G. L. Florant, B. A. Wunder, and B. Horwitz. Boulder, CO: Westview Press, pp. 175–183.

Heldmaier, G., R. Steiger, and T. Ruf 1993b. Suppression of metabolic rate in hibernation. In *Life in the Cold*, ed. C. Carey, G. L. Florant, B. A. Wunder, and B. Horwitz. Boulder, CO: Westview Press, pp. 545–548.

Henderson, J. A., and F. F. Gilbert 1978. Distribution and density of woodchuck burrow systems in relation to land-use. *Can. Field-Nat.* 92:128–136.

Herbers, J. 1981. Time resources and laziness in animals. *Oecologia* 49:252–262.

Heredia, R., and J. Herrero 1992. Bearded vulture (*Gypaetus barbatus*) and alpine marmot (*Marmota marmota*) interactions in the Southern Pyrenees. In *Proceedings of the 1st International*

*Symposium on Alpine Marmot (*Marmota marmota*) and Genus* Marmota, ed. B. Bassano, P. Durio, U. Gallo Orsi, and E. Macchi. Torino, Italy: Dipartimento di Produzioni animali, Epidemiologia ed Ecologia, pp. 227–229.

Herrero, J., J. Canut, D. Garcia-Ferre, R. Garcia-Gonzalez, and R. Hildalgo 1992. The alpine marmot (*Marmota marmota* L.) in the Spanish Pyrenees. *Z. Säugetierkunde* **57**:211–215.

Herrero, J., R. Garcia-Gonzalez, and A. Garcia-Serrano 1994. Altitudinal distribution of alpine marmot (*Marmota marmota*) in the Pyrenees, Spain/France. *Arct. Alp. Res.* **26**:328–331.

Herrero, J., R. Hidalgo, and R. Garcia-Gonzalez 1987. Colonization process of the alpine marmot (*Marmota marmota*) in Spanish Western Pyrenees. *Pirineos* **130**:87–94.

Hersey, J. 1953. *The Marmot Drive.* New York: Alfred A. Knopf, Inc.

Hill, V. L., and G. L. Florant 1999. Patterns of fatty acid composition in free-ranging yellow-bellied marmots (*Marmota flaviventris*) and their diet. *Can. J. Zool.* **77**:1494–1503.

Hock, R. J. 1967. Seasonal hematologic changes in high altitude hibernators. *Fed. Proc.* **26**:719.

Hoffmann, R. S., C. G. Anderson, R. W. Thorington, Jr., and L. R. Heaney 1993. Family Sciuridae. In *Mammal Species of the World,* 2nd edition, ed. D. E. Wilson and D. M. Reeder. Washington, DC: Smithsonian Institution Press, pp. 419–465.

Holmes, W. G. 1984a. The ecological basis of monogamy in Alaskan hoary marmots. In *Biology of Ground-dwelling Squirrels,* ed. J. O. Murie and G. R. Michener. Lincoln, NE: University of Nebraska Press, pp. 250–274.

Holmes, W. G. 1984b. Predation risk and foraging behavior of the hoary marmot in Alaska. *Behav. Ecol. Sociobiol.* **15**:293–301.

Holmes, W. G. 1984c. Sibling recognition in thirteen-lined ground squirrels: effects of genetic relatedness, rearing association, and olfaction. *Behav. Ecol. Sociobiol.* **14**:225–233.

Hoogland, J. L. 1979. Aggression, ectoparasitism, and other possible costs of prairie dog (Sciuridae, *Cynomys* spp.) coloniality. *Behaviour* **69**:1–35.

Hoogland, J. L. 2003. Black-tailed prairie dog. In *Wild Mammals of North America,* ed. G. A. Feldhamer, B. C. Thompson, and J. A. Chapman. Baltimore: Johns Hopkins University Press, pp. 232–247.

Hoogland, J. L. 2013. Prairie dogs disperse when all close kin have disappeared. *Science* **339**:1205–1207.

Huang, B., T. W. Wey, and D. T. Blumstein 2011. Correlates and consequences of dominance in a social rodent. *Ethology* **117**:1–13.

Huang, X., Z. Wang, J. Wu, and L. Liu. 1986. Breeding characteristics of Himalayan marmot at Reshui Bottomland and Wulannao Bottomland in Haiyan County, Qinhai. *Acta Theriol. Sinica* **6**:307–310.

Hugot, J.-P. 1980. Sur le genre *Citellina* Prendel, 1928 (Oxyuridae, Nematoda). *Annales de Parasitologie* **55**:97–109.

Hume, I. D. 2002. Digestive strategies of mammals. *Acta Zool. Sinica* **48**:1–19.

Hume, I. D. 2003. Aspects of digestive function in marmots. In *Adaptive Strategies and Diversity in Marmots,* ed. R. Ramousse, D. Allainé, and M. Le Berre. Lyon: International Marmot Network, pp. 111–116.

Hume, I. D., C. Beiglböck, T. Ruf, F. Frey-Roos, U. Bruns, and W. Arnold 2002. Seasonal changes in morphology and function of the gastrointestinal tract of free-living alpine marmots (*Marmota marmota*). *J. Comp. Physiol. B* **172**:197–207.

Hume, I. D., K. R. Morgan, and G. J. Kenagy 1993. Digesta retention and digestive performance in sciurid and microtine rodents: effects of hindgut morphology and body size. *Physiol. Zool.* **66**:396–411.

Humphreys, W. F. 1979. Production and respiration in animal populations. *J. Anim. Ecol.* **48**:427–451.

Inouye, D. W., B. Barr, K. B. Armitage, and B. D. Inouye 2000. Climate change is affecting altitudinal migrants and hibernating species. *PNAS* **97**:1630–1633.

Intergovernmental Panel on Climate Change 2007. *Summary for Policy Makers.* Geneva: IPCC Secretariat.

Jamieson, S. H., and K. B. Armitage 1987. Sex differences in the play behavior of yearling yellow-bellied marmots. *Ethology* **74**:237–253.

Johns, D. W., and K. B. Armitage 1979. Behavioral ecology of alpine yellow-bellied marmots. *Behav. Ecol. Sociobiol.* **5**:133–157.

Kalapos, G. 2006. *Fertility Goddesses, Groundhog Bellies and the Coca-cola Company. The Origins of Modern Holidays.* Toronto: Insomniac Press.

Kalthoff, D. C. 1999a. Ist *Marmota primigenia* (Kaup) eine eigenständige Art? Osteologische Variabiltät pleistozäner *Marmota*-Populationen (Rodentia: Sciuridae) in Neuwieder Becken (Rheinland-Pfalz, Deutschland) und benachbarter Gebiete. *Kaupia* **9**:127–186.

Kalthoff, D. C. 1999b. Jungpleistozäne Murmeltiere (Rodentia, Sciuridae) vom Mittelrhein (Deutschland) und ihre verwandtschaftlichen Beziehungen zu den beiden rezenten europäischen Arten. *Staphia* **63**:119–128.

Kapitonov, V. I. 1978. The Kamchatka marmot. In *Marmots. Biocenotic and Practical Significance*, ed. R. P. Zimina. Moscow: Nauka, pp. 178–209.

Karels, T. J., and D. S. Hik. 2003. Demographic responses of hoary marmots (*Marmota caligata*) to environmental variation. In *Adaptive Strategies and Diversity in Marmots*, ed. R. Ramousse, D. Allainé, and M. Le Berre. Lyon, France: International Marmot Network, pp. 167–168.

Kilgore, D. L., Jr., and K. B. Armitage 1978. Energetics of yellow-bellied marmots. *Ecology* **59**:78–88.

King, W. J., and D. Allainé 2002. Social, maternal, and environmental influences on reproductive success in female alpine marmots (*Marmota marmota*). *Can. J. Zool.* **80**:2137–2143.

Knight, R. L., and A. W. Erikson 1978. Marmots as a food source of golden eagles along the Columbia River. *Murrelet* **59**:28–29.

Koenig, L. 1957. Beobachtungen über Reviermarkierung sowie Drod-, Kampf- und Abwehrverhalten des Murmeltieres (*Marmota marmota* L.). *Zeitschrift für Tierpsychologie* **14**:510–521.

Kolesnikov, V. V., O. V. Brandler, and B. B. Badmaev 2009a. Folk use of marmots in Mongolia. *Ethol., Ecol., Evol.* **21**:285–287.

Kolesnikov, V. V., O. V. Brandler, B. B. Badmaev, D. Zoje, and Ya. Adiya 2009b. Factors that lead to a decline in numbers of Mongolian marmot populations. *Ethol., Ecol., Evol.* **21**:371–379.

Kortner, G., and G. Heldmaier 1995. Body weight cycles and energy balance in the alpine marmot (*Marmota marmota*). *Physiol. Zool.* **68**:149–163.

Krebs, C. J. 1985. *Ecology: The Experimental Analysis of Distribution and Abundance*, 3rd edition. New York: Harper and Row.

Krebs, C. J., X. Lambin, and J. O. Wolff 2007. Social behavior and self-regulation in murid rodents. In *Rodent Societies: An Ecological and Evolutionary Perspective*, ed. J. O. Wolff and P. W. Sherman. Chicago, IL: University of Chicago Press, pp. 173–184.

Krohne, D. T. 1997. Dynamics of metapopulations of small mammals. *J. Mammal.* **78**:1014–1026.

Kruckenhauser, L., A. A. Bryant, S. C. Griffin, S. J. Amish, and W. Pinsker 2009. Patterns of within and between-colony microsatellite variation in the endangered Vancouver Island marmot (*Marmota vancouverensis*): implications for conservation. *Cons. Gen.* **10**:1759–1772.

Kruckenhauser, L., W. Pinsker, E. Haring, and W. Arnold 1998. Marmot phylogeny revisited: molecular evidence for a diphyletic origin of sociality. *J. Zool. Syst. Evol. Research* **37**:49–56.

Kurtén, B., and E. Anderson 1980. *Pleistocene Mammals of North America*. New York: Columbia University Press.

Kwiecinski, G. G. 1998. *Marmota monax. Mammalian Species* **591**:1–8.

Kyle, C. J., T. J. Karels, C. S. Davis, S. Mebs, B. Clark, C. Strobeck, and D. S. Hik 2007. Social structure and facultative mating systems of hoary marmots (*Marmota caligata*). *Mol. Ecol.* **16**:1245–1255.

Labriola, M. C., C. Pasquaretta, G. Bogliani, and A. Von Hardenberg 2008. Home range size, foraging behaviour and risk of predation in alpine marmots (*Marmota marmota*). *Marmots in a Changing World*, Abstract: 22.

Lacey, E. A., and P. W. Sherman 1997. Cooperative breeding in naked mole-rats: implications for vertebrate and invertebrate sociality. In *Cooperative Breeding in Mammals*, ed. N. G. Solomon and J. A. French. Cambridge: Cambridge University Press, pp. 267–301.

Lambert, A. M., A. J. Miller-Rushing, and D. W. Inouye 2010. Changes in snowmelt and summer precipitation affect the flowering phenology of *Erythronium grandiflorum* (Glacier lily; Liliacae). *Am. J. Bot.* **97**:1431–1437.

Lange, A. L. 1956. Woodchuck remains in northern Arizona caves. *J. Mammal.* **37**:289–291.

Langenheim, J. H. 1955. Flora of the Crested Butte Quadrangle, Colorado. *Madroño* **13**:64–78.

Lardy, S., A. Cohas, E. Desouhant, M. Tafani, and D. Allainé 2012. Paternity and dominance loss in male breeders: the cost of helpers in a cooperatively breeding mammal. *PLoS ONE* **7**(1): e29508.

Lardy, S., A. Cohas, I. Figueroa, and D. Allainé 2011. Mate change in a socially monogamous mammal: evidences support the "forced divorce" hypothesis. *Behav. Ecol.* **22**:120–125.

Lea, A. J., and D. T. Blumstein 2011. Age and sex influence marmot antipredator behavior during periods of heightened risk. *Behav. Ecol. Sociobiol.* **65**:1525–1533.

Le Berre, M., D. Allainé, I. Rodrigue. G. V. Olenev, A. V. Lagunov, and V. D. Zakharov 1994. Some questions on the ecology of the steppe marmot in the southern Urals (analysis of the action of environmental factors). *Russian J. Ecol.* **25**:36–41.

Lee, J. N., B. M. Barnes, and C. L. Buck 2009. Body temperature patterns during hibernation in a free-living Alaska marmot (*Marmota broweri*). *Ethol., Ecol., Evol.* **21**:403–413.

Lehrer, E. W., and R. L. Schooley 2010. Space use of woodchucks across an urbanization gradient within an agricultural landscape. *J. Mammal.* **91**:1342–1349.

Lenihan, C., and D. Van Vuren 1996a. Growth and survival of juvenile yellow-bellied marmots (*Marmota flaviventris*). *Can. J. Zool.* **74**:297–302.

Lenihan, C. and D. Van Vuren 1996b. Costs and benefits of sociality in yellow-bellied marmots (*Marmota flaviventris*): do noncolonial females have lower fitness? *Ethol., Ecol., Evol.* **8**:177–189.

Lenti Boero, D. 1992. Alarm calling in alpine marmot (*Marmota marmota* L.): evidence for semantic communication. *Ethol., Ecol., Evol.* **4**:125–138.

Lenti Boero, D. 1994. Survivorship among young alpine marmots and their permanence in their natal territory in a high altitude colony. *IBEX J. M. E.* **2**:9–16.

Lenti Boero, D. 1995. Scent-deposition behaviour in alpine marmots (*Marmota marmota* L.): its role in territorial defense and social communication. *Ethology* **100**:26–38.

Lenti Boero, D. 1999. Population dynamics, mating system and philopatry in a high altitude colony of alpine marmots (*Marmota marmota* L.). *Ethol., Ecol., Evol.* **11**:105–122.

Lenti Boero, D. 2001. Occupation of hibernacula, seasonal activity, and body size in a high altitude colony of alpine marmots (*Marmota marmota*). *Ethol., Ecol., Evol.* **13**:209–223.

Lenti Boero, D. 2003a. Long-term dynamics of space and summer resource use in the alpine marmot (*Marmota marmota* L.). *Ethol., Ecol., Evol.* **15**:309–327.

Lenti Boero, D. 2003b. Spotting behaviour and daily activity cycle in the alpine marmots (*Marmota marmota* L.): a role for infant guarding. *Oecologia Montana* **12**:1–6.

Li, C., R. Monclus, T. J. Maul, Z. Jiang, and D. T. Blumstein 2011. Quantifying human disturbance on antipredator behavior and flush initiation distance in yellow-bellied marmots. *Appl. Anim. Behav. Sci.* **129**:146–152.

Lidicker, W. Z., Jr 1962. Emigration as a possible mechanism permitting the regulation of population density below carrying capacity. *Am. Nat.* **96**:29–33.

Liow, L. H., M. Fortelius, K. Lintulaakso, H. Mannila, and N. C. Stenseth 2009. Lower extinction risk in sleep-or-hide mammals. *Am. Nat.* **173**:264–272.

Lomnicki, A. 1988. *Population Ecology of Individuals*. Princeton, NJ: Princeton University Press.

Long, W. S. 1940. Notes on the life histories of some Utah mammals. *J. Mammal.* **21**:170–180.

Lopez, B. C., I. Figueroa, J. Pino, A. Lopez, and D. Potrony 2009. Potential distribution of the alpine marmot in Southern Pyrenees. *Ethol., Ecol., Evol.* **21**:225–235.

Lopez, B. C., J. Pino, and A. Lopez 2010. Explaining the successful introduction of the alpine marmot in the Pyrenees. *Biol. Invasions* **12**:3205–3217.

Louis, S., and M. Le Berre 2000. Ajustement des distances de fuite à l'homme chez *Marmota marmota*. *Can. J. Zool.* **78**:556–563.

Louis, S., and M. Le Berre 2002. Human disturbance and wildlife: preliminary results from the alpine marmot. In *Holarctic Marmots as a Factor of Biodiversity*, ed. K. B. Armitage and V. Yu. Rumiantsev. Moscow: ABF Publishing House, pp. 255–262.

Louis, S., O. Giboulet, and Yu. Semenov 2002. History of cohabitation of humans and marmots: the case of Aussois, Savoie, France. In *Holarctic Marmots as a Factor of Biodiversity*, ed. K. B. Armitage and V. Yu. Rumiantsev. Moscow: ABF Publishing House, pp. 249–254.

Lovegrove, B. G. 2000. The zoogeography of mammalian metabolic rate. *Am. Nat.* **156**:201–219.

Lovegrove, B. G. 2003. The influence of climate on the basal metabolic rate of small mammals: a slow-fast metabolic continuum. *J. Comp. Physiol B* **173**:87–112.

Lukovtsev, Yu. S., and V. N. Yasiliev 1992. The black-capped marmot (*Marmota camtschatica* Pall.) in Yakutia (The North-Eastern part of the USSR). In *Proceedings of the 1st International Symposium on Alpine Marmot (*Marmota marmota*) and Genus* Marmota, ed. B. Bassano, P. Durio, U. Gallo Orsi, and E. Macchi. Torino, Italy: Dipartimento di Produzioni animali, Epidemiologia ed Ecologia, pp. 231–232.

Lyapunova, E. A., G. G. Boyeskorov, and N. N. Vorontsov 1992. *Marmota camtschatica* Pall: Nearctic element in Palearctic marmota fauna. In *Proceedings of the 1st International Symposium on Alpine Marmot (*Marmota marmota*) and Genus* Marmota, ed. B. Bassano, P. Durio, U. Gallo Orsi, and E. Macchi. Torino, Italy: Dipartimento di Produzioni animali, Epidemiologia ed Ecologia, pp. 185–191.

Lyman, C. P., J. S. Willis, A. Malan, and L. C. H Wang 1982. *Hibernation and Torpor in Mammals and Birds*. New York: Academic Press.

Macchi, E., B. Bassano, P. Durio, M. Tarantola, and A. Vita 1992. Ecological parameters affecting the settlement's choice in alpine marmot (*Marmota marmota*). In *Proceedings of the 1st International Symposium on Alpine Marmot (*Marmota marmota*) and Genus* Marmota, ed. B. Bassano, P. Durio, U. Gallo Orsi, and E. Macchi. Torino, Italy: Dipartimento di Produzioni animali, Epidemiologia ed Ecologia, pp. 123–127.

Magnolon, S., J. Coulon, and D. Allainé 2002. Social interactions as a proximate factor in natal dispersal of the alpine marmot (*Marmota marmota*). In *Holarctic Marmots as a Factor of Biodiversity*, ed. K. B. Armitage and V. Yu. Rumiantsev. Moscow: ABF Publishing House, pp. 263–266.

Maher, C. R. 2004. Intrasexual territoriality in woodchucks (*Marmota monax*). *J. Mammal.* **85**:1087–1094.

Maher, C. R. 2006. Social organization in woodchucks (*Marmota monax*) and its relationship to growing season. *Ethology* **112**:313–324.

Maher, C. R. 2009a. Genetic relatedness and space use in a behaviorally flexible species of marmot, the woodchuck (*Marmota monax*). *Behav. Ecol. Sociobiol.* **63**:857–868.

Maher, C. R. 2009b. Effects of relatedness on social interaction rates in a solitary marmot. *Anim. Behav.* **78**:925–933.

Maher, C. R., and M. Duron 2010. Mating system and paternity in woodchucks (*Marmota monax*). *J. Mammal.* **91**:628–635.

Mainini, B., P. Neuhaus, and P. Ingold 1993. Behaviour of marmots *Marmota marmota* under the influence of different hiking activities. *Biol. Cons.* **64**:161–164.

Manfredi, M. T., E. Zanin, and A. P. Rizzoli 1992. Helminth community on alpine marmots. In *Proceedings of the 1st International Symposium on Alpine Marmot (*Marmota marmota*) and Genus* Marmota, ed. B. Bassano, P. Durio, U. Gallo Orsi, and E. Macchi. Torino, Italy: Dipartimento di Produzioni animali, Epidemiologia ed Ecologia, pp. 203–207.

Mann, C. S., E. Macchi, and G. Janeau 1993. Alpine marmot (*Marmota marmota* L.). *IBEX J. M. E.* **1**:17–30.

Marr, N. V., and R. L. Knight 1983. Food habits of golden eagles in eastern Washington. *Murrelet* **64**:73–77.

Martell, A. M., and R. J. Milko 1986. Seasonal diets of Vancouver Island marmots, *Marmota vancouverensis*. *Can. Field-Nat.* **100**:241–245.

Martin, L. C. 1994. *Wildlife Folklore*. Old Saybrook, CT: The Globe Pequot Press.

Mashkin, V. I. 1991. Hunting press and bobac population structure. In *Population Structure of the Marmot*, ed. D. I. Bibikov, A. A. Nikol'ski, V. Yu. Rumiantsev, and T. A. Seredneva. Moscow: USSR Theriological Society, pp. 119–147.

Mashkin, V. I. 2003. Interfamily regrouping of Eurasian marmots. In *Adaptive Strategies and Diversity in Marmots*, ed. R. Ramousse, D. Allainé, and M. Le Berre. Lyon: International Marmot Network, pp. 183–188.

Mashkin, V. I., V. V. Kolesnikov, and B. E. Zarubin 1994. Resources of a steppe marmot in the Ukraine. In *Actual Problems of Marmots Investigation*, ed. V. Yu. Rumiantsev. Moscow: ABF Publishing House, pp. 86–97.

Massemin, S., C. Gibault, R. Ramousse, and A. Butet 1996. Premiéres données sur le régime alimentaire de la marmotte alpine (*Marmota marmota*) en France. *Mammalia* **60**:351–361.

Matrosova, V. A., D. T. Blumstein, I. A. Volodin, and E. V. Volodina 2011. The potential to encode sex, age, and individual identity in the alarm calls of three species of Marmotinae. *Naturwissenschaften* **98**:181–192.

McBirnie, J. E., F. G. Pearson, G. A. Trusler, H. H. Karachi, and W. G. Bigelow 1953. Physiologic studies of the groundhog (*Marmota monax*). *Can. J. Med. Sci.* **31**:421–430.

McTaggart-Cowan, I. 1929. Notes on yellow-bellied marmot. *Murrelet* **10**:64.

McTaggart-Cowan, I. 1933. The British Columbian woodchuck *Marmota monax petrensis* Howell. *Can. Field-Nat.* **47**:57.

Meier, P. T. 1991. Response of adult woodchucks (*Marmota monax*) to oral-gland scents. *J. Mammal.* **72**:622–624.

Meier, P. T. 1992. Social organization of woodchucks (*Marmota monax*). *Behav. Ecol. Sociobiol.* **31**:393–400.

Mein, P. 1992. Taxonomy. In *Proceedings of the 1st International Symposium on Alpine Marmot (*Marmota marmota*) and Genus* Marmota, ed. B. Bassano, P. Durio, U. Gallo Orsi, and

E. Macchi. Torino, Italy: Dipartimento di Produzioni animali, Epidemiologia ed Ecologia, pp. 6–12.

Melcher, J. C. 1987. The influence of thermal energy exchange on the activity and energetics of yellow-bellied marmots. PhD dissertation, University of Kansas, Lawrence.

Melcher, J. C., K. B. Armitage, and W. P. Porter 1989. Energy allocation by yellow-bellied marmots. *Physiol. Zool.* **62**:429–448.

Melcher, J. C., K. B. Armitage, and W. P. Porter 1990. Thermal influences on the activity and energetics of yellow-bellied marmots (*Marmota flaviventris*). *Physiol. Zool.* **63**:803–820.

Mercer, J. M., and V. L. Roth 2003. The effects of Cenozoic global change on squirrel phylogeny. *Science* **299**:1568–1572.

Merriam, H. G., and A. Merriam 1965. Vegetation zones around woodchuck burrows. *Can. Field-Nat.* **79**:177–180.

Michener, G. R. 1974. Development of adult-young identification in Richardson's ground squirrel. *Dev. Psychobiol.* **7**:375–384.

Michener, G. R. 1983. Kin identification, matriarchies, and the evolution of sociality in ground-dwelling sciurids. In *Advances in the Study of Mammalian Behavior*, ed. J. F. Eisenberg and D. G. Kleinman. Am. Soc. Mammal. Sp. Publ. **7**, 528–572.

Mikhailuta, A. A. 1991. Family structure in grey marmots. In *Population Structure of the Marmot*, ed. D. L. Bibikov, A. A. Nikolski, V. Yu. Rumiantsev, and T. A. Seredneva. Moscow: USSR Theriological Society, pp. 172–187.

Milko, R. J. 1984. Vegetation and foraging ecology of the Vancouver Island marmot (*Marmota vancouverensis*). MS Thesis, University of Victoria, BC.

Mitchell, D., A. Fuller, R. S. Hetem, and S. K. Maloney 2008. Climate change physiology: the challenge of the decades. In *Molecules to Migration: The Pressures of Life*, ed. S. Morris and A. Vosloo. Bologna: Medimond, pp. 383–394.

Moehlman, P. D., and H. Hofer 1997. Cooperative breeding, reproductive suppression, and body mass in canids. In *Cooperative Breeding in Mammals*, ed. N. G. Solomon and J. A. French. Cambridge: Cambridge University Press, pp. 76–128.

Monclus, R., and D. T. Blumstein 2012. Litter sex composition affects life-history traits in yellow-bellied marmots. *J. Anim. Ecol.* **81**:80–86.

Monclus, R., T. Cool, and D. T. Blumstein 2012. Masculinized female yellow-bellied marmots initiate more social interactions. *Biol. Lett.* **8**:208–210.

Monclus, R., J. Tiulim, and D. T. Blumstein 2011. Older mothers follow conservative strategies under predator pressusre: the adaptive role of maternal glucocorticoids in yellow-bellied marmots. *Hormones Behav.* **60**:660–665.

Moore, D., and R. Ali 1984. Are dispersal and inbreeding avoidance related? *Anim. Behav.* **32**:94–112.

Morrison, P. 1960. Some interrelations between weight and hibernation function. In *Mammalian Hibernation*, ed. C. P. Lyman and A. R. Dave. *Bull. Mus. Comp. Zool.* **124**:75–91.

Mosolov, V. I., and V. A. Tokarsky 1994. The black-capped marmot (*Marmota camtschatica* Pall.) in the Kronotsky Reserve. In *Actual Problems of Marmots Investigation*, ed. V. Yu. Rumiantsev. Moscow: ABF Publishing House, pp. 98–110.

Moss, A. E. 1940. The woodchuck as a soil expert. *J. Wildl. Manage.* **4**:441–443.

Müller-Using, D. 1955. Vom "Pfeifen" des Murmeltieres. *Z. Jagdwissensch* **1**:32–33.

Müller-Using, D. 1957. Die Paarungsbiologie des Murmeltieres. *Z. Jagdwissensch.* **3**:24–28.

Münch, H. 1958. Zur Okologie und Psychologie von *Marmota m. marmota. Z. Säugetierk.* **23**:129–138.

Murdoch, J. D., T. Munkhzul, S. Buyandelger, R. P. Reading, and C. Sillero-Zubiri 2009. The endangered Siberian marmot *Marmota sibirica* as a keystone species? Observations and implications of burrow use by corsac foxes *Vulpes corsac* in Mongolia. *Oryx* **43**:431–434.

Murie, A. 1940. *Ecology of the Coyote in the Yellowstone. Fauna of the National Parks of the United States Bull #4.* Washington DC: U.S. Government Printing Office.

Murie, A. 1944. *The Wolves of Mt. Mckinley.* Washington DC: U.S. Government Printing Office.

Murie, J. O., and G. R. Michener, eds. 1984. *The Biology of Ground-dwelling Squirrels.* Lincoln, NE: University of Nebraska Press.

Murray, B. G., Jr. 1999. Can the population regulation controversy be buried and forgotten? *Oikos* **84**:148–152.

Nagorsen, D. W. 1987. *Marmota vancouverensis. Mammalian Species* **270**:1–5.

Nagorsen, D. W., and A. Cardini 2009. Tempo and mode of evolutionary divergence in modern and Holocene Vancouver Island marmots (*Marmota vancouverensis*) (Mammalia, Rodentia). *J. Zool. Syst. Evol. Res.* **47**:258–267.

Nagorsen, D. W., G. Keddie, and T. Luszcy 1996. Vancouver Island marmot bones from subalpine caves: archaeological and biological significance. *Occas. Paper* **4**:1–56. BC Ministry of Environment, Lands and Parks.

Nee, J. A. 1969. Reproduction in a population of yellow-bellied marmots (*Marmota flaviventris*). *J. Mammal.* **50**:756–765.

Nelson, B. C. 1980. Plague studies in California: the roles of various species of sylvatic rodents in plague ecology in California. *Proceedings of the 9th Vertebrate Pest Conference*, 89–96.

Nesterova, N. L. 1996. Age-dependent alarm behaviour and response to alarm call in bobac marmots (*Marmota bobak* Mull.). In *Biodiversity in Marmots*, ed. M. Le Berre, R. Ramousse, and L. LeGuelte. Lyon, France: International Marmot Network, pp. 181–186.

Neuhaus, P., and B. Mainini 1998. Reactions and adjustment of adult and young alpine marmots *Marmota marmota* to intense hiking activities. *Wildl. Biol.* **4**:119–123.

Neuhaus, P., B. Manini, and P. Ingold 1992. Human impact on marmot behaviour. In *Proceedings of the 1st International Symposium on Alpine Marmot (Marmota marmota) and Genus Marmota*, ed. B. Bassano, P. Durio, U. Gallo Orsi, and E. Macchi. Torino, Italy: Dipartimento di Produzioni animali, Epidemiologia ed Ecologia, pp. 165–199.

Nikol'skii, A. A. 2002a. Relative effects of soil and surface air on mammal burrow temperature: a study of the bobac burrow as an example. *Doklady Biol. Sci.* **382**:25–27.

Nikol'skii, A. A. 2002b. The geographical populations of the steppe marmot, *Marmota bobak* (a bioacoustical analysis). In *Holarctic Marmots as a Factor of Biodiversity*, ed. K. B. Armitage and V. Yu. Rumiantsev. Moscow: ABF Publishing House, pp. 290–298.

Nikol'skii, A. A. 2002c. Topographic relief as a factor in the geographical variation of the rhythmical structure of alarm calls of the steppe marmot (*Marmota bobak*). In *Holarctic Marmots as a Factor of Biodiversity*, ed. K. B. Armitage and V. Yu. Rumiantsev. Moscow: ABF Publishing House, pp. 299–307.

Nikol'skii, A. A. 2009. The hibernation temperature niche of the steppe marmot *Marmota bobak* Muller 1776. *Ethol., Ecol., Evol.* **21**:393–401.

Nikol'skii, A. A., and M. D. Khutorskoi 2001. Thermal characteristics of mammalian burrows in summer (using a burrow of the steppe marmot as an example). *Doklady Biol. Sci.* **378**:240–243.

Nikol'skii, A. A., V. M. Kotlyakov, and D. T. Blumstein 1999. Glaciation as a factor of geographic variation in the long-tailed marmot (bioacoustical analysis). *Doklady Biol. Sci.* **368**:509–513.

Nikol'skii, A. A., N. L. Nesterova, and M. V. Suchanova 1994. Situational variations in spectral structure in *Marmota bobak* Müll. alarm signal. In *Actual Problems of Marmots Investigation*, ed. V. Yu. Rumiantsev. Moscow: ABF Publishing House, pp. 127–148.

Nikol'skii, A. A., and V. Yu. Rumiantsev 2012. Center of species diversity of Eurasian marmots (Marmota, Rodentia) in an epi-platformal orogeny area. *Doklady Biol. Sci.* **445**:261–264.

Nikol'skii, A. A., and G. A. Savchenko 1999. Structure of family groups and space use by steppe marmots (*Marmota bobak*): preliminary results. *Vestnik zoologii* **33**:67–72.

Nikol'skii, A. A., and G. A. Savchenko 2002a. Structure of family groups and space use by steppe marmots: preliminary results. In *Holarctic Marmots as a Factor of Biodiversity*, ed. K. B. Armitage and V. Yu. Rumiantsev. Moscow: ABF Publishing House, pp. 308–316.

Nikol'skii, A. A., and G. A. Savchenko 2002b. Air temperature changes in a steppe marmot burrow in the summer–autumn period. *Russian J. Ecol.* **33**:109–114.

Nikol'skii, A. A., and G. A. Savchenko 2002c. Effect of atmosphere temperature near the soil surface on air convection in mammalian burrows (as exemplified by a steppe marmot burrow). *Doklady Biol. Sci.* **384**:242–245.

Nikol'skii, A. A., and A. Ulak 2006. Key factors determining the ecological niche of the Himalyan marmot, *Marmota himalayana*, Hodgson (1841). *Russian J. Ecol.* **37**:46–52.

Nowicki, S., and K. B. Armitage 1979. Behavior of juvenile yellow-bellied marmots: play and social integration. *Z. Tierpsychol.* **51**:85–105.

Noyes, D. H., and W. G. Holmes 1979. Behavioral responses of free-living hoary marmots to a model golden eagle. *J. Mammal.* **60**:408–411.

Nuckolls, K. R. 2010. Determinants of annual and lifetime reproductive success in female yellow-bellied marmots: a cross-generational study. PhD dissertation, University of Kansas, Lawrence.

Oli, M. K. 2003. Hamilton goes empirical: estimation of inclusive fitness from life-history data. *Proc. R. Soc. Lond. Ser. B* **270**:307–311.

Oli, M. K., and K. B. Armitage 2003. Sociality and individual fitness in yellow-bellied marmots: insights from a long-term study (1962–2001). *Oecologia* **136**:543–550.

Oli, M. K., and K. B. Armitage 2004. Yellow-bellied marmot population dynamics: demographic mechanisms of growth and decline. *Ecology* **85**:2446–2455.

Oli, M. K., and K. B. Armitage 2008. Indirect fitness benefits do not compensate for the loss of direct fitness in yellow-bellied marmots. *J. Mammal.* **89**:874–881.

Olivier, P. I., R. J. Van Arde, and S. M. Ferreira 2009. Support for a metapopulation structure among mammals. *Mammal Review* **39**:178–192.

Olson, L. E., and D. T. Blumstein 2010. Applying the coalitionary-traits metric: sociality without cooperation in male yellow-bellied marmots. *Behav. Ecol.* **21**:957–965.

Olson, L. E., D. T. Blumstein, J. R. Pollinger, and R. K. Wayne 2012. No evidence of inbreeding avoidance despite demonstrated survival costs in a polygynous rodent. *Mol. Ecol.* **21**:562–571.

Olson, L. E., V. Yovovich, and D. T. Blumstein. 2003. Early season arboreal behaviour in yellow-bellied marmots (*Marmota flaviventris*). *Oecologia Montana* **12**:12–14.

Ortmann, S., and G. Heldmaier 1992. Energetics of hibernating and normothermic alpine marmots. In *Proceedings of the 1st International Symposium on Alpine Marmot (Marmota marmota) and Genus Marmota*, ed. B. Bassano, P. Durio, U. Gallo Orsi, and E. Macchi. Torino, Italy: Dipartimento di Produzioni animali, Epidemiologia ed Ecologia, pp. 221–226.

Ortmann, S., and G. Heldmaier 2000. Regulation of body temperature and energy requirements in hibernating alpine marmots (*Marmota marmota*). *Am. J. Physiol. Reg. Integr. Comp. Physiol.* **278**:R698–R704.

Ouellet, J.-P., and J. Ferron 1988. Scent-marking behavior by woodchucks (*Marmota monax*). *J. Mammal.* **69**:365–368.

Ozgul, A., K. B. Armitage, D. T. Blumstein, and M. K. Oli 2006b. Spatiotemporal variation in survival rates: implications for population dynamics of yellow-bellied marmots. *Ecology* **87**:1027–1037.

Ozgul, A., K. B. Armitage, D. T. Blumstein, D. H. Van Vuren, and M. K. Oli 2006a. Effects of patch quality and network structure on patch occupancy dynamics of a yellow-bellied marmot metapopulation. *J. Anim. Ecol.* **75**:191–202.

Ozgul, A., D. Z. Childs, M. K. Oli, *et al.* 2010. Coupled dynamics of body mass and population growth in response to environmental change. *Nature* **466**:482–485.

Ozgul, A., M. K. Oli, K. B. Armitage, D. T. Blumstein, and D. H. Van Vuren 2009. Influence of local demography on asymptotic and transient dynamics of a yellow-bellied marmot meta-population. *Am. Nat.* **173**:517–530.

Ozgul, A., M. K. Oli, L. E. Olson, D. T. Blumstein, and K. B. Armitage 2007. Spatiotemporal variation in reproductive parameters of yellow-bellied marmots. *Oecologia* **154**:95–106.

Palmer, W. R. 1973. *Why the North Star Stands Still and Other Indian Legends*. Springdale, UT: Zion Natural History Association.

Panseri, M. 1992. The alpine marmot (*Marmota marmota*) on "Orobie" Alps and "Bergamasche" Prealps preliminary observations for an analysis of the population. In *Proceedings of the 1st International Symposium on Alpine Marmot (*Marmota marmota*) and Genus* Marmota, ed. B. Bassano, P. Durio, U. Gallo Orsi, and E. Macchi. Torino, Italy: Dipartimento di Produzioni animali, Epidemiologia ed Ecologia, pp. 235–237.

Panseri, M., and D. Frigerio 1996. Some considerations on marmot population expansion in Brembana Valley (Orobie Alps). In *Biodiversity in Marmots*, ed. M. Le Berre, R. Ramousse, and L. Le Guelte. Lyon, France: International Marmot Network, pp. 243–244.

Pattie, D. L. 1967. Observations on an alpine population of yellow-bellied marmots (*Marmota flaviventris*). *Northwest Sci.* **41**:96–102.

Pellis, S. M., and V. C. Pellis 1998. The structure-function interface in the analysis of play fighting. In *Animal Play: Evolutionary, Comparative, and Ecological Perspectives*, ed. M. Bekoff and J. A. Byers. Cambridge: Cambridge University Press, pp. 115–140.

Pengelley, E. T., and S. J. Asmundson 1974. Circannual rhythmicity in hibernating mammals. In *Circannual Clocks*, ed. E. T. Pengelley. New York: Academic Press, pp. 95–160.

Perrin, C., D. Allainé, and M. Le Berre 1993a. Socio-spatial organization and activity distribution of the alpine marmot *Marmota marmota*: preliminary results. *Ethology* **93**:21–30.

Perrin, C., D. Allainé, and M. Le Berre 1994. Instrusion de males et possibilité d'infanticide chez la marmotte alpine. *Mammalia* **58**:150–153.

Perrin, C., J. Coulon, and M. Le Berre 1993b. Social behavior of alpine marmots (*Marmota marmota*): seasonal, group, and individual variability. *Can. J. Zool.* **71**:1945–1953.

Perrin, C., J. Coulon, and M. Le Berre 1996. Social behaviour of alpine marmots (*Marmota marmota*): group and individual variability. In *Biodiversity in Marmots*, ed. M. Le Berre, R. Ramousse, and L. Le Guelte. Lyon, France: International Marmot Network, pp. 193–198.

Perrin, C., L. Le Guelte, and M. Le Berre 1992. Temporal and spatial distribution of activities during summer in the alpine marmot. In *Proceedings of the 1st International Symposium on Alpine Marmot (*Marmota marmota*) and Genus* Marmota, ed. B. Bassano, P. Durio, U. Gallo Orsi, and E. Macchi. Torino, Italy: Dipartimento di Produzioni animali, Epidemiologia ed Ecologia, pp. 101–108.

Perrin, N., and J. Goudet 2001. Inbreeding, kinship, and the evolution of natal dispersal. In *Dispersal*, ed. J. Clobert, E. Danchin, A. A. Dhondt, and J. D. Nichols. Oxford: Oxford University Press, pp. 123–142.

Pigozzi, G. 1984. The den system of the alpine marmot (*Marmota marmota marmota*) in the National Park of Stelvio, Northern Italy. *Z. Saugetierk.* **49**:13–21.

Pigozzi, G. 1989. Predation on alpine marmot. *Marmota marmota* (L) by a golden eagle, *Aquila chrysaetos* (L). *Atti della Ital. di Sci. Nat. e del Mus. Civ. di Storia Nat. de Milano* **130**:93–96.

Pole, S. B. 1992. Dynamics of the grey marmot population structure and size and the mechanisms of their restoration in the Tien-Shan natural plague focus. In *Proceedings of the 1st International Symposium on Alpine Marmot (*Marmota marmota*) and Genus* Marmota, ed. B. Bassano, P. Durio, U. Gallo Orsi, and E. Macchi. Torino, Italy: Dipartimento di Produzioni animali, Epidemiologia ed Ecologia, pp. 129–133.

Pole, S. B. 1996. Population heterogeneity in Tien Shan *Marmota baibacina*. In *Biodiversity in Marmots*, ed. M. Le Berre, R. Ramousse, and L. Le Guelte. Lyon, France: International Marmot Network, pp. 182–192.

Pole, S. B. 2003. Marmots and zoonotic infections in CIS. In *Adaptive Strategies and Diversity in Marmots*, ed. R. Ramousse, D. Allainé, and M. Le Berre. Lyon: International Marmot Network, pp. 13–18.

Pole, S. B., and D. I. Bibikov 1991. Dynamics of population structure and mechanisms of maintaining optimal population density in grey marmots. In *Population Structure of the Marmot*, ed. D. I. Bibikov, A. A. Nikolski, V. Yu. Rumiantsev, and T. A. Seredneva. Moscow: USSR Theriological Society, pp. 148–171.

Pole, D. S., V. N. Davydova, V. S. Ageev, and S. B. Pole 2003. Restoration of flea numbers after disinfection of grey marmot burrows in the Kokpak plague mesofocus. In *Adaptive Strategies and Diversity in Marmots*, ed. R. Ramousse, D. Allainé, and M. Le Berre. Lyon, France: International Marmot Network, pp. 19–22.

Polly, P. D. 2003. Paleophylogeography: the tempo of geographic differentiation in marmots (*Marmota*). *J. Mammal.* **84**:369–384.

Preleuthner, M., W. Pinsker, L. Kruckenhauser, W. J. Miller, and H. Prosl 1995. Alpine marmots in Austria. The present population structure as a result of postglacial distribution history. *Acta Theriol.* **40**(Suppl 3):87–100.

Preleuthner, M., H. Prosl, A. Bergmann, and W. Pinsker 1996. Infestation by endoparasites in different Austrian populations. In *Biodiversity in Marmots*, ed. M. Le Berre, R. Ramousse, and L. Le Guelte. Lyon, France: International Marmot Network, pp. 111–112.

Prosl, H., M. Preleuthner, and A. Bergmann 1992. Endoparasites of *Marmota marmota* in the Tyrolian Alps. In *Proceedings of the 1st International Symposium on Alpine Marmot (*Marmota marmota*) and Genus* Marmota, ed. B. Bassano, P. Durio, U. Gallo Orsi, and E. Macchi. Torino, Italy: Dipartimento di Produzioni animali, Epidemiologia ed Ecologia, pp. 215–216.

Pusey, A. E. 1987. Sex-biased dispersal and inbreeding avoidance in birds and mammals. *Trends Ecol. Evol.* **2**:295–299.

Pusey, A., and M. Wolf 1996. Inbreeding avoidance in animals. *Trends Ecol. Evol.* **11**:201–206.

Rains, D. L. 1979. Behavior of juvenile yellow-bellied marmots in two social environments. MS Thesis, The Ohio State University, Ohio.

Ramousse, R., and O. Giboulet 2002. In the name of marmot and chamois: value of trade use of their names. In *Holarctic Marmots as a Factor of Biodiversity*, ed. K. B. Armitage and V. Yu. Rumiantsev. Moscow: ABF Publishing House, pp. 323–328.

Ramousse, R., and M. Le Berre 1993. Management of alpine marmot populations. *Oecologia Montana* 2:23–29.

Ramousse, R., and M. Le Berre 2007. From mountain rat to alpine marmot, a tentative history. In *The Marmots of Eurasia: Origin and Current Status*, ed. A. V. Esipov, E. A. Bykova, O. V. Brandler, R. Ramousse, and E. V. Vashatko. Tashkent: International Marmot Network, pp. 108–117.

Ramousse, R., J. Metral, and M. Le Berre 2009. Twenty-seventh year of the alpine marmot introduction in the agricultural landscape of the Central Massif (France). *Ethol., Ecol., Evol.* 21:243–250.

Rankin, D. J., and H. Kokko 2007. Do males matter? The role of males in population dynamics. *Oikos* 116:335–348.

Rausch, R. L. 1980. Redescription of *Diandrya composita* Darrah, 1930 (Cestoda: Anoplocephalidae) from Nearctic marmots (Rodentia: Sciuridae) and the relationships of the genus *Diandrya* emend. *Proc. Helminthol. Soc. Wash.* 47:157–164.

Rausch, R. L., and J. G. Bridgens 1989. Structure and function of sudoriferous facial glands in Nearctic marmots, *Marmota* spp. (Rodential: Sciuridae). *Zool. Anz.* 223:265–282.

Rausch, R. L., and V. R. Rausch 1971. The somatic chromosomes of some North American marmots (Sciuridae), with remarks on the relationships of *Marmota broweri* Hall and Gilmore. *Mammalia* 35:85–101.

Rayor, L. S. 1985. Dynamics of a plague outbreak in Gunnison's prairie dog. *J. Mammal.* 66:194–196.

Rayor, L. S., and K. B. Armitage 1991. Social behavior and space-use of young of ground-dwelling squirrel species with different levels of sociality. *Ethol., Ecol., Evol.* 3:185–205.

Ricankova, V. P., J. Riegert, E. Semancikova, M. Hais, A. Cejkova, and K. Prach 2013. Habitat preferences in gray marmots *(Marmota baibacina)*. *Acta Theriologica* 2013:1–8.

Rodrigue, I., D. Allainé, R. Ramousse, and M. Le Berre 1992. Space occupation strategy related to ecological factors in alpine marmot (*Marmota marmota*). In *Proceedings of the 1st International Symposium on Alpine Marmot* (Marmota marmota) *and Genus* Marmota, ed. B. Bassano, P. Durio, U. Gallo Orsi, and E. Macchi. Torino, Italy: Dipartimento di Produzioni animali, Epidemiologia ed Ecologia. Lyon, France: International Marmot Network, pp. 135–141.

Roehrs, Z. P., and H. H. Genoways 2004. Historical biogeography of the woodchuck (*Marmota monax bunkeri*) in Nebraska and northern Kansas. *West. N.A. Natural.* 64:396–402.

Rogovin, K. A. 1992. Habitat use by two species of Mongolian marmots (*Marmota sibirica* and *M. baibacina*) in a zone of sympatry. *Acta Theriol.* 37:345–350.

Ronkin, V., G. Savchenko, and V. Tokarsky 2009. The place of the steppe marmot in steppe ecosystems of Ukraine: an historical approach. *Ethol., Ecol., Evol.* 21:277–284.

Ronkin, V. I., and V. A. Tokarsky 1993. Qualitative and quantitative assessment of feeding habits of bobak, *Marmota bobak*, and long-tailed marmot, *M. caudata* (Rodentia, Sciuridae) in captivity. *Zool. J.* 72:93–100.

Rubenstein, D. J., and R. W. Wrangham 1980. Why is altruism towards kin so rare? *Zeitschrift für Tierpsychologie* 54:381–387.

Rudi, V. N., N. N. Shevlyuk, and V. P. Soustin. 1994. Ecology and morphology of the bobac (*Marmota bobak*) in Orenburg Province. In *Actual Problems of Marmots Investigation*, ed. V. Yu. Rumiantsev. Moscow: ABF Publishing House, pp. 182–192.

Ruf, T., and W. Arnold 2000. Mechanisms of social thermoregulation in hibernating alpine marmots (*Marmota marmota*). In *Life in the Cold*, ed. G. Heldmaier and M. Klingenspor. Heidelberg: Springer-Verlag Berlin, pp. 81–94.

Ruf, T., and W. Arnold 2008. Effects of polyunsaturated fatty acids on hibernation and torpor: a review and hypothesis. *Am. J. Physiol. Regul. Integr. Comp. Physiol.* **294**:R1044–R1052.

Rumiantsev, V. Yu. 1991. Cartographic analysis of *Marmota bobak* distribution in Kazakhstan. In *Population Structure of the Marmot*, ed. D. I. Bibikov, A. A. Nikolski, V. Yu. Rumiantsev, and T. A. Seredneva. Moscow: USSR Theriological Society, pp. 71–97.

Rumiantsev, V. Yu. 1992. Marmot's impact on soil of solonetz complexes in northern Kazakhstan. In *Proceedings of the 1st International Symposium on Alpine Marmot (*Marmota marmota*) and Genus* Marmota, ed. B. Bassano, P. Durio, U. Gallo Orsi, and E. Macchi. Torino, Italy: Dipartimento di Produzioni animali, Epidemiologia ed Ecologia, pp. 241–243.

Rumyantsev, V. Yu., O. A. Ermakov, V. Yu. Il'in, A. N. Dobrolyubov, M. S. Soldatov, and E. A. Danilenko 2012. On the history and modern state of the steppe marmot (*Marmota marmota* Mull.) in Penza Oblast. *Arid Ecosystems* **2**:111–119.

Rymalov, I. V. 1994. On social structure and behaviour of steppe marmots during breeding period. In *Actual Problems of Marmots Investigation*, ed. V. Yu. Rumiantsev. Moscow: ABF Publishing House, pp. 225–235.

Sala, L., C. Sola, A. Spampanato, M. Magnanini, and P. Tongiorgi 1996. Space and time use in a population of *Marmota marmota* of the northern Apennines. In *Biodiversity in Marmots*, ed. M. Le Berre, R. Ramousse, and L. Le Guelte. Lyon, France: International Marmot Network, pp. 209–216.

Sala, L., C. Sola, A. Spampanato, and P. Tongiorgi 1992. The marmot population of the Tuscan-Emilian Apennine ridge. In *Proceedings of the 1st International Symposium on Alpine Marmot (*Marmota marmota*) and Genus* Marmota, ed. B. Bassano, P. Durio, U. Gallo Orsi, and E. Macchi. Torino, Italy: Dipartimento di Produzioni animali, Epidemiologia ed Ecologia, pp. 143–149.

Sala, L., C. Sola, A. Spampanato, P. Tongiorgi, and M. Maganini 1993. Capture and identification techniques of marmot on Mount Cimone (Northern Apennines). *IBEX J. M. E.* **1**:14–16.

Salsbury, C. M., and K. B. Armitage 1994a. Home-range size and exploratory excursions of adult, male yellow-bellied marmots. *J. Mammal.* **75**:648–656.

Salsbury, C. M., and K. B. Armitage 1994b. Resting and field metabolic rates of adult male yellow-bellied marmots, *Marmota flaviventris*. *Comp. Biochem. Physiol.* **108A**:579–588.

Salsbury, C. M., and K. B. Armitage 1995. Reproductive energetics of adult male yellow-bellied marmots (*Marmota flaviventris*). *Can. J. Zool.* **73**:1791–1797.

Salsbury, C. M., and K. B. Armitage 2003. Variation in growth rates of yellow-bellied marmots (*Marmota flaviventris*). In *Adaptive Strategies and Diversity in Marmots*, ed. R. Ramousse, D. Allainé, and M. Le Berre. Lyon: International Marmot Network, pp. 197–206.

Schmeltz, L. L., and J. O. Whitaker, Jr. 1977. Use of woodchuck burrows by woodchucks and other mammals. *Trans. Kentucky Acad. Sci.* **38**:79–82.

Schwartz, O. A., and K. B. Armitage 1980. Genetic variation in social mammals: the marmot model. *Science* **207**:665–667.

Schwartz, O. A., and K. B. Armitage 1981. Social substructure and dispersion of genetic variation in the yellow-bellied marmot. In *Mammalian Population Genetics*, ed. M. H. Smith and J. Soule. Athens: University of Georgia Press, pp. 139–159.

Schwartz, O. A., and K. B. Armitage 1998. Empirical considerations on the stable age distribution. *Oecologia Montana* **7**:1–6.

Schwartz, O. A., and K. B. Armitage 2002. Correlations between weather factors and life-history traits of yellow-bellied marmots. In *Holarctic Marmots as a Factor of Biodiversity*, ed. K. B. Armitage and V. Yu. Rumiantsev. Moscow: ABF Publishing House, pp. 345–351.

Schwartz, O. A., and K. B. Armitage 2003. Population biology of the yellow-bellied marmot: a 40 year perspective. In *Adaptive Strategies and Diversity in Marmots*, ed. R. Ramousse, D. Allainé, and M. Le Berre. Lyon, France: International Marmot Network, pp. 207–212.

Schwartz, O. A., and K. B. Armitage 2005. Weather influences on demography of the yellow-bellied marmot (*Marmota flaviventris*) *J. Zool., Lond.* **265**:73–79.

Schwartz, O. A., K. B. Armitage, and D. Van Vuren 1998. A 32-year demography of yellow-bellied marmots. *J. Zool., Lond.* **246**:337–346.

Semenov, Yu., S. Louis, O. Giboulet, and R. Ramousse 2002. Accommodation behaviour of alpine marmot (*Marmota marmota*, Linn. 1758) under direct anthropogenic influence. In *Holarctic Marmots as a Factor of Biodiversity*, ed. K. B. Armitage and V. Yu. Rumiantsev. Moscow: ABF Publishing House, pp. 358–365.

Semenov, Y., R. Ramousse, and M. Le Berre 2000. Effet de la lumiere de la temperature sur le rythme d'activite de la marmotte alpine (*Marmota marmota* Linne, 1758) en milieu naturel. *Can. J. Zool.* **78**:1980–1986.

Semenov, Y., R. Ramousse, M. Le Berre, V. Vassiliev, and N. Solomonov 2001a. Aboveground activity rhythm in Arctic black-capped marmot (*Marmota camtschatica bungei*, Katschenko 1901) under polar day conditions. *Acta Oecologia* **22**:99–107.

Semenov, Y., R. Ramousse, M. Le Berre, and Y. Tutukarov 2001. Impact of the black-capped marmot (*Marmota camtschatica bungei*) on floristic diversity of the arctic tundra in Northern Siberia. *Arct. Antarct. Alp. Res.* **33**:204–210.

Seredneva, T. A. 1991. Consistent pattern of the tarbagan spatial distribution in the eastern Khangai (Mongolia). In *Population Structure of the Marmot*, ed. D. I. Bibikov, A. A. Nikolski, V. Yu. Rumiantsev, and T. A. Seredneva. Moscow: USSR Theriological Society, pp. 233–274.

Shelley, E. L., and D. T. Blumstein 2004. The evolution of vocal alarm communication in rodents. *Behav. Ecol.* **16**:169–177.

Sherman, P. W. 1977. Nepotism and the evolution of alarm calls. *Science* **197**:1246–1253.

Shriner, W. M. 1998. Yellow-bellied marmot and golden-mantled ground squirrel responses to heterospecific alarm calls. *Anim. Behav.* **55**:529–536.

Shubin, V. I. 1991. Population structure and bobac reproduction in the northern part of the Kazakh Melkosopotchnik (= low hill area). In *Population Structure of the Marmot*, ed. D. I. Bibikov, A. A. Nikolski, V. Yu. Rumiantsev, and T. A Seredneva. Moscow: USSR Theriological Society, pp. 98–118.

Shubin, V. I., and L. V. Spivakova 1993. Cutaneous glands and scent marking of marmots (*Marmota*, Sciuridae). *Selevinia* **1**:69–80.

Silk, J. B. 2007. The adaptive value of sociality in mammalian groups. *Phil. Trans. R. Soc. B* **362**:539–559.

Sinha Hikim, A. P., A. Woolf, A. Bartke, and A. G. Amador 1991. The estrous cycle of captive woodchucks (*Marmota monax*). *Biol. Reprod.* **44**:733–738.

Sinha Hikim, A. P., A. Woolf, A. Bartke, and A. G. Amador 1992. Further observations on estrus and ovulation in woodchucks (*Marmota monax*) in captivity. *Biol. Reprod.* **46**:10–16.

Smith, J. E., R. Monclus, D. Wantuck, G. L. Florant, and D. T. Blumstein 2012. Fecal glucocorticoid metabolites in wild yellow-bellied marmots: Experimental validation, individual differences and ecological correlates. *Gen. Comp. Endocrinol.* **178**:417–426.

Snyder, R. L. 1962. Reproductive performance of a population of woodchucks after a change in sex ratio. *Ecology* **43**:506–515.

Snyder, R. L., and J. J. Christian 1960. Reproductive cycle and litter size of the woodchuck. *Ecology* **41**:647–656.

Snyder, R. L., D. E. Davis, and J. J. Christian 1961. Seasonal changes in the weights of wood-chucks. *J. Mammal.* **42**:297–312.

Solomon, N. G., and J. A. French 1997a. The study of mammalian cooperative breeding. In *Cooperative Breeding in Mammals*, ed. N. G. Solomon and J. A. French. Cambridge: Cambridge University Press, pp. 1–10.

Solomon, N. G., and J. A. French, eds. 1997b. *Cooperative Breeding in Mammals*. Cambridge: Cambridge University Press.

Solomonov, N. G., Yu. S. Lukovtsev, V. N. Vasiliev, and Yu. Semenov 1996. The black-capped marmot (*Marmota camtschatica* Pall.) in Yakutia. In *Biodiversity in Marmots*, ed. M. Le Berre, R. Ramousse, and L. Le Guelte. Lyon, France: International Marmot Network, pp. 251–265.

Soroka, O. V. 2000. How the environment affects the dynamics of *Marmota bobak* season activity (*Marmota bobak* Mull. 1776). In *Biology of Palaearctic Marmots*, ed. O. V. Brandler, and A. A. Nikol'skii. Moscow: BBF, pp. 145–158. (In Russian.)

Speakman, T. R. 2000. The cost of living: field metabolic rates of small mammals. *Adv. Ecol. Res.* **30**:177–297.

Stallman, E. L., and W. G. Holmes 2002. Selective foraging and food distribution of high-elevation yellow-bellied marmots (*Marmota flaviventris*). *J. Mammal.* **83**:576–584.

Stenseth, N. C., and W. Z. Lidicker Jr, eds. 1992. *Animal Dispersal: Small Mammals as a Model*. New York: Chapman and Hall.

Stephens, P. A., F. Frey-Roos, W. Arnold, and W. J. Sutherland 2002a. Model complexity and population predictions. The alpine marmot as a case study. *J. Anim. Ecol.* **71**:343–361.

Stephens, P. A., F. Frey-Roos, W. Arnold, and W. J. Sutherland 2002b. Sustainable exploitation of social species: a test and comparison of models. *J. Appl. Ecol.* **39**:629–642.

Steppan, S. J., M. R. Akhverdyan, E. A. Lyapunova, *et al.* 1999. Molecular phylogeny of the marmots (Rodentia: Sciuridae): tests of evolutionary and biogeographic hypotheses. *Syst. Biol.* **48**:715–734.

Steppan, S. J., G. J. Kenagy, C. Zawadzki, *et al.* 2011. Molecular data resolve placement of the Olympic marmot and estimate dates of trans-Beringian interchange. *J. Mammal.* **92**:1028–1037.

Steppan, S. J., B. L. Storz, and R. S. Hoffmann 2004. Nuclear DNA phylogeny of the squirrels (Mammalia: Rodentia) and the evolution of arboreality from c-myc and RAGI. *Mol. Phylogenet. Evol.* **30**:703–719.

Stone, R. 2010. Race to contain plague in quake zone. *Science* **328**:559.

Suntsov, V. V., and N. I. Suntsova 1991. Spatial structure of the tarbagan in Tuva. In *Population Structure of the Marmot*, ed. D. I. Bibikov, A. A. Nikolski, V. Yu. Rumiantsev, and T. A. Seredneva. Moscow: USSR Theriological Society, pp. 217–232.

Svendsen, G. E. 1974. Behavioral and environmental factors in the spatial distribution and pop-ulation dynamics of a yellow-bellied marmot population. *Ecology* **55**:760–771.

Svendsen, G. E. 1976. Structure and location of burrows of yellow-bellied marmot. *Southwest. Nat.* **20**:487–494.

Svendsen, G. E., and K. B. Armitage 1973. Mirror-image stimulation applied to field behavioral studies. *Ecology* **54**:623–627.

Swihart, R. K. 1990. Common components of orchard ground cover selected as food by captive woodchucks. *J. Wildl. Manage.* **54**:412–417.

Swihart, R. K. 1991a. Modifying scent-marking behavior to reduce woodchuck damage to fruit trees. *Ecol. Appl.* **1**:98–103.

Swihart, R. K. 1991b. Influence of *Marmota monax* on vegetation in hayfields. *J. Mammal.* **72**:791–795.

Swihart, R. K. 1992. Home-range attributes and spatial structure of woodchuck populations. *J. Mammal.* **73**:604–618.

Swihart, R. K., and P. M. Picone 1991a. Arboreal foraging and palatability of tree leaves to woodchucks. *Am. Midl. Nat.* **125**:372–374.

Swihart, R. K., and P. M. Picone 1991b. Effects of woodchuck activity on woody plants near burrows. *J. Mammal.* **72**:607–611.

Swihart, R. K., and P. M. Picone 1994. Damage to apple trees associated with woodchuck burrows in orchards. *J. Wildl. Manage.* **58**:357–360.

Tafani, M., A. Cohas, C. Bonenfant, J.-M. Gaillard, and D. Allainé 2013. Decreasing litter size of marmots over time: a life history response to climate change? *Ecology* **94**:580–586.

Tankersley, K. B., and B. G. Redmond 2000. Ice age Ohio. *Archaeology* **53**:42–46.

Tardif, S. D. 1997. The bioenergetics of parental behavior and the evolution of alloparental care in marmosets and tamarins. In *Cooperative Breeding in Mammals*, ed. N. G. Solomon and J. A. French. Cambridge: Cambridge University Press, pp. 11–33.

Taulman, J. F. 1977. Vocalizations of the hoary marmot, *Marmota caligata*. *J. Mammal.* **58**:681–683.

Taulman, J. F. 1990a. Late summer activity patterns in hoary marmots. *Northwest. Nat.* **71**:21–26.

Taulman, J. F. 1990b. Observation on scent marking in hoary marmots, *Marmota caligata*. *Can. Field-Nat.* **104**:479–482.

Terekhina, A., and P. Panteleyev 1994. Geographical variability of body size in marmots. *Abstracts 2nd International Conference on Marmots*, 146.

Thomas, W. K., and S. L. Martin 1993. A recent origin of marmots. *Mol. Phylogenet. Evol.* **2**:330–336.

Thompson, K. V. 1998. Self assessment in juvenile play. In *Animal Play: Evolutionary, Comparative, and Ecological Perspectives*, ed. M. Bekoff and J. A. Byers. Cambridge: Cambridge University Press, pp. 183–204.

Thompson, S. E. 1979. Socioecology of the yellow-bellied marmot (*Marmota flaviventris*) in central Oregon. PhD dissertation, University of California, Berkeley.

Thorp, C. R., P. K. Ran, and G. L. Florant 1994. Diet alters metabolic rate in the yellow-bellied marmot (*Marmota flaviventris*) during hibernation. *Physiol. Zool.* **67**:1213–1229.

Thunberg, C. P. 1796. *Voyages au Japon par le Cap de Bonne-Espérance, les Isles de la Sonde, &c. Traduits, rédigés et augmentes. . .par L. Langles*. Paris: Benoit Dandré Garnery et Obré, An 14.

Tokarski, V. A., O. V. Brandler, and A. V. Zavgorudko 1991. Spatial structure of the bobac population in the Ukraine. In *Population Structure of the Marmot*, ed. D. I. Bibikov, A. A. Nikolski, V. Yu. Rumiantsev, and T. A. Seredneva. Moscow: USSR Theriological Society, pp. 45–70.

Tokarsky, V. A. 1996. Biology, behaviour, and breeding black-capped marmots (*Marmota camtschatica* Pall.) in captivity. In *Biodiversity in Marmots*, ed. M. Le Berre, R. Ramousse, and L. Le Guelte. Lyon: International Marmot Network, pp. 257–260.

Tokarsky, V. A., and A. S. Vasiljev 1991. Distribution and number of the black-capped marmot in Kamchatka region. In *Population Structure of the Marmot*, ed. D. I. Bibikov, A. A. Nikolski, V. Yu. Rumiantsev, and L. A. Seredneva. Moscow: USSR Theriological Society, pp. 290–299.

Tomé, C., and L. Chaix 2003. Marmot's hunting and exploitation in the western Alps and the southern Jura from Late Pleistocene to Holocence. In *Adaptive Strategies and Diversity in Marmots*, ed. R. Ramousse, D. Allainé, and M. Le Berre. Lyon, France: International Marmot Network, pp. 77–84.

Townsend, S. E. 2006. Burrow cluster as a sampling unit: an approach to estimate marmot activity in the eastern steppe of Mongolia. *Mongolian J. Biol. Sci.* **41**:31–36.

Townsend, S. E. 2009. Estimating Siberian marmot (*Marmota sibirica*) densities in the eastern steppe of Mongolia. *Ethol., Ecol., Evol.* **21**:325–328.

Townsend, S. E., and P. Zahler 2006. Mongolian marmot crisis: status of the Siberian marmot in the Eastern Steppe. *Mongolian J. Biol. Sci.* **4**:37–44.

Travis, S. E., and K. B. Armitage 1972. Some quantitative aspects of the behavior of marmots. *Trans. Kans. Acad. Sci.* **75**:308–321.

Trivers, R. L. 1972. Parental investment and sexual selection. In *Sexual Selection and the Descent of Man, 1871–1971*, ed. B. Campbell. Chicago, IL: Aldine, pp. 136–179.

Turbill, C., C. Bieber, and T. Ruf 2011. Hibernation is associated with increased survival and the evolution of slow life histories among mammals. *Proc. R. Soc. B* **278**:3355–3363.

Turk, A., and W. Arnold 1988. Thermoregulation as a limit to habitat use in alpine marmots (*Marmota marmota*). *Oecologia* **76**:544–548.

Twitchell, A. R. 1939. Notes on the southern woodchuck in Missouri. *J. Mammal.* **20**:71–74.

Tyler, G. V., J. W. Summers, and R. L. Snyder 1981. Woodchuck hepatitis virus in natural woodchuck populations. *J. Wildl. Dis.* **17**:297–301.

Tyser, R. W., and T. C. Moermond 1983. Foraging behavior in two species of different-sized sciurids. *Am. Midl. Nat.* **109**:240–245.

Valentsev, A. S., V. A. Tokarski, and V. I. Mosolov 1996. The current status of black-headed marmot population on Kamtchatka. In *Biodiversity in Marmots*, ed. M. Le Berre, R. Ramousse, and L. Le Guelte. Lyon: International Marmot Network, pp. 261–264.

Van Vuren, D. 1990. Dispersal of yellow-bellied marmots. PhD dissertation, University of Kansas, Lawrence.

Van Vuren, D. 1991. Yellow-bellied marmots as prey of coyotes. *Am. Midl. Nat.* **125**:135–139.

Van Vuren, D. 1996. Ectoparasites, fitness, and social behavior of yellow-bellied marmots. *Ethology* **102**:686–694.

Van Vuren, D. H. 2001. Predation on yellow-bellied marmots (*Marmota flaviventris*). *Am. Midl. Nat.* **145**:94–100.

Van Vuren, D., and K. B. Armitage 1991. Duration of snow cover and its influence on life-history variation in yellow-bellied marmots. *Can. J. Zool.* **69**:1755–1758.

Van Vuren, D., and K. B. Armitage 1994a. Survival of dispersing and philopatric yellow-bellied marmots: what is the cost of dispersal? *Oikos* **69**:179–181.

Van Vuren, D., and K. B. Armitage 1994b. Reproductive success of colonial and noncolonial female yellow-bellied marmots (*Marmota flaviventris*). *J. Mammal.* **75**:950–955.

Van Wormer, J. 1974. *There's a Marmot on the Telephone*. Caldwell, ID: The Caxton Printers, Ltd.

Vasiliev, V. N. 1992. Hibernation in black-capped marmot (*Marmota camtschatica* Pallas, 1811). In *Proceedings of the 1st International Symposium on Alpine Marmot (*Marmota marmota*) and Genus* Marmota, ed. B. Bassano, P. Durio, U. Gallo Orsi, and E. Macchi. Torino, Italy: Dipartimento di Produzioni animali, Epidemiologia ed Ecologia, pp. 67–68.

Vasiliev, V. N., and N. G. Solomonov 1996. Body temperature and metabolism in hibernating marmots (*Marmota camtshatica* Pallas, 1811). In *Biodiversity in Marmots*, ed. M. Le Berre, R. Ramousse, and L. Le Guelte. Lyon, France: International Marmot Network, pp. 265–266.

Vita, A. 1992. The population of marmot (*Marmota marmota*) in the natural park Orsiera Rocciavré. In *Proceedings of the 1st International Symposium on Alpine Marmot (*Marmota marmota*) and Genus* Marmota, ed. B. Bassano, P. Durio, U. Gallo Orsi, and E. Macchi. Torino, Italy: Dipartimento di Produzioni animali, Epidemiologia ed Ecologia, pp. 265–268.

Van Staalduinen, M. A., and M. J. A. Werger 2007. Marmot disturbances in a Mongolian steppe vegetation. *J. Arid Environ.* **69**:344–351.

vom Saul, F. S. 1989. Sexual differentiation in litter-bearing mammals: influence of sex of adjacent fetuses in utero. *J. Anim. Sci.* **67**:1824–1840.

Wallens, K. 1970. The behavior of isolates of *Marmota flaviventris*. Unpublished research report. Rocky Mountain Biological Laboratory, Colorado.

Walro, J. M., P. T. Meier, and G. E. Svendsen 1983. Anatomy and histology of the scent glands associated with the oral angle in woodchucks. *J. Mammal.* **64**:701–703.

Waltari, E., and R. P. Guralnick 2009. Ecological niche modeling of montane mammals in the Great Basin, North America: examining past and present connectivity of species across basins and ranges. *J. Biogeogr.* **36**:148–161.

Ward, J. M., Jr., and K. B. Armitage 1981a. Circannual rhythms of food consumption, body mass, and metabolism in yellow-bellied marmots. *Comp. Biochem. Physiol.* **69A**:621–626.

Ward, J. M., Jr., and K. B. Armitage 1981b. Water budgets of montane-mesic and lowland-xeric populations of yellow-bellied marmots. *Comp. Biochem. Physiol.* **69A**:627–630.

Waring, G. H. 1965. Behavior of a marmot toward a marten. *J. Mammal.* **46**:631.

Waring, G. H. 1966. Sounds and communications of the yellow-bellied marmot (*Marmota flaviventris*). *Anim. Behav.* **14**:177–183.

Warren, E. R. 1916. Notes on the birds of the Elk Mountain region, Gunnison County, Colorado. *The Auk* **33**:292–317.

Wasser, S. K., and D. P. Barash 1983. Reproductive suppression among female mammals: implications for biomedicine and sexual selection theory. *Quart. Rev. Biol.* **58**:513–538.

Waterman, J. 2007. Male mating strategies in rodents. In *Rodent Societies: An Ecological and Evolutionary Perspective*, ed. J. O. Wolff and R. W. Sherman. Chicago, IL: University of Chicago Press, pp. 17–41.

Webb, D. R. 1980. Environmental harshness, heat stress, and *Marmota flaviventris*. *Oecologia* **44**:390–395.

Webb, D. R. 1981. Macro-habitat patch structure, environmental harshness, and *Marmota flaviventris*. *Behav. Ecol. Sociobiol.* **8**:175–182.

Weeks, H. P., Jr., and G. M. Kirkpatrick 1978. Salt preferences and sodium drive phenology in fox squirrels and woodchucks. *J. Mammal.* **59**:531–542.

Weigel, R. M. 1981. The distribution of altruism among kin: a mathematical model. *Am. Nat.* **118**:191–201.

Weiner, J. 1989. Metabolic constraints to mammalian energy budgets. *Acta Theriol.* **34**:3–35.

Weiner, J. 1992. Physiological limits to sustainable energy budgets in birds and mammals: ecological implications. *Trends Ecol. Evol.* **7**:384–388.

Werner, J. R. 2012. Vancouver Island marmots (*Marmota vancouverensis*) consume plants containing toxic secondary compounds. *Can. Field-Nat.* **126**:55–58.

West, S. A., I. Pen, and A. S. Griffinn 2002. Cooperation and competition between relatives. *Science* **296**:72–75.

West-Eberhard, M. J. 1989. Phenotypic plasticity and the origins of diversity. *Annu. Rev. Ecol. Syst.* **20**:249–278.

Wey, T. W., and D. T. Blumstein 2010. Social cohesion in yellow-bellied marmots is established through age and kin structure. *Anim. Behav.* **79**:1343–1352.

Whitehouse, M. E. A., and Y. Lubin. 2005. The function of societies and the evolution of group living: spider societies as a test case, *Biol. Rev.* **80**:347–361.

Williams, D. D., and R. L. Rausch 1973. Seasonal carbon dioxide and oxygen concentrations in the dens of hibernating mammals (Sciuridae). *Comp. Biochem. Physiol.* **44A**:1227–1235.

Williams, G. C. 1966. *Adaptation and Natural Selection. A Critique of Some Current Evolutionary Thought*. Princeton, NJ: Princeton University Press.

Wilson, E. O. 1975. *Sociobiology. The New Synthesis*. Cambridge, MA: The Belknap Press of Harvard University Press.

Winders, R. L., M. O. Farber, K. F. Atkinson, and F. Manfredi 1974. Parameters of oxygen delivery in the species *Marmota flaviventris* at sea level and 12 000 feet. *Comp. Biochem. Physiol.* **49A**:287–290.

Witczuk, J., S. Pagacz, and L. S. Mills 2013. Disproportionate predation on endemic marmots by invasive coyotes. *J. Mammal.* **94**:702–713.

Wolff, J. O. 1993. What is the role of adults in mammalian juvenile dispersal? *Oikos* **68**:173–176.

Wolff, J. O. 1997. Population regulation in mammals: an evolutionary perspective. *J. Anim. Ecol.* **66**:1–13.

Wood, D. L., and A. D. Barnosky 1994. Middle Pleistocene climate change in the Colorado Rocky Mountains indicated by fossil mammals from Porcupine Cave. *Quaternary Res.* **41**:366–375.

Wood, W. A. 1973. Habitat selection and energetics of the Olympic marmot. MS Thesis, Western Washington State College.

Woods, B. C. 2009. Using multivariate techniques to determine if yellow-bellied marmots feed selectively. *Ethol., Ecol., Evol.* **21**:261–276.

Woods, B. C., and K. B. Armitage 2003a. Effect of food supplementation on juvenile growth and survival in *Marmota flaviventris. J. Mammal.* **84**:903–914.

Woods, B. C., and K. B. Armitage 2003b. Effects of food addition on life history of yellow-bellied marmots. *Oecologia Montana* **12**:29–36.

Woods, B. C., K. B. Armitage, and D. T. Blumstein 2002a. Circadian rhythm is maintained during hibernation in yellow-bellied marmots. In *Holarctic Marmots as a Factor of Biodiversity*, ed. K. B. Armitage and V. Yu. Rumiantsev. Moscow: ABF Publishing House, pp. 389–395.

Woods, B. C., K. B. Armitage, and D. T. Blumstein 2002b. Yellow-bellied marmots depress metabolism to enter torpor. In *Holarctic Marmots as a Factor of Biodiversity*, ed. K. B. Armitage and V. Yu. Rumiantsev. Moscow: ABF Publishing House, pp. 400–404.

Woods, B. C., C. L. Brown, and M. A. Cobb 2009. Elevation variation in life-history characteristics of populations of yellow-bellied marmots (*Marmota flaviventris*). *Ethol., Ecol., Evol.* **21**:381–392.

Wright, J., B. C. Tennant, and B. May 1987. Genetic variation between woodchuck populations with high and low prevalence rates of woodchuck hepatitis virus infection. *J. Widl. Dis.* **23**:186–191.

Wright, P. L., and C. H. Conaway 1950. Golden-mantled marmot in northwestern Montana. *Murrelet* **31**:32.

Wynne-Edwards, V. C. 1962. *Animal Dispersion in Relation to Social Behaviour*. London: Oliver and Boyd.

Wynne-Edwards, V. C. 1965. Self-regulating systems in populations of animals. *Science* **147**:1543–1548.

Yakovlev, F. G., and E. G. Shadrina 1996. Density and demographic structure of black-capped marmot (*Marmota camtschatica* Pall.) in northeastern Yakutia. In *Biodiversity in Marmots*, ed. M. Le Berre, R. Ramousse, and L. Le Guelte. Lyon: International Marmot Network, pp. 267–268.

Yensen, E., and P. W. Sherman 2003. Ground squirrels (*Spermophilus* species and *Ammospermophilus* species). In *Wild Mammals of North America: Biology, Management, and Conservation*, 2nd edition, ed. G. A. Feldhamer, B. C. Thompson, and J. A. Chapman. Baltimore, MD: The Johns Hopkins University Press, pp. 211–231.

Yoshihara, Yu., T. Okuro, B. Buuveibaatar, J. Undarmaa, and K. Takeuchi 2010a. Responses of vegetation to soil disturbance by Siberian marmots within a landscape and between landscape positions in Hustai National Park, Mongolia. *Jap. Soc. Grassland Sci.* **56**:42–50.

Yoshihara, Yu., T. Okuro, B. Buuveibaatar, J. Undarmaa, and K. Takeuchi 2010b. Complementary effects of disturbance by livestock and marmots on the spatial heterogeneity of vegetation and soil in a Mongolian steppe system. *Agri., Ecosystems and Environ.* **135**:155–159.

Young, R. A., and E. A. H. Sims 1979. The woodchuck, *Marmota monax*, as a laboratory animal. *Lab. An. Sci.* **29**:770–780.

Zegers, D. A. 1984. *Spermophilus elegans. Mammalian Species* **214**:1–7.

Zeidman, J. 2000. *The Marmot Chronicles: A Crested Butte Fable.* Columbus, OH: Greyden Press.

Zervanos, S. M., C. R. Maher, J. A. Waldvogel, and G. L. Florant 2010. Latitudinal differences in the hibernation characteristics of woodchucks (*Marmota monax*). *Physiol. Biochem. Zool.* **83**:135–141.

Zervanos, S. M., and C. M. Salsbury 2003. Seasonal body temperature fluctuations and energetic strategies in free-ranging eastern woodchucks (*Marmota monax*). *J. Mammal.* **84**:299–310.

Zervanos, S. M., C. M. Salsbury, and J. K. Brown 2009. Maintenance of biological rhythms during hibernation in Eastern woodchucks (*Marmota monax*). *J. Comp. Physiol. B* **179**:411–418.

Zhaltsanova, D. S. D., and N. M. Shalaeva 1990. Ecological pecularities of helminthofauna of rodents of squirrels family (Rodentia: Sciuridae) in Zabaikalie. *Helminthologia* **27**:217–223.

Zheleznov, N. K. 1991. Pattern of the black-capped marmot distribution. In *Population Structure of the Marmot*, ed. D. I. Bibikov, A. A. Nikolski, V. Yu. Rumiantsev, and T. A. Seredneva. Moscow: USSR Theriological Society, pp. 275–289.

Zheleznov, N. 1996. Spatial structure of the Kamtchatka marmot (*Marmota camtschatica*) in the far northeast of Russia. In *Biodiversity in Marmots*, ed. M. Le Berre, R. Ramousse, and L. Le Guelte. Lyon: International Marmot Network, pp. 269–270.

Zholnerovskaya, E. I., D. I. Bibikov, and V. I. Ermolaev 1992. Immunogenetical analysis of systematical relations among marmots. In *Proceedings of the 1st International Symposium on Alpine Marmot (*Marmota marmota*) and Genus* Marmota, ed. B. Bassano, P. Durio, U. Gallo Orsi, and E. Macchi. Torino, Italy: Dipartimento di Produzioni animali, Epidemiologia ed Ecologia, pp. 193–196.

Zholnerovskaya, E. I., and V. I. Ermolaev 1996. Immunogenetic differences between *Marmota menzbieri* and other Palearctic marmots. In *Biodiversity in Marmots*, ed. M. Le Berre, R. Ramousse, and L. Le Guelte. Lyon: International Marmot Network, pp. 217–222.

Zimina, R. P. 1996. Role of marmots in landscape transformations since Pleistocene. In *Biodiversity in Marmots*, ed. M. Le Berre, R. Ramousse, and L. Le Guelte. Lyon: International Marmot Network, pp. 59–62.

Zimina, R. P., and I. P. Gerasimov 1973. The periglacial expansion of marmots (Marmota) in Middle Europe during Late Pleistocene. *J. Mammal.* **54**:327–340.

Index

Printed in the United States
By Bookmasters